Backgrounds of Arithmetic and Geometry
An Introduction

SERIES IN PURE MATHEMATICS

Editor: C C Hsiung
Associate Editors: S S Chern, S Kobayashi, I Satake, Y-T Siu, W-T Wu
and M Yamaguti

Series in Pure Mathematics – Volume 23

BACKGROUNDS OF ARITHMETIC AND GEOMETRY
An Introduction

Radu Miron
Dan Brânzei
University Al. I. Cuza
Romania

World Scientific
Singapore • New Jersey • London • Hong Kong

Published by

World Scientific Publishing Co. Pte. Ltd.
P O Box 128, Farrer Road, Singapore 9128
USA office: Suite 1B, 1060 Main Street, River Edge, NJ 07661
UK office: 57 Shelton Street, Covent Garden, London WC2H 9HE

Sci
QA
155
M57
1995

ISBN 981-02-2210-6

Printed in Singapore.

PREFACE

There is a general opinion that the foundations of Mathematics make up a part of Mathematics completely from the remaining ones, which only a small number of researchers will be interested in. In supporting this opinion, some *arguments* have also been put up: natural numbers have been efficiently used for millieniums before being defined on the basis of Peano's Axiomatic system. Real numbers had been achieved but in the Middle Ages, sets had taken up the role of foundation of Mathematics on the basis of Cantor's naïve set theory long before the edification of a rigorous logical theory.

This opinion is profoundly mistaken. First of all, in present-day Mathematics it is not possible to make a delimitation, not even a vague one, of its foundations. Secondly, the efforts meant to found Mathematics have appeared very early, Euclid's monumental work constituting an undeniable example. We can say that Mathematics evolved in close relations with its foundations, and to know this fact is an essential element of general culture. The Romanian Mathematician Dan Barbilian, also known as an outstanding poet under the name of Ion Barbu, synthetized this truth in a remarkable assertion for Mathematics, History and Philosophy of Culture: *Homer is not the obligatory gate through which one can approach the Greek world. Greek Geometry is a wider gate, from which the eye can look upon an austere, but essential landscape.*

School's aim is not that of acquiring knowledge, but that of achieving creative thinking capacities, a non-dogmatic and non-prejudiced one. If Mathematics could be torn from its foundations, it would became a series of formulae, receipts and tautologies that could not be applied any longer to the objective reality, but only to some rigid, mortified schemes of this reality.

The teaching of Mathematics expresses a remarkable balance between Mathematical concrete constructions and their rigorous logical founding. But balance is dynamic, variable in time and space; the teacher presents not only the contents of a theory, but its relevance in Mathematics and life, as well.

Likewise, theoretical Informatics can not be thought separately from the foundations of Mathematics.

The foundations of Mathematics constitute an ample domain too for only a book. In order to achieve a relevant prospect we chose Arithmetic for the constructive method and Geometry for the axiomatic one. The alternation of the two methods ensures a unitary understanding of *elementary Geometry* and of its organic link to the study of real numbers.

Here is a description of the contents of the book.

The first chapter presents the naïve set theory with the constructions and structurings of the families of cardinal and ordinal numbers.

The second chapter presents the natural numbers by means of the Frege-Russell constructions, then successively extending N to Z, Q, R – through the factorization of the ring of Cauchy sequences by means of the ideal of null sequences – and C.

The third chapter presents the axiomatic Theories differentiated according to their degree of formalization and the central metatheoretical problems. The subject is exemplified by the reconstruction of N on the basis of Peano's semiformalized axiomatics.

The fourth chapter presents the *algebraic backgrounds* of Geometry. The problems of this chapter constitute a recent and important contribution of Romanian Mathematics, especially of the first of the authors of this book. A minimal Weyl type axiomatic system introduces the linear, affine and Euclidean spaces.

The fifth chapter edifies Geometry on the basis of Hilbert's axiomatic system. Here, the becoming attention is paid to the deduction of some first consequences of axioms; the way from these to the main Euclidean results is assumed to be know. The metatheoretical analysis points out the arithmetical model i.e. the leading ideas of the analytical Geometry of the space.

In the sixth chapter, Hilbert's axiomatics is correlated to that of Birkhoff, quite recently adopted to the didactics of Geometry in many countries.

The seventh chapter presents in brief the mean types of geometrical transformations, their actions upon the fundamental geometric figures and the results of certain geometrical transformations compounding.

The eighth chapter exposes in a modern manner Felix Klein's conception on Geometries and exemplifies it through the study of affine, projective and plane hyperbolic Geometries.

Francisc Radó, a specialist of international rate in Geometries algebraic foundations, Barbilian spaces, finite Geometries, etc, has attentively studied the manuscript and made pertinent observations of great usefulness. Appreciating the quality of the material, he outlined a new unitary prospect to it by the ninth chapter in which he presents the construction of Geometry through an axiomatics of Bachmann type, with Isometry as a primary notion.

The first eight chapters contain exercises or problems.

The book benefits from a beautiful tradition of the city Iaşi (Jassy) materialized by Professor Izu Vaisman's **The Foundations of Mathematics** (in Romanian, EDP, Bucharest, 1968), as well. The didactic experience of the authors, who have been teaching this discipline for over two decades at the oldest University in Romania has also benefited from an ample dialogue with specialists in this field who studied their work **The Foundations of Mathematics and Geometry** (in Romanian, Ed. Academiei, Bucharest, 1983). This work turns to account this experience, possessing a personal hint by its structure and conception. The original feature is at its top in chapters IV, VIII and IX, but it is to be found in each paragraph, as well.

The unity of the work, as a mirroring of the unitary character of Mathematics in general and especially of Geometry is evident. The book is quite accessible to high school pupils in the last forms, but its being taken for a handbook is out of question; first of all, its reading is supposed to follow a first systematic study of Geometry on the basis of an axiomatic system. Secondly, details of a technical character meant to point out their ideas and connections have often been omitted.

The book is useful to the students in Mathematics faculties (for getting the *master* or in drawing up a doctorship thesis included), to the Mathematics teacher, to the researchers in Mathematics (even for those in very specialized domains), and to-other researchers that use the mathematical apparatus, to computer sciences, for instance.

The authors consider that the reader will be convinced not only by the essential aspects ensuring the unity and vigour of Mathematics, but of the unaltered value of the geometrical measure with its remarkable elegance, as well.

Acknowledgments. We express appreciation to Mr. *Mihai Eugeniu Avădanei* and Miss *Oana Brânzei* for their essential contribution in the improvements of English version.

We would like to extend special thanks to our colleagues *Mihai Anastasiei* and *Adrian Albu*, who brought along valuable comments and suggestions in the evolution of this text.

We are also grateful to *Mihai Postolache, Valentin Clocotici, Radu Negrescu, Daniel Rameder, Bogdan Balica* and *Cornelia Ivaşc* who realised the computer-typed version of the manuscript.

Last but not least we are thankful to all our students who have been carefully studying our courses and conferences. Their opinions decisively contributed to a better crystallized conception on the subjects didactics.

THE AUTHORS

CONTENTS

CHAPTER I

ELEMENTS OF SET THEORY

The set theory has a fundamental importance in present-day Mathematics. From the logicians' point of view, Mathematics is the theory of sets and of their consequences. In the following lines, the construction of natural numbers and of the other known numbers is based on the concept of set, and the notion of mathematical structure is considered as a set of elements subjected to a system of axioms.

It is not easy to achieve a rigorous presentation of the notion of set. Its axiomatic description is made through a complicated process. That is why we place ourselves within the naive point of view here, according to which the notion of set is supposed to be known.

§1. The Set Algebra

We shall denote the sets by A, B, C, ..., the elements of a set by a, b, c, ..., x, y, z, ... etc., and will use the logical symbols $=$, $\Rightarrow$, $\forall$, $\exists$, etc., whose meanings we consider to be known.

A set will often be characterized by certain properties of its elements. One writes $\{x: P(x)\}$, where $P(x)$ is a proposition referring to x, this meaning the set of all those x-es, so that $P(x)$ is true. From this point of view, the terms "property" and "set" are synonymous ones.

If x is an element of the set A, we will write $x \in A$, while $x \notin A$ means that "x is not an element of the set A".

We shall admit the existence of a set, denoted by $\varnothing$, which has no elements.

For any x, $\{x\}$ denote the set containing only the element x. Analogously, $\{x_1, x_2, ..., x_n\}$ will be the set containing only the elements x_1, x_2, ..., x_n (supposed to be distinct).

Definition 1.1. *Let be A and B two sets. We shall say that A is a subset of B or that A is included into B or A is a part of B, if $x \in A \Rightarrow x \in B$, and we shall write $A \subset B$. If $A \subset B$ and $B \subset A$, we shall say that A and B are equal, and write $A = B$; to deny that we write $A \neq B$.*

If $A \subset B$, $A \neq B$, then A will be called a *proper subset* of B. This property is denoted by $A \subsetneq B$. It follows that $A \subset B$ means that $A \subsetneq B$, or $A = B$.

It is easy to deduce from the definition of inclusion that:

1. $A \subset A$ (reflexivity);
2. $A \subset B$, $B \subset C \Rightarrow A \subset C$ (transitivity);
3. $A \subset B$, $B \subset A \Rightarrow A = B$ (antisymmetry).

The antisymmetry property expresses the fact that a set A is completely characterized through the totality of elements belonging to it; for example:

$$\{x_1, x_2, x_3, x_4\} = \{x_2, x_4, x_3, x_1\}.$$

Of course, for any set A it occurs that $\varnothing \subset A$, and it immediately follows that the void set is unique.

Definition 1.2. *The set $A \bigcup B = \{x : x \in A$ or $x \in B\}$ is called the union of the sets A and B.*

This definition extends to a certain family of sets $\{A_i\}_{i \in I}$ and is written $\bigcup_{i \in I} A_i$. We draw the attention to the fact that we are using the term of "family" synonymous to that of sets in order to avoid annoying repetitions.

Definition 1.3. *The set $A \bigcap B = \{x : X \in A$ and $x \in B\}$ is called the intersection of the sets A and B.*

By extending that definition to a certain family of sets $\{A_i\}_{i \in I}$, $\bigcap_{i \in I} A_i = \{x : \forall i \in I, x_i \in A_i\}$ holds.

One denotes by $\mathcal{P}(A)$ the set of all A's subsets; one prefers for $\mathcal{P}(A)$ the name *power set* of A better than the set of parts of A. Obviously:

$$A \in \mathcal{P}(A), \quad \varnothing \in \mathcal{P}(A), \quad M \subset A \Rightarrow M \in \mathcal{P}(A).$$

Theorem 1.1. *For any sets A, B, C the following properties hold:*

$A \bigcup B = B \bigcup A$	$A \bigcap B = B \bigcap A$	*(commutativity)*;
$A \bigcup A = A$	$A \bigcap A = A$	*(idempotence)*;
$A \bigcup \varnothing = A$	$A \bigcap \varnothing = \varnothing$;	
$A \bigcup (B \bigcup C) = (A \bigcup B) \bigcup C$	$A \bigcap (B \bigcap C) = (A \bigcap B) \bigcap C$	*(associativity)*;
$A \subset A \bigcup B$	$A \bigcap B \subset A$;	
$A \subset B \Rightarrow A \bigcup B = B$	$A \subset B \Rightarrow A \bigcap B = A$.	

The proof of this theorem is immediate; we leave it to the reader. It is also easy to show the following:

Theorem 1.2 (The laws of distributivity).

$$A\cap(B\cup C) = (A\cap B)\cup(A\cap C);$$
$$A\cup(B\cap C) = (A\cup B)\cap(A\cup C).$$

Definition 1.4. *The set $A\backslash B = \{x : x \in A, x \notin B\}$ is called the difference of the sets A and B. The set $C_A B \equiv A\backslash B$, for $B \subset A$ is called the complement of B with respect to A. If A is fixed, it is called universal set. $C_A B$ is denoted by $\bar{B}$ and it is simply called the complement of B.*

One immediately deduces $A \subset B \Rightarrow \bar{B} \subset \bar{A}, \bar{\bar{A}} = A, C_A \varnothing = A, C_A A = \varnothing$, and

Theorem 1.3 (Rules of Pierce and De Morgan).

$$\overline{A\cup B} = \bar{A}\cap\bar{B}; \quad \overline{A\cap B} = \bar{A}\cup\bar{B};$$
$$\overline{\bigcup_{i\in I} A_i} = \bigcap_{i\in I} \bar{A}_i; \quad \overline{\bigcap_{i\in I} A_i} = \bigcup_{i\in I} \bar{A}_i.$$

The proof does not entail difficulties.

From the last three theorems it follows *the principle of duality* in the set theory.

From any general relationship between sets that makes the operation $\cup$, $\cap$, $\subset$, $\supset$ *intervene, a new true relationship is obtained, replacing respectively the previous operations by* $\cap$, $\cup$, $\supset$, $\subset$ *and the void set* $\varnothing$ *by the universal set A and vice versa.*

If the sets A and B have the property $A\cap B = \varnothing$ we say that A and B are *disjoint sets* or *non-overlapping*.

Let be a and b distinct elements; the set $\{a, b\}$ enables us to know this pair of elements without specifying which is "the first" and which "the second". We can favour one of these elements, a, for example, in order to consider it as being the first one, giving the set $\{\{a, b\}, a\}$ which we denote more briefly by (a, b) and call it *ordered pair* or *couple*. The symbol (a, b) is considered in the case $a = b$, too, but now it notes the set $\{a, \{a\}\}$.

The couple $(a, (b, c))$ permits the identification of the elements a, b, c, but an ordering of them, as well; it is more simply denoted by (a, b, c) and it is called *ordered triple*.

For a certain set A, the set $\widetilde{A} = \{(x, A): x \in A\}$ is called *associated* with A. Let us notice that $A = B \Rightarrow \widetilde{A} = \widetilde{B}$ and $A \neq B \Rightarrow \widetilde{A} \cap \widetilde{B} = \varnothing$.

Definition 1.5. *Let* $\left\{A_i\right\}_{i \in I}$ *be a certain family of sets. The set* $\bigvee_{i \in I} A_i = \bigcup_{i \in I} \widetilde{A}_i$ *is called* disjoint union *of the sets of the family under consideration. If the family contains only two (distinct) sets A, B, for their disjoint union the notation $A \vee B$ is also used. This notation is also used when $A = B$, but its meaning being now specified by $A \vee B = A \cup \widetilde{A}$.*

§2. Binary Relations and Functions

The notions of relation and that of function are of a general interest. We will briefly expose them.

If X and Y are two sets, then $X \times Y = \{(x, y): x \in X, y \in Y\}$ is called *their Cartesian product*.

Definition 2.1. *A binary relation ρ is an ordered triple $\rho = (X, Y, G)$, where X and Y are sets (called base sets of ρ), and $G \subset X \times Y$. (The set G is called the graph of the relation ρ.)* Instead of the notation $(x, y) \in G$, the notation $x \rho y$ (which is read "x is in the relation ρ with y") is preferred.

Of course, when the base sets, X, Y are implied in the context, the binary relation ρ is specified by its graph, G, but that does not mean that a binary relation would coincide with its graph.

For the binary relation $\rho = (X, Y, G)$, the dom $\rho = \{x: \exists_y \ x \rho y\}$ and codom $\rho = \{y: \exists_x \ x \rho y\}$ are considered, sets which are called the *domain* and *codomain*, respectively, of the binary relation ρ.

The binary relation $\rho^{-1} = (Y, X, G')$ characterized by $y \rho^{-1} x \Leftrightarrow x \rho y$ is called the *reverse* or the *inverse* of the relation ρ.

Obviously, dom ρ^{-1} = codom ρ and codom ρ^{-1} = dom ρ.

Definition 2.2. *Let* $\rho = (X, Y, G)$ *and* $\sigma = (Y, Z, H)$ *be two binary relations; we call composition of these binary relations the binary relation* $\sigma \circ \rho = (X, Z, K)$, *where*

$$x \sigma \circ \rho z \Leftrightarrow \exists y, \ x\rho y \ and \ y\sigma z.$$

An important particular case of binary relations consists of the homogeneous relations, when the two base sets coincide. Although for the binary relation $\rho = (X, X, G)$, $G \subset X^2$ occurs it is said that ρ is a binary relation on X.

Among the homogeneous binary relations the relations of equivalence and order distinguish themselves as most important ones.

Relations of Equivalence

Definition 2.3. *The binary relation* $\rho = (X, X, G)$ *is an equivalence (on the set* X), *if it satisfies the following conditions:*

1. *it is reflexive, that is* $x \in X \Rightarrow x\rho x$;
2. *it is symmetric, that is* $x\rho y \Rightarrow y\rho x$;
3. *it is transitive, that is* $x\rho y$ *and* $y\rho z \Rightarrow x\rho z$.

For such a binary relation, to an arbitrary element $x \in X$ its *class of equivalence* $\rho_x = \{y: y\rho x\}$ is associated. Frequently ρ_x is named *coset of x modulo* ρ. For the distinct elements x, y, the classes of equivalence ρ_x, ρ_y are disjoint or coincide. Due to the reflexivity property, any element x from X belongs at least to a class of equivalence

$$(x \in \rho_x), \text{ hence } \bigcup_{x \in X} \rho_x = X.$$

The set $X/\rho = \{\rho_x : x \in X\}$ is called *factor set* or *quotient set of X through the relation* ρ; its elements are the classes of equivalence and make up a *partitioning* of X. The partitioning X/ρ is obviously uniquely determining the equivalence ρ. The map $\pi: X \to X/\rho$ defined through $\pi(x) = \rho_x$ is called *the canonical projection of the equivalence* ρ or the *quotient map* of ρ.

Relation of (Partial) Ordering

Definition 2.4. *The binary relation* $\leq = (X, X, G)$ *is called (partial) ordering of the set* X, *if it is reflexive, transitive and "antisymmetric", that is* $(x \leq y$ *and* $y \leq x) \Rightarrow x = y$. *The adjective "partial" may be omitted and will be replaced by "total" when the property of "trichotomy" occurs:*

$$\forall x, y \in X \quad x \le y \text{ or } y \le x.$$

For order relations, symbols of the following forms $\le$, $\ge$, $\preceq$, $\succeq$ are generally used, since they have the advantage of permitting suggestive and handy notations for the reverse ones ($\ge$ instead of $\le^{-1}$ etc.), and for "strict orderings" (for example, $x < y$ means that $x \le y$ and $x \ne y$).

Functions

Definition 2.5. *The binary relation* $\rho = (X, Y, G)$ *is called univocal, univalent or partial map, if there occurs*

$$\forall x \in X \ (xfy \text{ and } xfz \Rightarrow y = z).$$

Definition 2.6. *We call function (from X to Y) a binary relation* $f = (X, Y, G)$ *which is univocal and "full", that is* $\mathrm{dom} f = X$. *We specify that* f *is a function using the notation* $f \colon X \to Y$.

When a binary relation ρ is a function f, the notation $\mathrm{ran} f$ (to be read "range of f") is preferred to co$-\mathrm{dom} f$.

Several synonyms with *"function"* are sometime preferred in some branches of Mathematics: map, mapping, transformation, correspondence, operator, etc.

The reverse f^{-1} of a function $f = (X, Y, G)$ is a binary relation, but, in general, it is not a function; if f^{-1} is univocal, it is said that f is *injective*, or *one-one*, if f^{-1} is full it is said that f is *surjective* or *onto*, if f^{-1} is a function it is said that f is *bijective* or *bi-univocal* (correspondence). For these peculiar functions f we use the words injection, surjection respectively bijection, too.

When f^{-1} is a function one prefers to call it the *inverse* of the function f.

§3. Cardinal Numbers

Let us suppose as fixed a set U (a "Universe") and any sets X which we shall subsequently consider will be supposed to be elements of U. Although the choice of such a universe does not essentially influence the constructions to follow, such a choice is necessary to avoid the apparition of certain antinomies.

An equivalence $\sim$ on the universe U is defined by saying that $A \sim B$ holds if there exists a bijection $f \colon A \to B$. The relation is called *equipotence* and, obviously, it is an equivalence.

Definition 3.1. *We call cardinal number (in the universe U) a class of equivalence of* $\sim$. *The class of equipotence of a set A will be called the cardinal of A (in U) and it will benefit from the notation* $\bar{\bar{A}}$.

For certain particular sets (supposed to be in U) there appear special notations such as $\bar{\bar{\varnothing}} = 0$, $\overline{\overline{\{a\}}} = 1$, $\overline{\overline{\{a, \{a\}\}}} = 2$, etc.

We draw the attention upon the fact that for the following theory it is not necessary to know the numbers and the sets of numbers N, Z, Q, R which will be defined in the next chapter just on the basis of a theory of the cardinals. Supposing these sets to be intuitively known and familiar, we will use them in certain *examples* which can not create a vicious circle for the theory.

For example, we will consider the cardinal $\aleph_0$ (it is read *"aleph zero"*) of N and the cardinal c of R. If n is an arbitrary natural number, we will also denote by n the cardinal of the set $\{x: x \in N \text{ and } x < n\}$. Although this double significance of a notation may create confusions, we appreciate it for its merit of suggesting the interpretation of natural number as cardinals of certain special sets (*"finite"*), suggestion that will be made explicit in the next chapter.

To ensure the possibility of certain constructions that are to follow and the rigour of certain propositions, we will suppose that the universe U is *"ample enough"*; for example, for $A \in U$ and $B \in U$, we will suppose that $\tilde{A}$, $\tilde{B}$, $A \vee B$, $A \times B$, $\mathcal{P}(A)$ and $\{f: B \to A\}$ are also elements of U.

Theorem 3.1. *If $A_1 \sim A_2$ and $B_1 \sim B_2$, then:*

1. $\tilde{A}_1 \sim \tilde{A}_2$, $\tilde{B}_1 \sim \tilde{B}_2$;
2. $A_1 \vee B_1 \sim A_2 \vee B_2$;
3. $A_1 \times B_1 \sim A_2 \times B_2$;
4. $\{f: B_1 \to A_1\} \sim \{g: B_2 \to A_2\}$.

Proof. 1. Let $f: A_1 \to A_2$ be a bijection; then $\tilde{f}((a, A_1)) = (f(a), A_2)$ defines a bijection $\tilde{f}: \tilde{A}_1 \to \tilde{A}_2$.

2. We do not restrict the generality by supposing that for $i = 1, 2: A_i \neq B_i$, hence $\tilde{A}_i \cap \tilde{B}_i = \varnothing$. The bijection $\tilde{f}: \tilde{A}_1 \to \tilde{A}_2$ and $\tilde{g}: \tilde{B}_1 \to \tilde{B}_2$ determine a bijection $F: \tilde{A}_1 \cup \tilde{B}_1 \to \tilde{A}_2 \cup \tilde{B}_2$ defined by $F((a, A_1)) = \tilde{f}((a, A_1))$ and $F((b, B_1)) = \tilde{g}((b, B_1))$.

3. If $f: A_1 \to A_2$ and $g: B_1 \to B_2$ are bijections, then $f \times g: A_1 \times B_1 \to A_2 \times B_2$ given by $(f \times g)(a, b) = (f(a), f(b))$ is a bijection, too.

4. Let $\alpha: A_1 \to A_2$ and $\beta: B_2 \to B_1$ be two bijections. For any $f: B_1 \to A_1$ one

defines $F(t) = \alpha \circ f \circ \beta : B_2 \to A_2$ and it is easy to prove that F is a bijection from $\{f: B_1 \to A_1\}$ to $\{g: B_2 \to A_2\}$.

Definition 3.2. *Let be* $a = \overline{\overline{A}}$, $b = \overline{\overline{B}}$.

1. *We call sum of cardinal numbers* a *and* b *the cardinal number (denoted by* $a+b$) *of the set* $A \vee B$.

2. *We call product of cardinal numbers* a *and* b *the cardinal number (denoted by* ab) *of the set* $A \times B$.

3. *By* a *raised to the power* b *we understand the cardinal number* a^b *of the set of the maps from* B *to* A, $\{f: B \to A\}$.

According to the theorem 3.1 it is easy to prove the independence of $a+b, ab$ and a^b on the representatives A, B chosen for the cardinals a, b. It is also easy to prove the following theorem:

Theorem 3.2. *For any cardinal numbers* a, b, c:
1. $a+(b+c) = (a+b)+c$;
2. $a+b = b+a$;
3. $a(b+c) = ab+ac$;
4. $a(bc) = (ab)c$;
5. $ab = ba$;
6. $a^b a^c = a^{b+c}$;
7. $a^c b^c = (ab)^c$;
8. $(a^b)^c = a^{b \cdot c}$.

Definition 3.3. *One says that the cardinal number* $a = \overline{\overline{A}}$ *is smaller or equal to the cardinal number* $b = \overline{\overline{B}}$ *and one writes* $a \leq b$ *if there is a subset* B' *within* B *so that* $A \sim B'$. *(We will write* $a < b$ *if* $a \leq b$, $a \neq b$.) *A binary relation* $\leq$ *was thus defined on the set* $U/\sim$ *of the cardinals of the universe* U.

Proposition 3.1. a, b, c *being cardinal number:*
1. $a \leq a$;
2. $a \leq b$ *and* $b \leq c$ *imply* $a \leq c$.
Indeed, $A \sim A$ occurs, therefore 1. is checked. From $A \sim B'$, $B' \subset B$,

$B \sim C'$, $C' \subset C$ it results $A \sim C''$, $C'' \subset C$ and 2. is proved.

By means of this proposition we assure ourselves that the binary relation $\leq$ we introduced above is reflexive and transitive; the theorem to follow will also ensure the anti-symmetry of this relation.

Theorem 3.3 (Bernstein). *Let be* $a = \overline{\overline{A}}$, $b = \overline{\overline{B}}$. *If* $a \leq b$ *and* $b \leq a$, *then* $a = b$.

Proof. According to the hypothesis there exist the bijections $f: A \to B'$ and $g: B \to A'$ where $B' \subset B$ and $A' \subset A$. We define a function $\varphi: \mathcal{P}(A) \to \mathcal{P}(B)$ by taking

(1) $\varphi(E) = A \cap \overline{[g(B \cap \overline{f(E)})]}$.

(Naturally, the complements refer to the set B and A.) It is easy to state the following property of *"monotony"*:

(2) $E \subset F \Rightarrow \varphi(E) \subset \varphi(F)$.

Let us consider the set $M \subset \mathcal{P}(A)$:

$$M = \{E: E \in \mathcal{P}(A) \text{ and } E \subset \varphi(E)\}$$

(We can say that M is the family of those subsets E of A which are not restricted by φ.) Obviously $\varnothing \in M$, hence $M \neq \varnothing$. Let us consider the union D of the unvoid family M. Since for any $E \in M$, $E \subset D$, (2) implies $\forall E \in M$, $E \subset \varphi(E) \subset \varphi(D)$. Taking the union and applying (2) we get

(3) $D \subset \varphi(D)$, $\varphi(D) \subset \varphi(\varphi(D))$.

It results that $\varphi(D) \in M$, hence $\varphi(D) \subset D$. Therefore, $\varphi(D) = D$; by applying (1) one finds:

$$D = A \cap \overline{[g(B \cap \overline{f(D)})]}.$$

It follows that the function f defined by

$$h(x) = \begin{cases} f(x), & \text{for } x \in D \\ g^{-1}(x), & \text{for } x \in A \backslash D \end{cases}$$

achieves a bijection from A to B, q.e.d.

Corollary 3.1. *On the set of cardinal numbers, $\leq$ is a relation of (partial) ordering.*

Example. We have $0 < n < \aleph_0 < c$.

The sets that do not have the cardinal 0, n are called infinite. The sets $\mathbf{N}$, $\mathbf{Z}$, $\mathbf{Q}$ have the cardinal $\aleph$. An infinite set has a characteristic property proved by Dedekind, that is it has the same cardinal with a proper subset of it.

For example $\overline{\overline{(0, 1)}} = \overline{\overline{[0, 1]}} = \overline{\overline{\mathbf{R}}} = c$.

Indeed, the function $f: (0, 1) \to \mathbf{R}$, defined by $f(x) = \frac{1-2x}{x(1-x)}$ is bijective. Therefore $\overline{\overline{[0, 1]}} = c$. The other equalities are analogously checked, too.

Theorem 3.4 (Cantor). *For any cardinal number a, we have $a < 2^a$.*

Proof. Let be $\overline{\overline{A}} = a$ and $\mathcal{P}_1(A) \subset \mathcal{P}(A)$ the family of the subsets of A, each of them made up of a single element. Then we have $A \sim \mathcal{P}_1(A)$. There follows that $a = \overline{\overline{A}} = \overline{\overline{\mathcal{P}_1(A)}} \leq \overline{\overline{\mathcal{P}(A)}}$. But $a \neq \overline{\overline{\mathcal{P}(A)}}$. Indeed, let us admit, by reductio ad absurdum, that $a = \overline{\overline{\mathcal{P}(A)}}$. Then there exists a bijections $h: A \to \mathcal{P}(A)$. We are defining $S = \{x: x \in A \text{ and } x \notin h(x)\}$. Since $S \subset A$ we have $S \in \mathcal{P}(A)$ and, because h is a bijection there exists $x_0 \in A$ such that $h(x_0) = S$. If $x_0 \in S$, then, from the definition of the set S, we have $x_0 \notin h(x_0) = S$. Therefore $x_0 \notin S$. But S is the set $h(x_0)$ and $x_0 \notin h(x_0)$ implies $x_0 \in S$. This contradiction shows that $a \neq \mathcal{P}(A)$. Therefore, $a < \mathcal{P}(A)$. Now let us show that $\mathcal{P}(A) = 2^a = \{f: A \to \{0, 1\}\}$. For this purpose, let us associate to each subset B of A the function $\xi_B: A \to \{0, 1\}$ defined by

$$\xi_B(x) = \begin{cases} 1, & \text{if } x \in B \\ 0, & \text{if } x \in C_A B \end{cases}$$

called *the characteristic function* of the subset B of A.

Let us consider now the map $\varphi: \mathcal{P}(A) \to \{f: A \to \{0, 1\}\}$, given by $\varphi(B) = \xi_B$. Then φ is a bijection. Then follows that $\overline{\overline{\mathcal{P}}} = 2^a$, q.e.d.

Corollary 3.2. *For any cardinal number* a:

$$a < 2^a < 2^{2^a} < \ldots$$

In particular:

$$\aleph_0 < 2^{\aleph_0} < 2^{2^{\aleph_0}} < \ldots$$

Theorem 3.5. $2^{\aleph} = c$.

Proof. Let us denote with A the set of the maps φ with the domain N and the range $\{0, 1\}$. The cardinal number of A is $2^{\aleph_0}$. Since the set of real numbers $[0, 1)$ has the cardinal number c, we will show that $A \sim [0, 1)$. We are considering the map $f: A \to [0, 1)$ given by $f(\varphi) = \sum_{n=1}^{\infty} \frac{\varphi(n)}{3^n}$. It is clear that this series is convergent and that f is injective. It follows that $2^{\aleph_0} \leq c$. Let be now a number $x \in [0, 1]$. It can be written uniquely under the form of $x = \sum_{n=1}^{\infty} \frac{x_n}{2^n}$, where x_n is 0 or 1 and $x_n = 0$ for an infinity of $n \in N$. We are defining the map $g: [0, 1] \to A$ by $g(x) = \varphi$ with $\varphi: N \to [0, 1]$ given by $\varphi(n) = x_n$; obviously, g is injective, which implies $c \leq 2^{\aleph_0}$. Now, by applying Bernstein's theorem, the assertion follows, q.e.d.

Some other proprieties of cardinal numbers and of the operations with them (some of them implying the axiom of choice) will be stated without proves.

Let A be an infinite set and $a = \overline{\overline{A}}$. Then:

1. $n + a = a$ (Particularly $n + \aleph_0 = \aleph_0$, $n + c = c$);
2. $a + a = a$ (Particularly $\aleph_0 + \aleph_0 = \aleph_0$, $c + c = c$);
3. $b \leq a \Rightarrow a + b = a$ (Particularly $\aleph_0 + c = c$);
4. $a^2 = a \cdot a = a$ (Particularly $\aleph_0^2 = \aleph_0$, $c^2 = c$);
5. $b \leq a \Rightarrow ab = a$ (Particularly $n\aleph_0 = \aleph_0$, $nc = c$, $\aleph_0 c = c$).

Cantor put forward *the hypothesis of alephs*: between a and 2^a there exist no other cardinal numbers; the particular case of $a = \aleph_0$ constitutes the famous *hypothesis of the continuum*. In 1963, I.P. Cohen succeeded in proving, [567], that this hypothesis of the continuum is an independent proposition from the Zermelo-Fraenkel axioms (the axiom of choice included) that are taken for the basis of the rigorous theory of sets. For this outstanding achievement, I.P. Cohen was awarded the Field prize (the equivalent of the Nobel prize in Mathematics), in Moscow, in 1966.

§4. Ordinal Numbers

Definition 4.1. *Let* $(A, \leq)$ *be a totally ordered set. We say that* $(A, \leq)$ *is well ordered if for any non-void subset* B *of* A *there exists an element* $b \in B$ *with the property* $\forall x \in B$, $b \leq x$. *The element* $b \in B$ *with the property* $\forall x \in B$, $b \leq x$, *is called the smallest element or the first element of* B, *or the prime element in* B. *Any well ordered set* $(A, \leq)$ *has a prime element and it is a unique one.*

Any totally ordered finite set is obviously well ordered. The set $\mathbf{N}$ with usual ordering is well ordered; the sets $\mathbf{Q}$ and $\mathbf{R}$ are totally ordered, but not well ordered.

Let $(A, \leq)$ be a totally ordered set and $a \in A$. The subset $A_a = \{x \in A : x < a\}$ is called *segment of A determined by the element* a. As we can see, $a \notin A_a$.

For the well ordered sets the following theorem is particularly important in applications.

Theorem 4.1 (The Principle of Transfinite Induction). *Let* $(A, \leq)$ *be a well ordered set. If* A' *is a subset of* A *having the properties:*

1. A' *contains the prime element of* A*;*

2. For any segment $A_a \subset A'$, *we have* $a \in A'$, *then the set* A' *coincides with the set* A.

Proof. Let us suppose by reductio ad absurdum that the subset A' of A has the properties 1. and 2. from the enunciation of the theorem, but that $A' \neq A$. Then $A \setminus A'$ is a non-void subset of the well ordered set A. It has a prime element b which is not to be found in A'. The segment A_b is included into A'. According to 2., it results that $b \in A'$. That is absurd, q.e.d.

Conversely, we have:

Theorem 4.2. *Let* $(A, \leq)$ *be a totally ordered set having a prime element. If the only subset* A' *of a which satisfies conditions* $1°$ *and* $2°$ *in theorem* 4.1 *is* A, *then* $(A, \leq)$ *is well ordered.*

Proof. Let us suppose that $B \neq \varnothing$ is a subset of A which has no prime element. Then $A' = A \setminus B$ is a subset of A containing the prime element of A. If A_a is a segment of A', then $a \in A'$. In the contrary case, a would be the prime element of B. By applying the principle of transfinite induction, it results that $A' = A$. That is $A \setminus B = A$, therefore $B = \varnothing$, in contradiction with $B \neq \varnothing$, q.e.d.

The two last theorems show that the principle of transfinite induction is equivalent to the well ordering. We will say theorems proved with the help of this principle are proved "by recurrence".

Definition 4.2. *Two totally ordered sets* (A, ρ) *and* (B, σ) *are called similar and we write* $A \approx B$ *if there exists a bijection* $f : A \to B$ *with the property*

$$\forall \, x, y \in A \quad x \rho y \Rightarrow f(x) \sigma f(y).$$

For such a bijection we shall use the name similitude or similarity.

Let us consider now a "universe" V of totally ordered sets. Within this universe, $\approx$ is obviously an equivalence. The class of equivalence with respect to $\approx$ of a totally ordered set $(A, \leq)$ is denoted by ord A and is called *the type of order* of $(A, \leq)$. If $(A, \leq)$ is well ordered, any other element $(B, \leq)$ in ord A will have the same property, and it is said that ord A constitutes *the ordinal number* of $(A, \leq)$.

We may accept that the void set is well ordered; instead of the notation ord $\varnothing$, the notation 0 is preferred. If n is a natural number, the set $\{x : x \in \mathbf{N}$ and $x < n\}$ usually ordered, is well ordered and no confusion appear if we denote its ordinal number by n, too. If $\mathbf{N}$, $\mathbf{Q}$, $\mathbf{R}$, with their usual orderings are elements of the universe V, we renote: ord $\mathbf{N} = \omega$, ord $\mathbf{Q} = \eta$, ord $\mathbf{R} = \lambda$; ω is an ordinal number, but η and λ are only types of order.

On the set $V / \approx$ of the types of order, particularly on the set $\mathbf{Z}$ of the ordinal numbers, the operations of addition and multiplication can naturally be defined. The Arithmetic of the types of order is complicated enough due to the fact that the operation of addition and multiplication are not commutative.

We will define on $\mathbf{Z}$ a relation of partial order, $\prec$, as follows. Let (A, ρ) and (B, σ) be two well ordered sets and $\alpha = \text{ord } A$, $\beta = \text{ord } B$. We will write down $\alpha < \beta$ if there exists $x \in B$ with the property $A \approx B_x$. We put $\alpha \leq \beta$ if $\alpha < \beta$ or $\alpha = \beta$. It is easy to check that $\alpha \preceq \beta$ depends only on the ordinal numbers α and β and not on their representatives (A, ρ), (B, σ).

Certain properties in applications are given by the following two theorems:

Theorem 4.3. *If $(A, \leq)$ is a well ordered set and $f: A \to A$ is a similitude, then $\forall x \in a$, $x \leq f(x)$.*

Proof. Let us suppose, by reductio ad absurdum, that there exists $x \in A$ with the property $f(x_0) < x_0$. Let a be the smallest element x_0 with this property. We have $f(a) < a$. By applying the similitude f we have $f(f(a)) < f(a)$. Therefore, $f(a) = b$ has the property $f(b) < b$. And now $f(a) < a$ that is $b < a$; it follows that a is not the smallest element with this property, which contradicts the property of minimality of a, q.e.d.

Theorem 4.4. *If $(A, \leq)$ and $(B, \preceq)$ are two well ordered sets, then:*
1. *A is not similar to any segment A_a of it;*
2. *$A_x \approx A_y \Rightarrow x = y$;*
3. *$A \approx B$ implies the existence of a unique similarity $f: A \to B$.*

Proof. 1. Let us admit by reductio ad absurdum that $f: A \to A_x$ achieves a similarity for an $x \in A$. According to the theorem 4.3, $x \leq f(x)$. But $f(x) \in A_x$. Therefore, $f(x) < x$. These last relations are contradictory.

2. From $A_x \approx A_y$, $x \neq y$ we can have $y < x$, then A_y is similar to a segment A_x of it, in contradiction with 1.

3. Let f, $g: A \to B$ be two similitudes. It results that $f^{-1} \circ g: A \to A$ is a similitude. According to the theorem 4.3, $\forall x \in A$, $x \leq f^{-1} \circ g(x)$. Then $f(x) \preceq g(x)$. By replacing the role of f with g we also have $g(x) \leq f(x)$. Therefore, for any x in A, $f(x) = g(x)$, q.e.d.

By admitting the axiom of choice we can establish the property of trichotomy of the ordered set $(Z, \preceq)$.

The axiom of choice. *For any non-void family $\mathscr{F}$ of non-void sets disjoint two by two, there exists a set A which has an element and only one in common with each set in $\mathscr{F}$.*

Let (M, μ) be a partially ordered set. An element $m \in M$ is called *maximal* if $m \mu x$

implies $m = x$. An element $n \in M$ is called *minimal* if $x\mu n$ implies $x = n$. A *chain* in (M, μ) is a subset $A \subset M$ together with the induced order α, so that (A, α) is totally ordered.

Some other auxiliary matters. A family $\mathcal{F}$ of sets is called of a *finite character* if for each set A we have $A \in \mathcal{F}$ if and only if any finite part of A belongs to $\mathcal{F}$. These notion may be applied to the families of sets on which the inclusion relation was chosen as relation of partial order.

It is proved that the axiom of choice is equivalent to each of the following propositions:

Tukey's Lemma. *Any non-void family of finite character has a maximal element.*

Hausdorff's Principle of Maximality. *Any partially ordered non-void set contains a maximal chain.*

Zorn's Lemma. *Any partially ordered non-void set in which each chain is upper bounded has a maximal element.*

Theorem of Well Ordering (Zermelo). *Any set can be well ordered.*

But we draw attention to the fact that this theorem ensures only *the existence* of a well ordering on an arbitrary set A, without indicating an effective *construction* of such a relation; the theorem ascertains no *uniqueness* of a well ordering on A, and such an assertion is not correct, either. A cardinal number a being given we can put in evidence a representative of it A (so $\overline{\overline{A}} = a$); further on, different well orderings on A able to lead to *diverse* ordinal numbers benefitting from the notation ord$A$ can be imagined. We keep then in mind that the notation ord$A$ is an ambiguous one since it does not refer to a certain well ordering on A, but this ambiguity does not create confusions.

Theorem of the Cartesian Product. *The Cartesian product of a family of non-void sets is non-void.*

On the basis of the axiom of choice, we can also prove:

Theorem 4.5. *For any pair of ordinal numbers α and β, one and only one of the relations $\alpha < \beta$, $\alpha = \beta$, $\beta < \alpha$ holds.*

Proof. Theorem 4.4 states that at most one of these relations occurs. Let us show that at least one of them holds. If $\alpha = \text{ord}A$, $\beta = \text{ord}B$, (A, ρ) and (B, σ) being well ordered, let $\mathscr{F}$ be the family of all similitudes from the segments of A or A to the segments of B or B. We have $\{(a, b)\} \in \mathscr{F}$, a being the first element in A and b the first element in B. Therefore, $\mathscr{F} \neq \varnothing$. According to Hausdorff's principle of maximality there exists a maximal chain $\mathscr{L} \subset \mathscr{F}$. Let be $h = U_{\mathscr{L}}$. It can be easily proved that h belongs to $\mathscr{F}$, If $\text{dom}h$ and $\text{ran}h$ are the segments A_x and B_y of A and B, respectively, then $h \cup \{(x, y)\}$ may be added to $\mathscr{L}$ and this contradicts the maximality of $\mathscr{L}$. The situation $\text{dom}h \neq A$ and $\text{ran}h \neq B$ is excluded. From $\text{dom}h = A$ and $\text{ran}h = B$ it follows $\alpha = \beta$; from $\text{dom}h = A$ and $\text{ran}b = B_y$ it results $\alpha \prec \beta$; the last eventuality $\text{dom}h = A_x$ and $\text{ran}h = B$ leads to $\beta \prec \alpha$, q.e.d.

We deduce:

Theorem 4.6. *The set* $(Z, \preceq)$ *is totally ordered; for any* a *in* Z, $\text{ord}Z_a = a$.
Proof. Let be $a = \text{ord}A$. We will construct a map $f: Z_a \to A$ that will be a theorem 4.4 there exists a unique y in A, so that $Z_x = A_y$; we define $y = f(x)$.

Theorem 4.7. *For any cardinal* a *there exists at least an ordinal number* α, *so that* $a = \overline{\overline{Z}}_\alpha$.

Proof. Let be $\overline{\overline{A}} = a$; on the basis of the axiom of choice there exists a well ordering ρ on A. We note $\alpha = \text{ord}A$, and by means of the previous theorem we deduce that (A, ρ) is similar to $(Z_\alpha, \preceq)$. We deduce $A \sim Z_\alpha$ and, subsequently, $a = \overline{\overline{Z}}_\alpha$.

But we once again draw attention to the fact that, in general, α is not unique.

Remark. Let us admit that the universe V is ample enough to contain at least a countable set and at least an uncountable one (and, of course, all the segments of these sets). On the basis of the well ordering of Z there will exists the smallest ordinal numbers ω, W, so that $Z_{\overline{\omega}}$ may be countable, and Z_W may be uncountable, respectively. The continuum hypothesis is now expressed by $\overline{\overline{Z}}_W = c$ and is equivalent to the following proposition: for $A \subset \mathbb{R}$, A is finite, countable or equipotent to $\mathbb{R}$.

The following theorem illustrates the role of ordinal numbers for the study of cardinal numbers.

Theorem 4.8. *Let be a universe* U *and the family* $U/\sim$ *of its cardinals. Then*

$(U/\sim, \leq)$ *is a totally and well ordered set.*

Proof. By specifying (according to Zermelo's theorem) good orderings $\leq_A$ for each element A in U, we get the universe $U' = \{(A, \leq_A): A \in U\}$ of well ordered sets. Let $\mathcal{Z}$ be the set of the ordinals of U'. Now, for a cardinal number a there exists, according to the previous theorem, an ordinal α, so that $a = \overline{\overline{\mathcal{Z}_\alpha}}$. If the ordinal β corresponds to another cardinal b, it is obvious that $a \leq b \Leftrightarrow \alpha \preceq \beta$ holds. According to the Theorems 4.6 and 4.7 $(U/\sim, \leq)$ is totally and well ordered.

Exercises

1. Prove that, if for binary relations α, β, γ there exists one of the compositions $(\alpha \circ \beta) \circ \gamma$ or $\alpha \circ (\beta \circ \gamma)$ or , then there also exists the other one, and that these compositions coincide.

2. Prove that $\overline{\overline{\mathbf{Z}}} = \overline{\overline{\mathbf{N}}}$ and $\overline{\overline{\mathbf{Q}}} = \overline{\overline{\mathbf{N}}}$.

3. Knowing that $\overline{\overline{A}} = n \in \mathbf{N}$, define $\overline{\overline{\mathcal{C}(A)}}$, where $\mathcal{C}(A)$ is the set of equivalence relations on A.

4. On a set A, an equivalence relation ε and a binary operation $\perp : A^2 \to A$ are precised. In what conditions does the equality $\varepsilon_x T \varepsilon_y = \varepsilon_{x \perp y}$ precise a binary operation $T : (A/\varepsilon)^2 \to (A/\varepsilon)$?

5. For $\omega = \text{ord } \mathbf{N}$, prove that:

$$1 + \omega = \omega \neq \omega + 1 \text{ and } \omega \cdot 2 \neq 2 \cdot \omega.$$

6. A set X being fixed, let us take the universe U_X of the pairs (A, α), where $\alpha : A \to X$. Draw up a theory of the "*cardinals*" of this universe, taking into account the equivalence ε given by $(A, \alpha) \varepsilon (B, \beta)$ when a bijection $f : A \to B$ so that $\beta \circ f = \alpha$ and $\alpha \circ f^{-1} = \beta$ hold.

CHAPTER II

ARITHMETIC

Contemporary Mathematics uses two methods for its edification: the constructive method and the axiomatic one. The first method, the constructive one, consists in defining the notions involved in mathematical theories in a direct, effective manner, as it has been done in the set theory in the preceding chapter. We will apply this idea to the construction of some set of numbers. The axiomatic method will be used in the other chapters.

§1. The set N of Natural Numbers

The problems discussed in the previous chapter regarding cardinal numbers find their first application in defining the set N of natural numbers. The basic idea, in constructing this set, has been suggested by Frege and Russel, and that is why it is called the *Frege -Russell theory of natural numbers.*

By using the void set $\varnothing$, whose uniqueness has already been remarked, let us consider the sequence:

(1.1) $\varnothing, \{\varnothing\}, \{\varnothing, \{\varnothing\}\}, \{\varnothing, \{\varnothing\}, \{\varnothing, \{\varnothing\}\}\}, \ldots$

in which each term, beginning with the second one, is the set whose elements are all the preceding terms in the sequence.

Definition 1.1 *We call natural numbers the cardinal numbers of the sets in the sequence (1.1).*

We note the natural numbers with $0, 1, 2, \ldots$ and their set with N, respectively.

From the way of constructing the sequence (1.1) it results

$$n = \overline{\overline{\{0, 1, 2, \ldots, n-1\}}}$$

We recall that the cardinal n sets are called finite, those of cardinal $\aleph_0 = \overline{\overline{N}}$ being called countable. On the basis of the definition of order relation between cardinal numbers it follows that $0 < n < \aleph_0$. Therefore: the set N is not finite and any element $n \in N$ is smaller than $\overline{\overline{N}}$.

It is obvious that the order relation on N, determined by the order between cardinal numbers is equivalent to the relation of natural order given by the succession in the sequence (1.1). Since the ordering of cardinal numbers, according to the Theorem 4.9 (Chapter I), is good, there follows:

Theorem 1.1. *The set* $(N, \leq)$ *is well ordered.*

Particularizing the Theorem 4.1. in the preceding chapter, we deduce :

Theorem 1.2. *For the set* $(N, \leq)$ *the principle of transfinite induction holds. Namely: if* $M \subset N$ *has the following properties:*
1. $0 \in M$;
2. *For any segment* $M_n \subset M$, *we have* $n \in M$,
then $M = N$.

In this case, it will be also called *the principle of complete induction.*

For the set $(N, \leq)$ it is convenient for us to give another form to this principle. In order to reach this target let us call *successive* two natural numbers n and n' which are the cardinal numbers of two successive terms in the sequence (1.1); it is also said that n' is *the successor* of n. From their construction way it results that between n and n' there exist no other natural numbers. Then, if n is given in N, its successor n' is uniquely determined and belongs to the set N.

Theorem 1.3. *The principle of complete induction is equivalent to the following assertion: "If for a set* $M \subset N$ *we have*
1. $0 \in M$ *and*
$2'$. $n \in M \Rightarrow n' \in M$,
then $M = N$".

Proof. It is easy to see that set M with the properties 1. and $2'$ has the properties 1. and 2., as well, and therefore the theorem 1.2 implies the theorem 1.3. The converse implication results on the basis of the observation that any non-null natural number has a predecessor and (1. and 2.) $\Rightarrow$ (1. and $2'$).

Now, let us notice that we have:

Theorem 1.4. *Addition and multiplication of cardinal numbers induce operations of addition and multiplication on the set* N *of natural numbers.*
Proof. We have to show that the sum of two natural numbers m and n, defined

as a sum of two cardinal numbers $m + n$, gives us a natural number for any m and n. The same property must be proved for the product.

Since the addition of cardinal numbers is associative and $m + 0 = 0 + m = m$, $m \in \mathbf{N}$, let us observe that we have

$$m + n' = m + (n + 1) = (m + n) + 1 = (m + n)'.$$

By fixing any natural number m, let M be the set of those natural numbers n for which the sum of cardinal numbers $m + n$ is a natural number. From $m + 0 = m$ there results $0 \in M$. Then, if $n \in M$, there follows that $m + n$ is in $\mathbf{N}$ and $m + n' = (m + n)'$ is in $\mathbf{N}$. Since m was "any", there results $m + n \in \mathbf{N}$ for any m and n in $\mathbf{N}$.

In the case of the product of cardinal numbers let us observe that we obviously have

$$m \cdot 0 = 0, \; m \cdot 1 = m, \; m \cdot n' = m \cdot (n + 1) = m \cdot n + m.$$

Let us fix any number m and let M be the set of those natural numbers n for which $m \cdot n$ is a natural number. From $m \cdot 0 = 0$, $0 \in M$ is deduced, and from $m \cdot n \in M$ we have $m \cdot n' = m \cdot n + m \in M$. There follows that $n \in M$ implies $n' \in M$. By applying the principle of induction there results $M = \mathbf{N}$. And since m was arbitrary, we can state that $m \cdot n \in \mathbf{N}$, $\forall m$, $n \in \mathbf{N}$, q.e.d.

The most important result as regards the set of natural numbers $\mathbf{N}$ is given by:

Theorem 1.5. *The set* $\mathbf{N}$ *of natural numbers together with the operation of addition and multiplication, given by theorem 1.4, is a semi-domain of integrity.*

Proof. We have to show that for any m, n, $p \in \mathbf{N}$ the following equalities hold: $m + (n + p) = (m + n) + p$; $\; m + 0 = m$; $\; m + n = n + m$; $\; m \cdot (n \cdot p) = (m \cdot n) \cdot p$; $m \cdot n = n \cdot m$; $(m + n) \cdot p = m \cdot p + n \cdot p$ and $m \cdot n = 0 \Rightarrow (m = 0$ or $n = 0)$.

Except the last proprieties, the other ones have been proved either with the operations with cardinal numbers, or in the theorem 1.4. In order to justify the property $m \cdot n = 0 \Rightarrow (m = 0$ or $n = 0)$ let us observe that the Cartesian product of two sets is void when at least one of the sets is the void one. By applying this observation to the Cartesian product of two terms in the sequence (1), there immediately results the assertion, q.e.d.

Further on, we intend to show that the addition and multiplication operations in N are compatible with the structure of well ordering of N.

Theorem 1.6. *We have $m \leq n$ if and only if there exists a natural number p so that $m + p = n$.*

Proof. Let be $m = \overline{M}$ and $n = \overline{N}$. The inequality of the cardinals returns to the existence of a bijection $f: M \to M' \subset N$ We evidence the set $P = N \setminus M'$ and its cardinal p. From $P \subset N$ there follows that p is a natural number and the definition of the addition of cardinals numbers ensures $m + p = n$.

Let us suppose that $m + p = n$. We will show by complete induction related to p that such an equality implies $m \leq n$. For $p = 0$ we deduce $n = m + 0 = n$, therefore $m = n$. If from $m + k = n$ there follows $m \leq n$, the equality $m + k' = r$ becomes $(m + k)' = r$, therefore $n' = r$, that is $n \leq r$. From $m \leq n$ and $n \leq r$ there follows $m \leq r$ etc.

Remark. The natural number p, which exists according to the theorem in the hypothesis $m \leq n$, is unique, is noted with $n - m$ and is called *the difference of natural numbers n and m.*

Theorem 1.7. *We have:*
1. $\forall p \in N$, $m \leq n \Leftrightarrow m + p \leq n + p$;
2. $\forall p \in N \setminus \{0\}$, $m \leq n \Rightarrow mp \leq np$;
3. $\forall n \in N \setminus \{0\}$, $m \leq mn$.

Proof. 1. From $m \leq n$ and the theorem 1.6 there results $n = m + q$, $q \in N$. Then $n + p = (m + p) + q$. By applying again the theorem 1.6, there results $m + p \leq n + p$. Conversely, $m + p \leq n + p \Rightarrow n + p = (m + r) + p$; $r \in N$. By induction, we get $n = m + r$. By applying once again the theorem 1.6 , we have $m \leq n$.

Point 2. is analogously proved.

3. If $n \neq 0$, p exists so that $n = p'$. It follows $mn = mn' = mp + m$ and therefore $m \leq mn$.

Remark. We have $m = mn$ if and only if $m = 0$ or $n = 1$.

Theorem 1.8 (Archimedes'axiom). *For any natural numbers m and p, with $m \neq 0$, there exists a natural number n so that $nm > p$.*

Proof. We apply the induction with respect to p. We remark that for $p = 0$ the property is checked with $n = 1$. If now $m \neq 0$ is "any", fixed, and $p \neq 0$, and there exists $n \in N$ so that $nm > p$, then $n'm > p'$, q.e.d.

Theorem 1.9 (The algorithm of dividing with remainder in N). *If n, m are natural numbers with $m \neq 0$, there exist unique natural numbers q (called quotient) and r (called remainder) so that $n = mq + r$ and $r < m$.*

Proof. Let be $A = \{x/x \in N$ and $mx \leq n\}$; by the preceding theorem, A is a proper subset of N. We have $0 \in A$ and in order not to reach by induction the false conclusion $A = N$, there must exist $q \in A$ so that $q' \notin A$. From $q \in A$ there results the existence of an $r \in N$ so that $mq + r = n$. Since $q' \notin A$, it results $m(q+1) > mq + r$, therefore $r < m$. To prove the uniqueness of (q, r) let us suppose that there also exists $(h, k) \neq (q, r)$ so that $n = mh + k$ and $k < m$. There follows that $h \in A$, $h' \notin A$. If, for example, $h < q$ it follows $h' \leq q$ and, therefore, $mh' \leq mq \leq n$ in contradiction with $h' \notin A$. Analogously from $q < h$ we would contradict $q' \notin A$. Therefore, $h = q$. There follows $n = mq + r = mq + k$, therefore $r = k$, q.e.d.

Remarks. The notion of natural number is obviously objective and non-contradictory. But here it is not a question of abstracting this notion from the objective reality but of constructing it within Mathematics. The fact that we can not prove the objective character of the natural number within Mathematics constitutes a secondary, non-edifying aspect. Frege's conviction is firm in this context:

"Within Mathematics, we do not deal with objects we know as something extraneous, exterior, by means of our senses, but with objects given directly to the reason, which can fully intuit them as something characteristic, of its own. In spite of the above mentioned fact, or, better said, just due to it, these objects are not subjective airy visions. There is nothing more objective than the laws of Arithmetic" (see [172]).

§2. The Set Z of Integers

Starting from the set N of natural numbers, we shall construct the set Z of the integer numbers with the idea of completing the additive semi-group $(N, +)$ so that we may get a group on the new set of numbers.

Let us consider $\mathcal{H} = (N \times N)$. We define a binary relation $\sim$ on $\mathcal{H}$ by:

$$(m, n) \sim (m_1, n_1) \Leftrightarrow m + n_1 = m_1 + n.$$

Proposition 2.1. *The relation* $\sim$ *is an equivalence on the set* $\mathcal{H}.$

Indeed, $(m, n) \sim (m, n)$ occurs because $m + n = n + m$. From $(m, n) \sim (m_1, n_1)$ it results $(m_1, n_1) \sim (m, n)$, because $m + n_1 = m_1 + n \Rightarrow m_1 + n = m + n_1$. We also have $[(m, n) \sim (m_1, n_1) \text{ and } (m_1, n_1) \sim (m_2, n_2)] \Rightarrow (m, n) \sim (m_2, n_2)$, q.e.d.

Proposition 2.2. *The following properties hold:*
1. $\forall m \in \mathbf{R}, (m, m) \sim (0, 0)$;
2. $(m, n) \sim (m - n, 0)$ for $m > n$;
3. $(m, n) \sim (0, n - m)$ for $n > m$.
The proofs of these properties are immediate.

Let be $\mathbf{Z} = \mathcal{H}/\sim;$; an element of this set is called *integer number*, or simply *integer*, and $\mathbf{Z}$ is called the set of the integer numbers. As $\mathbf{Z}$ is the set of the classes of equivalence related to $\sim$, we will denote by $[m, n]$ the integer number determined by (m, n).

Proposition 2.3. *If* $(m, n) \sim (m_1, n_1)$ *and* $(p, q) \sim (p_1, q_1)$, *then*
1. $(m + p, n + q) \sim (m_1 + p_1, n_1 + q_1)$;
2. $(mp + nq, mq + np) \sim (m_1 p_1 + n_1 q_1, m_1 q_1 + n_1 p_1)$.
Proof. 1. From $m + n_1 = m_1 + n$ and $p + q_1 = p_1 + q$ are immediately results $m + p + n_1 + q_1 = m_1 + p_1 + n + q$. One analogously checks 2.

Proposition 2.4. *The applications* $+ : Z^2 \to Z$ *and* $\cdot : Z^2 \to Z$ *defined by* $[m, n] + [p, q] = [m + p, n + q]$ *and* $[m, n] \cdot [p, q] = [mp + nq, mq + np]$, *respectively, do not depend on the representatives* (m, n) *and* (p, q) *of the integers* $[m, n]$ *and* $[p, q]$, *respectively.*

Indeed, the preceding equalities are immediate consequences of Proposition 2.3. Therefore, $+$ and $\cdot$ are binary operations on the set Z; we will call them *addition* and *multiplication* of the integers.

Theorem 2.1. *The set* Z *of the integers together with the operations of addition and multiplication is a domain of integrity.*

Proof. 1) $(Z, +)$ is an *Abelian group.* Indeed,

$$[m, n] + [[p, q] + [r, s]] = [m, n] + [p + r, q + s] = [m + q + r, n + q + s] =$$
$$= [[m, n] + [p, q]] + [r, s].$$

The neutral element is $[0, 0]$, $[m, n] + [0, 0] = [m, n]$. The opposite $-[m, n]$ of the number $[m, n]$ is the integer $[n, m]$ since the following equality holds $[m, n] + [n, m] = [m + n, n + m] = [0, 0]$. Finally, $[m, n] + [p, q] = [p, q] + [m, n]$.

2) $(Z, \cdot)$ is a *unitary, commutative semi-group*.

$$[[m, n] \cdot [p, q]] \cdot [r, s] = [mp + nq, mq + np] \cdot [r, s] =$$
$$= [mpr + nqr + mqs + nps, mps + nqs + mqr + npr] = [m, n] \cdot [[p, q] \cdot [r, s]]$$

$[1, 0]$ is the unit: $[m, n] \cdot [1, 0] = [m, n]$.

Finally, $[m, n] \cdot [p, q] = [mp + nq, mq + np] = = [p, q] \cdot [m, n]$.

3) Multiplication is *distributive* in relation with addition.

$$[m, n] \cdot [[p, q] + [r, s]] = [m, n] \cdot [p + r, q + s] =$$
$$= [mp + mr + nq + ns, mq + ms + np + nr] = [m, n] \cdot [p, q] + [m, n] \cdot [r, s].$$

4) The absence of the divisors of zero. Let be $[m, n] \cdot [p, q] = [0, 0]$; if $[m, n] \neq [0, 0]$ occurs, $m \neq n$. We are fixing the idea supposing, for instance, $m > n$; therefore there exists $r \neq 0$ such that $m = n + r$ and so $[m, n] = [r, 0]$. The initial equality becomes $[r, 0] \cdot [p, q] = [0, 0]$ that is $rp = rq$. There follows that $p = q$, that is $[p, q] = [0, 0]$, q.e.d.

We will denote again the element $[0, 0]$ in Z by 0; we will subsequently find that this "identification" does not generate confusions. Let be $Z_+ = \{[m, n]: m > n\}$ and $Z_- = \{[m, n]: m < n\}$.

Proposition 2.5. 1. $Z = Z_+ \cup \{0\} \cup Z_-$.

2. *Any integer $[m, n]$ belongs to one and only to one of the sets Z_-, $\{0\}$, Z_+.*

Indeed, 1. is obvious on the basis of the definitions of the sets Z_-, $\{0\}$ and Z_+.

2. Let $[m, n]$ be any integer, and (m, n) one of its representatives. If $[m, n] \sim [m', n']$ and $m > n$, then $m' > n'$ because $m + n = n + m'$ and $m + n' > n + n' \Rightarrow n + m' > n + n' \Rightarrow m' > n'$. Therefore, the property $m > n$ does not depend on the representatives, but only on the integer $[m, n]$. Then by applying the trichotomy of natural numbers, we either have $m > n$ and then $[m, n] \in Z_+$ or we have $m = n$ and then $[m, n] = 0$, or $m < n$ and then $[m, n] \in Z_-$, but only one of the situation can be presented, q.e.d.

The set Z_+ will be called *the set of the positive integers,* and Z_- *the set of the negative integers.*

Proposition 2.6. *The operation of addition,* $+$, *in* Z *induces on* $Z_+ \cup \{0\}$, *an operation also denoted by* $+$, *and the operation of multiplication on* Z *induces on* $Z_+ \cup \{0\}$ *an operation also denoted by* $\cdot$; *the triple* $(Z_+ \cup \{0\}, +, \cdot)$ *is a semi-domain of integrity.*

Indeed, $Z_+ \cup \{0\}$ is a stable part related to $+$ and $\cdot$. Then, the restrictions of the operation $+$ and $\cdot$ to this subset give a structure of semi-domain of integrity, as we may immediately see.

Definition 2.1. *We say that the integer* $[m, n]$ *is smaller than the integer* $[p, q]$ *if the difference* $[p, q] - [m, n] \in Z_+$. *We denote this relation with the sign* $<$ *and will put* $[m, n] \leq [p, q]$ *if* $[m, n] = [p, q]$ *or* $[m, n] < [p, q]$.

Theorem 2.2. *The set* $(Z, \leq)$ *is totally ordered.*

Proof. We obviously have $[m, n] \leq [m, n]$. Let now $[m, n]$ and $[p, q]$ be two integers. According to the proposition 2.5, the integer $[p, q] - [m, n]$ belongs either to Z_+, to $\{0\}$, or to Z_-, but only to one of these disjoint sets; if this number belongs to Z_+, then $[m, n] < [p, q]$. If it is zero, then $[m, n] = [p, q]$. Finally, if it is in Z_-, then $[p, q] < [m, n]$.

The set $Z_+ \cup \{0\}$ is well ordered by means of $\leq$. Then:

Theorem 2.3. 1. *There exists a unique similitude* f *of the ordered sets* $(N, \leq)$ *and* $(Z_+ \cup \{0\}, \leq)$;

2. f *is an isomorphism of the integrity semi-domains* $(N, +, \cdot)$ *and* $(Z_+ \cup \{0\}, +, \cdot)$.

Proof. 1. Let us show first of all that $[m, n] \in Z_+ \cup \{0\}$ if and only there exists $p \in N$ so that $[m, n] = [p, 0]$. Indeed,

$$[m, n] \in Z_+ \cup \{0\} \Rightarrow m \geq n \Rightarrow \exists p \in N, \ m + p = n.$$

Then $[m, n] = [p, 0]$ holds. Conversely, $[p, 0] \in Z_+ \cup \{0\}$ and $[m, n] = [p, 0]$ belongs to this set.

We have now $[m, 0] \leq [p, 0] \Leftrightarrow m \leq p$.

The application $f: \mathbf{N} \rightarrow \mathbf{Z}_+ \cup \{0\}$, $f(m) = [m, 0]$ is a bijection possessing the property $m \leq n \Rightarrow f(m) \leq f(n)$. Therefore, f is a similitude. According to the Theorem 4.4 in Chapter I, this similitude is unique.

2. We notice that $f(m + n) = [m + n, 0] = [m, 0] + [n, 0] = f(m) + f(n)$ and $f(m \cdot n) = [m \cdot n, 0] = [m, 0] \cdot [n, 0] = f(m) \cdot f(n)$, q.e.d.

By means of this isomorphism f, we identify the elements in $\mathbf{Z}_+ \cup \{0\}$ with the natural numbers to which they correspond. After this identification, $\mathbf{N}$ appears as a subset of $\mathbf{Z}$ therefore we conceive $\mathbf{Z}$ as an extension of $\mathbf{N}$. By the following theorem we will prove that this extension is "*minimal*".

Theorem 2.4. *If $V = (V, \oplus, \circ)$ is a ring, $\mathbf{N} \subset V$, and the restrictions of the operations $\oplus$ and $\circ$ to $\mathbf{N}$ coincide with the operations of addition and multiplication of natural numbers, then there exists an injective homomorphism h from the ring $(\mathbf{Z}, +, \cdot)$ of the integers to V.*

The proof is immediate by putting $h(m, n) = m \oplus (-n)$, where $(-n)$ is the opposite of n in V.

Theorem 2.5. *The ring $(\mathbf{Z}, +, \cdot)$ ordered by $\leq$ is "Archimedean", that is, for any $a \in \mathbf{Z}_+$ and $b \in \mathbf{Z}$, there exists a natural number n so that $n \cdot a > b$.*

Proof. We may consider $a = [m, 0]$, with $m \neq 0$. If $b = [p, q]$, the inequality $n \cdot a > b$ turns to $n \cdot m + q > p$. If $p \leq q$, we can take $n = 0$. Otherwise, there exists h so that $p = q + h$ and, since $\mathbf{N}$ is Archimedean, we find n so that $n \cdot m > h$, q.e.d.

It is observed that, regarding $\mathbf{Z}$ as an extension of $\mathbf{N}$, we note its elements with small letters of Latin alphabet, as well.

§3. Divisibility in the Ring of Integers

We are considering as known the function of absolute value $| \ |: \mathbf{Z} \rightarrow \mathbf{N}$, where $|x| = \max \{-x, +x\}$, and its main properties.

Theorem 3.1 (The algorithm of division with remainder in $\mathbf{Z}$). *For any integers a, b with $b \neq 0$, there exist unique integers q (quotient) and r (remainder) so that $a = b \cdot q + r$ and $0 \leq r < |b|$.*

Proof. According to the Theorem 1.9 there will exist the natural numbers h, k

so that $|a| = |b| \cdot h + k$ and $k < |b|$. There appear only the following possibilities and, corresponding to them, the expression of q and r:

$$
\begin{aligned}
&a \geq 0, \, b > 0 \qquad &&q = h, \, r = k \\
&a \geq 0, \, b = 0 \qquad &&q = -h, \, r = k \\
&a < 0, \, b < 0, \, k = 0 \qquad &&q = -h, \, r = 0 \\
&a < 0, \, b > 0, \, k \neq 0 \qquad &&q = -(1+h), \, r = b - k \\
&a < 0, \, b < 0, \, k = 0 \qquad &&q = h, \, r = 0 \\
&a < 0, \, b < 0, \, k \neq 0 \qquad &&q = h - 1, \, r = b - k.
\end{aligned}
$$

One considers the binary relation $|$ on $\mathbf{Z}$ defined by

$$x \,|\, y \Leftrightarrow \exists\, a \in \mathbf{Z}, \, y = a \cdot x.$$

The symbol $x \,|\, y$ is read *"x divides y"*. The converse relation, $|^{-1}$, benefits from the notation: $\vdots$; $x \vdots y$ means $y \,|\, x$ and is read *"x is divided by y"*. The notation $x \,\slash\!\!\!| \, y$ is also used, with the significance of: x does not divide y.

Proposition 3.1. *The binary relation $|$ is a "pre-order" on $\mathbf{Z}$, that is it is reflexive and transitive. From $x \,|\, y$ and $y \,|\, x$ there results that x and y are "associated integers", that is $x = y$ or $x = -y$.*

It is found that the restriction to $\mathbf{N}$ of the relation $|$ is a partial ordering for which 0 is a maximal element. The ordering $|$ of $\mathbf{N}$ is not total because, for example, neither $2 \,|\, 3$, nor $3 \,|\, 2$.

Theorem 3.2. *If $x \,|\, y$ then $y = 0$ or $|x| \leq |y|$.*
Indeed, from $x \,|\, y$ there follows $y = a \cdot x$ and $|y| = |a| \cdot |x|$. Point 3. of the Theorem 1.7 leads us quickly to the conclusion.

Theorem 3.3.
1. $x \,|\, y \Rightarrow x \,|\, (y \cdot z)$;
2. $x \,|\, y$ and $x \,|\, z \Rightarrow x \,|\, (y + z)$.
We also record the following formulae easy to prove:

$$
\begin{aligned}
&x \in \mathbf{Z} \text{ and } x \neq 0 \Rightarrow 1 \,|\, x \text{ and } (-1) \,|\, x \\
&x \in \mathbf{Z} \Rightarrow x \,|\, 0 \\
&x \,|\, y \Leftrightarrow (-x) \,|\, y \Leftrightarrow x \,|\, (-y) \Leftrightarrow (-x) \,|\, (-y)
\end{aligned}
$$

For any element a in Z one considers the set $D_a = \{x: x \mid a\}$, whose elements are called *divisors* of a. For $a = 0$, we get $D_0 = Z$. For any a we have $1 \in D_a$, $-1 \in D_a$, $a \in D_a$, $-a \in D_a$; numbers 1, -1, a, $-a$ are called *improper divisors* of a; the other elements in D_a are called *proper divisors*.

Definition 3.1. *The integer a is prime if set D_a has the cardinal* 4.

The condition that a be prime turns to $a \notin \{-1, 0, 1\}$ and to the fact that a does not admit proper divisors. An integer a satisfying $a \notin \{-1, 0, 1\}$ and not being prime is called *composite*.

Theorem 3.4. *Any composite number admits at least a prime divisor, which is a natural number.*

Proof. If n is composite, the cardinal of D_n is bigger than 4, therefore $D_n \backslash \{-1, +1, n, -n\}$ is a non-void set D_n'. The set $E = \{\mid x \mid : x \in D_n'\}$ admits a minimal element p which is obviously a natural number different from 0 and constitutes a proper divisor of n. Therefore, we found that there also exists an integer q so that $n = p \cdot q$. Obviously, $p \notin \{-1, 0, 1\}$ even if, by reductio ad absurdum, it were not prime, it would be composite and we could repeat for above phase in order to find $p = p_1 \cdot p_2$, p_1 being natural number, proper divisor of p. It would follow that p_1 is a proper divisor of n, therefore $p_1 \in E$, $p_1 < p_2$, contradicting the minimality of p.

Remark. It also results $p \leq \mid q \mid$, therefore $p^2 \leq \mid p \cdot q \mid = n$ that is the composite number n admits at least a prime divisor p so that $0 < p < \sqrt{\mid n \mid}$.

Theorem 3.5. *If p is a prime number and $p \mid (x \cdot y)$ then $p \mid x$ or $p \mid y$.*

Proof. We may suppose, without restricting the generality, that p is a natural number. If $p \nmid x$, there will exist a non-null natural number $x' < p$ so that $x = p \cdot q + x'$. Analogously, if (by reductio ad absurdum) $p \mid y$ does not occur either, we find s in Z and y' in N^* so that $y = p \cdot s + y'$ and $y' < p$. There follows $x \cdot y = p \cdot k + x' \cdot y'$, where $k = p \cdot q \cdot s + q \cdot y' + s \cdot x'$. There results $p \mid (x' \cdot y')$ and the above remark is contradicted.

Theorem 3.6 (of decomposition in prime factors). *For any integer number* *n* $\neq$ 0 *there exists an unique decomposition of the form*

(3.1) $n = \varepsilon(p_1)^{a_1}(p_2)^{a_2}\dots(p_k)^{a_k}$

where $\varepsilon \in \{-1, +1\}$ *are prime natural numbers,* $p_1 < p_2 < \dots < p_k$ *and* a_i *are non-null natural numbers.*

A decomposition (3.1) is easily achieved on the way suggested in the proof of the Theorem 3.4. For uniqueness, it is supposed by reductio ad absurdum that except (3.1) there exists another decomposition, as well:

$$n = \varepsilon'(q_1)^{b_1}(q_2)^{b_2}\dots(q_h)^{b_h}$$

satisfying the specified conditions. There immediately follows $\varepsilon' = \varepsilon$. We deduce $q_1 \mid n$ and there follows $q_1 = p_1$; by simplifying a_1 or b_1 times we find $a_1 = b_1$ etc.

Theorem 3.7. *There exists an infinity of prime numbers.*

Proof. The proof is achieved supposing by reductio ad absurdum that there would exist only a finite number of prime numbers, $p_1, p_2, \dots, p_k$. The number $n = p_1 p_2 \dots p_k + 1$ cannot be prime, and $n \in \{-1, 0, 1\}$ does not occur, either. If n were composite, it would admit a positive prime factor p for which $p = p_i$ cannot occur.

Let m be a non-null natural number. We introduce a binary relation $\equiv$ on Z, defining

$$x \equiv y \Leftrightarrow m \mid (x-y).$$

When the number m is not indicated by the contest, instead of $x \equiv y$ we write $x \equiv y (\bmod m)$; the binary relation $\equiv$ is read *"congruent with"*, and the specification $\bmod m$ is read *"modulo m"*. The factor set, $Z/\equiv$ will be denoted by Z_m and the equivalent classes are called *residue classes* $\bmod m$.

Theorem 3.8. *The congruence modulo m is an equivalence. The factor set Z_m is provided with operations + and $\cdot$ so that $(Z_m, +, \cdot)$ is a ring, and the canonical projection $\pi : Z \to Z_m$ is homomorphism of rings. If m is a prime number, $(Z_m, +, \cdot)$ is a field.*

The proof is immediate. We can suppose that m is a natural number without restricting the generality. The function $\pi : Z \to Z_m$ associates to an integer Z the remainder of its dividing by m. The elements of Z_m, the factor classes, are denoted by $\hat{0}$, $\hat{1}$, ..., $\widehat{m-1}$, where $\hat{n}$ is the equivalence class of n, that is

$$\hat{n} = \{n + x \cdot m : x \in Z\}.$$

We can see that $\{\hat{0}\}$ is an ideal of the ring $(Z, +, \cdot)$. If m is not prime, $(Z_m, +, \cdot)$ is not domain of integrity; for example, $2 \cdot 3 \equiv 0 \pmod 6$. For prime m, the quality of domain of integrity of $(Z_m, +, \cdot)$ results from the theorem 3.5. Taking into account that Z_m has a finite number of elements, there results now that it is a field. Indeed, for $\hat{a} \in Z_m \setminus \{\hat{0}\}$ the function $f_a: Z_m \to Z_m$ given by $f_a(\hat{x}) = (a\hat{x})$ is, for prime m, injective, and, due to the finitude of Z_m, surjective, too. Therefore there will exist a unique solution $\hat{x}$ of the equation $\hat{a} \cdot \hat{x} = \hat{1}$, which constitutes the inverse of the element $\hat{a}$ etc.

Theorem 3.9 (Fermat's Little Theorem). *If m is prime and a is not divisible by m, then*

$$a^{m-1} \equiv 1 \pmod m.$$

Proof. It easily comes out that in the sequence of numbers $a \cdot 1$, $a \cdot 2$, ..., $a \cdot (m-1)$ there are not two congruents modulo m and that none of them is congruent with zero. Therefore,

$$\{\hat{a}, 2\hat{a}, ..., (m-1)\hat{a}\} = \{\hat{1}, \hat{2}, ..., \widehat{m-1}\}.$$

Consequently, the products of the elements of the two sets coincide, that is

$$(m-1)! \cdot a^{m-1} \equiv (m-1)! \pmod m, \text{ or } m \,|\, (m-1)!(a^{m-1}-1).$$

But $m \nmid (m-1)!$ since (prime) m does not divide any of the numbers $1, 2, ..., m-1$ strictly inferior to it. There results $m \,|\, (a^{m-1}-1)$, q.e.d.

Theorem 3.10 (Lagrange). *If m is prime and*

$$f(x) = a_0 x^n + a_1 x^{n-1} + ... + a_{n-1} x + a_n$$

is a polynomial of $n \geq 1$ *degree with integer coefficients so that* $m \nmid a_0$, *then in the set* $N_m = \{0, 1, 2, \ldots, m-1\}$ *there exist at most n numbers x for which* $m \mid f(x)$.

The proof is made by induction with respect to the degree n. For $n = 1$, if there existed in N_m distinct elements x, y so that $m \mid f(x)$ and $m \mid f(y)$, then m would also divide the difference $f(x) - f(y) = a_0(x-y)$. It would follow $m \mid x - y$, absurd. Let be an arbitrary natural number n so that the conclusion of the theorem may be valid for any polynomials g of degrees at most equal with $(n-1)$ which have their dominant coefficient not divisible by m. We are supposing by reductio ad absurdum that for the polynomial f of degree n in the enounce of the theorem the conclusion is not true and in N_m there exist $(n+1)$ distinct elements $x_0, x_1, \ldots, x_n$ so that $m \mid f(x_k)$ for $0 \leq k < n$. It is easy to find that there exists a polynomial g of $(n-1)$ degree with integer coefficients so that $f(x) - f(x_0) = (x - x_0)g(x)$ and the dominant coefficient of g is a_0. We deduce that for $1 \leq k \leq m$ there occurs $m \mid f(x_k) - f(x_0)$ and because $m \nmid (x - x_0)$ there follows $m \mid g(x_k)$, in contradiction with the inductive hypothesis.

We immediately conclude on the validity of the conclusion of the theorem.

Corollary. *If m is prime and f is a polynomial of degree n with integer coefficients and there exist in* N_m *more than n numbers x so that* $m \mid f(x)$ *then m divides all the coefficients of f*.

Proof. We are supposing that f does not satisfy the conclusions of the corollary and a_k is its first coefficient that does not divide by m; according to the theorem, $k > 0$. At first, we suppose $k < n$. Let be $g(x) = a_k \cdot x^{n-k} + a_{k+1} \cdot x^{n-k+1} + \ldots + a_n$. Since $m \mid f(x)$ and $m \mid g(x)$ occur simultaneously, we deduce that the polynomial g of $(n-k)$ degree does not satisfy Lagrange's theorem. In order not to enter a contradiction, it is necessary that $k = n$ should occur. But in this case $f(x) = mx \cdot g(x) + a_n$ and because there exist integers x so that $m \mid f(x)$ there follows $m \mid a_n$. Therefore, the hypothesis that the conclusion of the corollary would not be satisfied leads to a contradiction.

Theorem 3.11 (Wilson). *For any prime natural number m we have*

$$m \mid [(m-1)! + 1].$$

Proof. We consider the polynomial

$$f(x) = (x-1)(x-2)\ldots(x-m+1) - x^{m-1} + 1$$

of $(m-2)$ degree and having obviously integer coefficients. For $x = 1, 2, \ldots, m-1$ there occurs $f(x) = -(x^{m-1}-1)$ and, according to Fermat's theorem, $m \mid f(x)$. According to the above corollary (considered for $n = m-2$) all the coefficients of f are divisible by m, particularly the free term, as well, $f(0) = (-1)^{m-1} \cdot 1 \cdot 2 \cdot \ldots \cdot (m-1) + 1$. If m is odd $f(0) = (m-1)! + 1$, therefore the conclusion will be satisfied for any prime m, with the eventual exception of 2. But for $m = 2$ we find directly that $(m-1)! + 1 = 1 + 1 = 2$, number divisible by 2, q.e.d.

§4. The Set of Rational Numbers

In general, in Z there are no inverses with respect to the multiplication; we will construct an extension Q of Z which should be a field.

Let be $Z^* = Z \setminus \{0\}$ and $D = Z \times Z^*$. We define a binary relation $\sim$ on D, demanding:

$$(a, b) \sim (c, d) \Leftrightarrow ad = bc.$$

One immediately deduces:

Proposition 4.1. $\sim$ *is an equivalence on* D.

Proposition 4.2. *The following hold:*
1. $\forall a \in Z; \forall b, c \in Z^*, (a, b) \sim (ac, bc)$;
2. $\forall a \in Z; \forall b, c, d \in Z^*, (a, b) \sim (ac, d) \Rightarrow d = bc$;
3. $\forall a \in Z^*, (a, a) \sim (1, 1)$ and $(0, a) \sim (0, 1)$;
4. $[(a, b) \sim (c, d)$ and $a \neq 0] \Rightarrow [c \neq 0$ and $(b, a) \sim (d, c)]$.

The justification is elementary. We prove 4. If $(a, b) \sim (c, d)$, $a \neq 0$, then $ad = cb$. If $c = 0$ we have the contradiction $a = 0$ or $b = 0$. Consequently, $c \neq 0$. There follows that $(b, a), (d, c) \in D$. But $ad = cb \Rightarrow (b, a) \sim (d, c)$, q.e.d.

Definition 4.1. *The set* $Q = D / \sim$ *is called the set of rational numbers.*

The classes of $D / \sim$ are *rational numbers* and will be denoted by: $< a, b >$, $(a, b) \in D$ being a representative of the rational number, $< a, b >$.
From the proposition 4.2 it is immediately deduced:

Proposition 4.3. *There occur the properties:*

1. $\forall c \in Z^*$, $<a, b> = <ac, bc>$;
2. $<a, b> = <ac, d> \Rightarrow d = bc$;
3. $\forall a \in Z^*$, $<a, a> = <1, 1>$ and $<0, a> = <0, 1>$;
4. $[<a, b> = <c, d>$ and $a \neq 0] \Rightarrow [c \neq 0$ and $<b, a> = <d, c>]$.

Proposition 4.4. *If there occur* $(a, b) \sim (a', b')$ *and* $(c, d) \sim (c', d')$, *then:*

1. $(ad+bc, bd) \sim (a'd'+b'c', b'd')$;
2. $(ac, bd) \sim (a'c', b'd')$.

Proof. 1. $(ab' = a'b$ and $cd' = c'd) \Rightarrow ab'dd'+cd'bb' =$
$= a'bdd'+c'dbb' \Rightarrow (ad+bc, bd) \sim (a'd'+b'c', b'd')$.

2. $(ab' = a'b$ and $cd' = c'd) \Rightarrow acb'd' = a'c'bd \Rightarrow$
$\Rightarrow (ac, bd) \sim (a'c', b'd')$.

Proposition 4.5. *The application* $+ : Q^2 \to Q$ *and* $\cdot : Q^2 \to Q$ *defined by*

$$<a, b> + <c, d> = <ad+bc, bd> \quad <a, b> \cdot <c, d> = <ac, bd>,$$

respectively, do not depend on the representatives (a, b), (c, d) *of the rational numbers* $<a, b>$ *and* $<c, d>$, *respectively.*

Applying the propositions 4.4, there immediately results the assertion.

Therefore, $+$ and $\cdot$, given by the last proposition, are operations on the set Q of rational numbers. We call them the *addition* and *multiplication* of rational numbers.

Theorem 4.1. *The addition* $+$ *and the multiplication* $\cdot$ *give to* Q *a structure of commutative field.*

Proof. 1) $(Q, +)$ is an *Abelian group*:

$$(<a, b> + <c, d>) + <e, f> = <a \cdot d+b \cdot c, b \cdot d> + <e, f> =$$
$$= <a \cdot d \cdot f+b \cdot c \cdot f+b \cdot d \cdot e, b \cdot d \cdot f> = <a, b> + (<c, d> + <e, f>)$$

$<a, b> + <0, 1> = <a, b>$, so $<0, 1>$ is the neutral element for the addition. The opposite of the element $<a, b>$ is the rational number $<-a, b>$. Finally,

$$<a, b> + <c, d> = <c, d> + <a, b>.$$

2) $(Q, \cdot)$ is an *Abelian semi-group with unity.*

$$(<a,b> \cdot <c,d>) \cdot <e,f> = <a \cdot c, b \cdot d> \cdot <e,f> =$$
$$= <a \cdot c \cdot e, b \cdot d \cdot f> = <a,b> \cdot (<c,d> \cdot <e,f>)$$

$<a,b> \cdot <1,1> = <a,b>$ and

$$<a,b> \cdot <c,d> = <a \cdot c, b \cdot d> = <c,d> \cdot <a,b>.$$

3) $Q^{\overset{*}{\cdot}} = Q\backslash\{0, 1\}$ and $\cdot$ is a *commutative group.*

We have to show that $<a,b> \neq <0,1>$ has an inverse with respect to multiplication. But $<a,b> \neq <0,1> \Rightarrow a \neq 0$. Then we obtain $<a,b> \cdot <b,a> = <a \cdot b, b \cdot a> = <1,1>$, (because $a \cdot b \neq 0$). Therefore, for $a \neq 0$, $<a,b>^{-1} = <b,a>$.

4) Multiplication is *distributive* with respect to addition.

$$(<a,b> + <c,d>) \cdot <e,f> = <a \cdot d + b \cdot c, b \cdot d> \cdot <e,f> =$$
$$= <a \cdot d \cdot e + b \cdot c \cdot e, b \cdot d \cdot f> = <a,b> \cdot <e,f> + <c,d> \cdot <e,f>.$$

Convention: *As in the following we shall consider only commutative fields we will renonce to specify the "commutativity" by supposing it implied.*

We will construct a relation of total order in Q and will show that it is compatible with the operations of addition and multiplication of rational numbers. Particularizing in proposition 4.3.1. $c = -1$, we get $<a,b> = <-a, -b>$ and it results that for any rational number $<a,b>$ we can determine a representative (a, b) or $(-a, -b)$ so that its second factor, b or $-b$, may be positive, that is $Q = \{<a,b>: a \in Z$ and $b \in Z_+\}$. Let then be

$Q_+ = \{<a,b>: a \in Z_+$ and $b \in Z_+\}$ and $Q_- = \{<a,b>: a \in Z_-$ and $b \in Z_+\}$.

Proposition 4.6. 1. *The sets Q_+, Q_- and $\{0\}$ make up a partition for Q.*

2. Q_+ *is stable part in $(Q, +)$.*

3. $(Q_+, \cdot)$ *is an Abelian group.*

Definition 4.2. *We say that the rational number $<a,b>$ is smaller than the* rational number $<c,d>$ and we write $<a,b> < <c,d>$ if $<c,d> - <a,b> \in Q_+$. We write $<a,b> \leq <c,d>$ if $<a,b> < <c,d>$ or $<a,b> = <c,d>$.

Theorem 4.2. $(Q, +, \cdot, \leq)$ *is a totally ordered field, that is: the set* $(Q, \leq)$ *is totally ordered and the relation of order is compatible with the operations of addition and multiplication in* Q.

Proof. According to the definition 4.2, $<a,b> \leq <a,b>$. Then $<a,b> \leq <c,d>$ and $<c,d> \leq <a,b>$, by Proposition 4.6.1., imply $<a,b> = <c,d>$. From $<a,b> \leq <c,d>$, $<c,d> \leq <e,f>$ we have $<c,d> - <a,b> \in Q_+ \cup \{0\}$, $<e,f> - <c,d> \in Q_+ \cup \{0\}$, which imply $<e,f> - <a,b> \in Q_+ \cup \{0\}$ that is $<a,b> \leq <c,d>$. The principle of trichotomy immediately results by applying the proposition 4.6.1.. The first part of the theorem is proved.

Let us show now:

(4.1.) $\quad \begin{aligned} &<a,b> < <c,d> \text{ and } <a_1,b_1> \leq <c_1,d_1> \Rightarrow \\ &\Rightarrow <a,b> + <a_1,b_1> < <c,d> + <c_1,d_1> \end{aligned}$

and

(4.2.) $\quad \begin{aligned} &<a,b> < <c,d> \text{ and } <e,f> \in Q_+ \Rightarrow \\ &\Rightarrow <a,b> \cdot <e,f> < <c,d> \cdot <e,f>. \end{aligned}$

Indeed, $<c,d> - <a,b> \in Q_+$, $<c_1,d_1> - <a_1,b_1> \in Q_+ \cup \{0\}$ imply $<c,d> + <c_1,d_1> - (<a,b> + <a_1,b_1>) \in Q_+$ and (4.1) occurs. Relation (4.2) can be proved in the same way.

Let be $\tilde{Q} = \{<a,1>; a \in Z\}$. Obviously, $\varnothing \neq \tilde{Q} \subset Q$. The elements of $\tilde{Q}$ are called *rational integers*. The restriction of the relation $\leq$ determines a total order on $\tilde{Q}$. Likewise, one can easily see that $\tilde{Q}$ is stable part with respect to the operations $+$ and $\cdot$ in Q. Then:

Proposition 4.7. $(\tilde{Q}, +, \cdot)$ *is a domain of integrity.*

Theorem 4.3. 1. *The sets* $(Z, \leq)$, $(\tilde{Q}, \leq)$ *are similar.*
2. *There exists an isomorphism between the domains of integrity* $(Z, +, \cdot)$ *and*

$(\tilde{Q}, +, \cdot)$, *which is a similitude, too.*

Proof. Let $f: \mathbf{Z} \to \tilde{Q}$ be the application given by $f(a) = <a, 1>$. From $f(a) = f(b)$ we deduce $<a, 1> = <b, 1>$ which implies $a = b$. Therefore f is injective. Let $<a, 1>$ be any of the elements in $\tilde{Q}$ and $<a, 1>$ one of its representatives; then $f(a) = <a, 1>$. It follows that f is surjective. Therefore f is bijective. Let us admit now that $a \leq b$; then $f(a) \leq f(b)$, since $<b, 1> - <a, 1> = <b, 1> + <-a, 1> = <b-a, 1>$ and $b - a \geq 0$.

2. The preceding similitude is also an isomorphism, because
$$f(a+b) = <a+b, 1> = <a, 1> + <b, 1> = f(a) + f(b)$$

and

$$f(a \cdot b) = <ab, 1> = <a, 1> \cdot <b, 1> = f(a) \cdot f(b).$$

On the basis of this result, the sets $\mathbf{Z}$ and $\tilde{Q}$ can be identified and Q appears as an extension of the domain of integrity $\mathbf{Z}$.

Theorem 4.4. *Any rational number $<a, b>$ can be written under the form of product of a rational integer by the inverse of a rational integer.*

Proof. Let be $<a, b> \in Q$, $b \in \mathbf{Z}_+$. If $<a, b> = 0$, then $<a, b> = <0, b> \cdot <b, 1>^{-1}$. If $<a, b> \neq 0$, then $<a, b> = <a, 1> \cdot <b, 1>^{-1}$, q.e.d.

Theorem 4.5. *The field of rational numbers Q is a minimal extension of the domain of integrity $\mathbf{Z}$.*

Proof. Any other field K containing a subset $\tilde{K}$ isomorphic with $\mathbf{Z}$ will also contain the inverses of the elements in $\tilde{K} \backslash \{0\}$. According to the theorem 4.4, K will contain a subfield isomorphic with Q and then Q identifies with a subfield in K, q.e.d.

Proposition 4.8. $(Q, \leq)$ *is a totally ordered set, without any prime or last element, and dense.*

The total character of the ordering $\leq$ was proved in the theorem 4.2. For any x in Q we find $x - <1, 1> < x$ and $x + <1, 1> > x$, therefore no prime or last element can exist. Let be $x = <a, b>$ and $y = <c, d>$ so that $x < y$, that is $ad < bc$. The property of density assumes the existence of at least one $z = <e,f>$ so that $x < z < y$. Indeed, we can take $z = \frac{1}{2}(x+y) = <1,2> \cdot <ad+bc, bd> = <ad+bc, 2bd>$ and $x < z$, $z < y$ is easy to be checked.

An ultimate result we are presenting, as regards the field Q, is given by:

Theorem 4.6. *The field Q of rational numbers is Archimedean.*

Proof. We have to show that in $(Q, \leq)$ the following property holds: If $<a,b> \in Q_+$ and $<c,d> \in Q$ then there exists a natural number n so that $n \cdot <a,b> = <n,1> \cdot <a,b> > <c,d>$.

To justify this property, let be the integers $ad > 0$ and cb. There exists $n \in N^*$ so that $n \cdot ad > cb$. Then

$$<n,1> \cdot <a,b> = <n,1> \cdot <ad, bd> > <cb, bd> = <c,d>,$$

q.e.d.

Remark. We identify the sets Z and $\tilde{Q}$: the rational integers $<a,1>$ with a, the inverse of $<b,1>$ with $\frac{1}{b}$. Then $<a,b>$ can be denoted by $\frac{a}{b}$. Thus we have

$$\frac{a}{b} + \frac{c}{d} = \frac{ad+bc}{bd}, \quad \frac{a}{b} \cdot \frac{c}{d} = \frac{ac}{bd}.$$

§5. The Set **R** of Real Numbers

The construction of the set of real numbers is complicated enough, but it constitutes an important problem for Arithmetic's backgrounds. That is why we will expose here one of the possible ways of constructing real numbers.

As we have seen, the field of rational numbers Q is ordered and Archimedean. It is convenient to note now with a, b,..., x, y, ... its elements.

We recall that $\leq$ is a relation of total order on Q and that it is compatible with the operations in Q.

Thus we have for any rational numbers a, b, c, d:

$$a < b, c \leq d \Rightarrow a + c < b + d,$$
$$a < b, c > 0 \Rightarrow ac < bc,$$
(5.1)
$$a < b, c < 0 \Rightarrow bc < ac,$$
$$a^2 \in \mathbf{Q}_+ \text{ for } a \neq 0; 1 \in \mathbf{Q}_+$$
$$ab \in \mathbf{Q}_+, a \in \mathbf{Q}_+ \Rightarrow b \in \mathbf{Q}_+.$$

With our notations, for $a \in \mathbf{Q}^*$ we have $a^{-1} = \dfrac{1}{a} \in \mathbf{Q}^*$. Then

(5.2) $0 < a < b \Rightarrow 0 < \dfrac{1}{b} < \dfrac{1}{a}.$

Let be the function $| \ | : \mathbf{Q} \to \mathbf{Q}_+ \cup \{0\}$, called absolute value, defined by

(5.3) $|x| = \begin{cases} x & \text{if } x \geq 0, \\ -x & \text{if } x < 0. \end{cases}$

Proposition 5.1. 1. $|x| = |-x|$; 2. $|x \cdot y| = |x| \cdot |y|$;
$3° \ |x + y| \leq |x| + |y|$; 4. $\|x| - |y\| \leq |x - y|$.
We prove 3. Since $x \leq |x|, y \leq |y| \Rightarrow x + y \leq |x| + |y|$. Analogously,
$-(x + y) \leq |x| + |y|$, hence $|x + y| \leq |x| + |y|$, q.e.d.
We further consider sequences (a_n) of elements in $\mathbf{Q}$,

$$(a_n) = (a_1, a_2, \dots, a_n, \dots).$$

Definition 5.1. 1. *The sequence (a_n) is called bounded if there exists $b \in \mathbf{Q}_+$ so that $|a_n| < b, \forall n \in \mathbf{N}^*$. Let $\underline{B}$ be the set of bounded sequences.*

2. (a_n) is called Cauchy sequence if for any $e \in \mathbf{Q}_+$ there exists a natural number $N(e)$ so that $|a_p - a_q| < e$, for all $p, q > N(e)$. We denote by $\mathcal{C}$ the set of Cauchy sequences.

3. (a_n) is called null sequence if, for any $e \in \mathbf{Q}_+$, there exists a natural number $L(e)$ so that $|a_p| < e$ for all $p > L(e)$. We denote by $\underline{N}$ the set of null sequences.

Theorem 5.1. $\underline{N} \subset \mathcal{C} \subset \underline{B}$.

Proof. It is clear that $\underline{N} \subset \mathcal{C}$. Let be $(a_n) \in \mathcal{C}$; then $p, q > N(1) \Rightarrow |a_p - a_q| < 1$. If b is the biggest of rational numbers $|a_1|, \dots, |a_{N(1)}|, |a_{N(1)} + 1|$, then $|a_p| < b$ for all $p \in \mathbf{N}^*$, q.e.d.

On the set S of the sequences of elements in Q, we define the operations:

(5.4)
$$(a_n) + (b_n) = (a_n + b_n),$$
$$(a_n) \cdot (b_n) = (a_n \cdot b_n)$$

and $(S, +, \cdot)$ is then a commutative ring with unity.

Theorem 5.2. 1. $\mathfrak{C}$ *is stable part in S with respect to the operations* "+" *and* "$\cdot$";

2. $(\mathfrak{C}, +, \cdot)$ *is a commutative ring with unity;*

3. $\underline{N} \neq \mathfrak{C}$;

4. $\underline{N}$ *is an ideal in* $\mathfrak{C}$.

Proof. 1. Let (a_n), (b_n) be elements of $\mathfrak{C}$ and $N(e)$ and $M(e)$ the integers in $\mathbf{Z}$, associated to these sequences, respectively, for any $e \in Q_+$. If $p, q \geq \max[N(\frac{e}{2}), M(\frac{e}{2})]$, then $|a_p + b_p - (a_q + b_q)| < e$. Therefore $(a_n) + (b_n) \in \mathfrak{C}$.

Let now be c and d with the property $\forall n \in N^*$, $|a_n| < c$, $|b_n| < d$. If

$$p, q > \max[N(\frac{e}{2d}), M(\frac{e}{2c})]$$

then $|a_p b_p - a_q b_q| = |a_p(b_p - b_q) + (a_p - a_q)b_q| \leq c \cdot (\frac{e}{2c}) + (\frac{e}{2d}) \cdot d = e$. Therefore $(a_n) \cdot (b_n) \in \mathfrak{C}$.

2. $(\mathfrak{C}, +, \cdot)$, according to 1. and to the observation $(a_n)(b_n) \in \mathfrak{C} \Rightarrow (a_n) - (b_n) \in \mathfrak{C}$ is a subring of the commutative ring S. The constant sequence $(1, 1, \dots)$ is element of $\mathfrak{C}$ and the unity with respect to the product in $\mathfrak{C}$.

Property 3. is obvious since $(1, 1, \dots) \in \mathfrak{C} \setminus \underline{N}$.

4. If (a_n), $(b_n) \in \underline{N}$ then it is obvious that $(a_n) - (b_n) \in \underline{N}$, therefore $(\underline{N}, +)$ is a subgroup in $(\mathfrak{C}, +)$. Let be $(a_n) \in \underline{N}$ and $(b_n) \in \mathfrak{C}$; let us show that $(a_n)(b_n) \in \underline{N}$. Since (b_n) is bounded, there exists $c \in Q_+$ so that $\forall n \in N$, $|b_n| < c$. Let then take any $e \in Q_+$. There exists $L(\frac{e}{c})$ so that for $\forall p > N(\frac{e}{2c})$ we may have $|a_p| < \frac{e}{c}$. Therefore, for $p > L(\frac{e}{c})$, we have $|a_p b_p| < |a_p| \cdot c < (\frac{e}{c}) \cdot c = e$. There follows that $(a_n)(b_n) \in \underline{N}$, q.e.d.

Definition 5.2. *The set* $\mathbf{R} = \mathfrak{C} / \underline{N}$ *is called the set of real numbers. We denote* the real numbers by α, β, γ. They are the elements of the factor set $\mathbb{C}/\underline{N}$, therefore they have the form $\alpha = (a_n) + \underline{N}$. We consider the constant sequences (0_n), (1_n) and we denote: $\overline{0} = (0_n) + \underline{N}$, $\overline{1} = (1_n) + \underline{N}$; $\overline{0}$ is called the zero real number, and $\overline{1}$, the unity in $\mathbb{R}$.

Proposition 5.2. *If* $\alpha = (a_n) + \underline{N}$ *and* $\beta = (b_n) + \underline{N}$ *are any real numbers, then* $\alpha + \beta = (a_n + b_n) + \underline{N}$ *and* $\alpha \cdot \beta = (a_n \cdot b_n) + \underline{N}$ *depend only on the numbers* α *and* β *and not on their representatives* (a_n) *and* (b_n).

Indeed, if $\alpha = (a_n) + \underline{N} = (a_n') + \underline{N}$, $\beta = (b_n) + \underline{N} = (b_n') + \underline{N}$, then let us consider $(c_n) = (a_n) - (a_n') \in \underline{N}$, $d_n = (b_n) - (b_n') \in \underline{N}$. There follows that

$$\alpha + \beta = (a_n + b_n) + \underline{N} = [(a_n' + c_n) + (b_n' + d_n)] + \underline{N} = (a_n' + b_n') + (c_n + d_n + \underline{N} = (a_n' + b_n') + \underline{N}$$

Also, $\alpha \cdot \beta = (a_n \cdot b_n) + \underline{N} = (a_n' \cdot b_n') + \underline{N}$, q.e.d.

We can consider then the operations $+: \mathbb{R}^2 \to \mathbb{R}$, $\cdot: \mathbb{R}^2 \to \mathbb{R}$, where $+: (\alpha, \beta) \to \alpha + \beta$, $\cdot:(\alpha, \beta) \to \alpha \cdot \beta$, in which $+$ and $\cdot$ are given by (5.4).

Theorem 5.3. $(\mathbb{R}, +, \cdot)$ *is a field.*

Proof. According to the theorem 5.2, $(\mathbb{R}, +, \cdot)$ is a commutative ring with unity. It remains to show that any α in $\mathbb{R}^* = \mathbb{R} \setminus \{\overline{0}\}$ has an inverse in $\mathbb{R}$ with respect to multiplication $\cdot$.

Let be $\alpha = (a_n) + \underline{N}$, that is $(a_n) \in \mathbb{C} \setminus \underline{N}$. We will have to determine a Cauchy sequence (b_n) so that $\beta = (b_n) + \underline{N}$ may have the property $\alpha \cdot \beta = 1$ that is $(a_n b_n) + \underline{N} = (1_n) + \underline{N}$ or even $(a_n b_n) - (1_n) \in \underline{N}$. Since (a_n) is not a null sequence, there exists a rational number $e \in Q_+$ so that for any $r \in \mathbb{N}^*$, natural numbers $s > r$ for which $|a| > e$ may exist. For all p, $q > N(\frac{\varepsilon}{2})$ we have $|a_p - a_q| < \frac{\varepsilon}{2}$. Let be $s > N(\frac{\varepsilon}{2})$ so that $|a_s| \geq e$. Then, for any arbitrary $p \geq N(\frac{\varepsilon}{2})$, we have $e \leq |a_r| = |a_r - a_p + a_p| \leq |a_r - a_p| + |a_p| < \frac{\varepsilon}{2} + |a_p|$, hence $|a_p| > \frac{\varepsilon}{2}$. We take $m = N(\frac{\varepsilon}{2})$, $b_1 = ... = b_{m-1} = 1$ and $b_p = \dfrac{1}{a_p}$ if $p > m$.

We have $(a_n b_n) = (a_1, ..., a_{m-1}, 1, ..., 1)$ and therefore

$(a_n b_n) = (a_1 - 1, \dots, a_{m-1} - 1, 0, \dots, 0) \in \underline{N}$. Let us show now that the constructed sequence (b_n) is a Cauchy one. If p, $q > N(\frac{\epsilon}{2})$, then

$$|b_p - b_q| = |\frac{1}{a_p} - \frac{1}{a_q}| = \frac{|a_q - a_p|}{|a_q| \cdot |a_p|} < \frac{2}{e} \cdot \frac{2}{e} \cdot |a_p - a_q|.$$

Consequently, if e' is arbitrary in Q_+, taking p, $q \geq \max[N(\frac{e}{2}), N(\frac{e^2 \cdot e'}{4})]$, we have $|b_p - b_q| < e'$, q.e.d.

Definition 5.3. *We say that the element α of $\mathbf{R}$ is positive and we set $\alpha \in \mathbf{R}_+$ if there exists $r \in Q_+$ and a natural number $P(r)$, so that $n > P(r) \Rightarrow a_n > r$. We say that the element α under consideration is negative if there exists $r \in Q_+$ and a natural number $P(r)$ so that $n > P(r) \Rightarrow a_n < -r$; we set now $\alpha \in \mathbf{R}_-$.*

It is easy to find that the above given definitions are independent from the representative (a_n) chosen for α. We conceive $\mathbf{R}_+$ and $\mathbf{R}_-$ as subsets in $\mathbf{R}$.

Theorem 5.4. 1. $\mathbf{R}_+$ *and* $\mathbf{R}_-$ *are disjoint sets;*
2. $\alpha \in \mathbf{R}_- \Rightarrow -\alpha \in \mathbf{R}_+;$
3. $\mathbf{R} \setminus (\mathbf{R}_+ \bigcup \mathbf{R}_-) = \{\bar{0}\};$
4. $\alpha, \beta \in \mathbf{R}_+ \Rightarrow (\alpha + \beta \in \mathbf{R}_+ \text{ and } \alpha \cdot \beta \in \mathbf{R}_+).$
The proofs of these assertions develop easily on the basis of Definition 5.3.

Proposition 5.3. 1. $\alpha \in \mathbf{R}^* = \mathbf{R} \setminus \{\bar{0}\} \Rightarrow \alpha^2 \in \mathbf{R}_+;$
2. $1 \in \mathbf{R}_+;$
3. $(\alpha \cdot \beta \in \mathbf{R}_+ \text{ and } \alpha \in \mathbf{R}_+) \Rightarrow \beta \in \mathbf{R}_+.$
Proof. 1. If $\alpha \in \mathbf{R}_+$, then $\alpha^2 = \alpha \cdot \alpha \in \mathbf{R}_+$. If $\alpha \in \mathbf{R}_-$, then $-\alpha \in \mathbf{R}_+$ and $\alpha^2 = (-\alpha)^2 \in \mathbf{R}_+;$
2. $\bar{1} = \bar{1} \cdot \bar{1} = (\bar{1})^2 \in \mathbf{R}_+;$
3. If $\beta \notin \mathbf{R}_+$ we can have: $\beta = \bar{0}$ but $\alpha \cdot \beta = \bar{0} \notin \mathbf{R}_+$ or $\beta \in \mathbf{R}_-$ which implies: $-\beta \in \mathbf{R}_+$, $\alpha \cdot (-\beta) = -(\alpha \cdot \beta) \in \mathbf{R}_+$, contrary to the hypothesis etc.

Definition 5.4. *For α and β in $\mathbf{R}$, we write $\alpha < \beta$, if $\beta - \alpha \in \mathbf{R}_+$ and $\alpha \leq \beta$, if $\alpha < \beta$ or $\alpha = \beta$. The binary relation $\leq$ on $\mathbf{R}$ is thus introduced.*

Theorem 5.5. *The set* $(\mathbf{R}, \leq)$ *is totally ordered and the relation of order* $\leq$ *is compatible with the operations in the field* $\mathbf{R}$.

Proof. On the basis of Definition 5.4, $\alpha \leq \alpha$. If $\alpha \leq \beta$, $\beta \leq \alpha$, we have $\alpha = \beta$; otherwise, $\alpha < \beta$, $\beta < \alpha$, hence $(\beta - \alpha) \in \mathbf{R}_+$ and $-(\beta - \alpha) \in \mathbf{R}_+$, which is absurd. Also, the property of transitivity is immediate. Let now α, β be two arbitrary real numbers; $\beta - \alpha$ is in $\mathbf{R}_+$, in $\{\bar{0}\}$ or in $\mathbf{R}_-$ and only in one of these sets, therefore we have $\alpha < \beta$ or $\alpha = \beta$ or $\beta < \alpha$ and only one of these relations. There follows that $(\mathbf{R}, \leq)$ is a totally ordered set.

Let us show now that

$$(\alpha < \beta \text{ and } \gamma \leq \delta) \Rightarrow \alpha + \gamma < \beta + \delta,$$
$$(\alpha < \beta \text{ and } \gamma > \bar{0}) \Rightarrow \alpha \cdot \gamma < \beta \cdot \gamma.$$

Indeed, from $\beta - \alpha \in \mathbf{R}_+$ and $\delta - \gamma \in \mathbf{R}_+ \cup \{\bar{0}\}$ there results $\beta + \delta - (\alpha + \gamma) \in \mathbf{R}_+$. Also, $(\beta - \alpha \in \mathbf{R}_+ \text{ and } \gamma \in \mathbf{R}_+) \Rightarrow (\beta - \alpha) \cdot \gamma = \beta \cdot \gamma - \alpha \cdot \gamma \in \mathbf{R}_+$, q.e.d.

Consequently, $(\mathbf{R}, +, \cdot, \leq)$ is an ordered field.

Theorem 5.6. *The field of real numbers* $\mathbf{R}$ *contains a subfield similar with the field of rational numbers* $\mathbf{Q}$.

Proof. Let us notice that $\forall n \in \mathbf{N}$, $n \cdot \bar{1} \neq \bar{0}$. Indeed, $\bar{1} + \bar{1} + \ldots + \bar{1} = n \cdot \bar{1} \in \mathbf{R}_+$. So $\mathbf{R}$ is of zero characteristic.

We consider the application $f \colon \mathbf{Q} \to \mathbf{R}$, defined by

$$f\left(\frac{m}{n}\right) = (m \cdot \bar{1}) \cdot [(n \cdot \bar{1})^{-1}];$$

f is a morphism of fields because

$$f\left(\frac{m}{n} + \frac{p}{q}\right) = f\left(\frac{mq + np}{nq}\right) = [(mq + pn) \cdot \bar{1}] \cdot [(nq \cdot \bar{1})^{-1}] =$$
$$= (m \cdot \bar{1}) \cdot (n \cdot \bar{1})^{-1} + (p \cdot \bar{1}) \cdot (q \cdot \bar{1})^{-1} = f\left(\frac{m}{n}\right) + f\left(\frac{p}{q}\right)$$

Analogously, $f\left(\frac{m}{n} \cdot \frac{p}{q}\right) = f\left(\frac{m}{n}\right) \cdot f\left(\frac{p}{q}\right)$. If we show that f is injective, it will result that the image of Q by f is a subfield in $\mathbb{R}$. For this purpose it is enough to prove $f(m) = f(n) \Rightarrow m = n$ only for m, n rational integers. But $f(m) = f(n) \Rightarrow \Rightarrow m \cdot \bar{1} = n \cdot \bar{1} \Rightarrow m = n$.

Summarizing, $f: Q \to f(Q) \subset \mathbb{R}$ is an isomorphism of fields. Let us show that f is a similitude. For this purpose we notice that for $\frac{m}{n} > 0$ we have $f(\frac{m}{n}) = (m \cdot \bar{1}) \cdot [(n \cdot \bar{1})^{-1}] \in \mathbb{R}_+$. If $a < b$ in Q, then $b - a \in Q_+$ and $f(b-a) = f(b) - f(a) \in \mathbb{R}_+$, therefore $f(a) < f(b)$, q.e.d.

On the basis of this theorem, it is convenient to set $\bar{a} = f(a)$, $\forall a \in Q$.

Theorem 5.7. *The field of real numbers $\mathbb{R}$ is Archimedean.*

Proof. Let $\alpha \in \mathbb{R}_+$ and β be two real numbers. Let us show that there exists $n \in N$ so that $n\alpha > \beta$ ($n\alpha$ being $\alpha + \alpha + \ldots + \alpha$, n-terms). If $\beta < \alpha$, then the existence of n is obvious. Let be $0 < \alpha < \beta$. It is easy to see that $e \in Q_+$ exists, so that $\bar{e} = (e, e, \ldots, e) + \underline{N}$ should have the property $0 < \bar{e} < \alpha$. Let be $\beta = (b_n) \in N$. Since the sequence (b_n) is bounded, there exists $d \in Q$ so that $\beta < \bar{d}$. We can apply Archimede's axiom to the rational numbers $e > 0$ and d: there exists $n \in N^*$ so that $ne > d$; there follows that $n\bar{e} > \bar{d}$. Then $n\alpha > n\bar{e} > \bar{d} > \beta$, q.e.d.

Definition 5.5. *We say that the sequence $(a_n) \in Q$ has the limit $b \in Q$ and write* $\lim_{n \to \infty} a_n = b$ *or* $a_n \to b$, *if for any* $e \in Q_+$ *there exists a natural number* $N(e)$ *so that* $|a_n - b| < e$ *for all* $n > N(e)$.

Proposition 5.4. *Any sequence of rational numbers having a limit is a Cauchy sequence.*

The proof is elementary.

Remark. Examples of Cauchy sequences of rational numbers without a limit are easy to construct. That is why it is said about Q that it is not "complete".

The notions of absolute value, Cauchy sequence and sequence with limit immediately extend to $\mathbb{R}$. Having the notions extended this way, an important theorem, whose proof we will just sketch, is formulated.

Theorem 5.8. *The field $\mathbb{R}$ of real numbers is "complete" that is any Cauchy* *sequence of real numbers admits a limit in* **R**.

Let be (α_n) a Cauchy sequence in **R**. For any index p fixed in **N***, α_p is then a real number for which we can adequately choose a representative $(\alpha_{p,n})$, Cauchy sequence in Q. Let us consider now for each n in **N*** the rational number $b_n = a_{n,n}$. On the basis of certain technical details which we are omitting, we find that the sequence of rational numbers (b_n) is a Cauchy one and, therefore, represents a real number β. In the end it is proved that $\lim_{p \to \infty} \alpha_p = \beta$ holds.

Remark. By applying to the field **R** the same construction we have applied to Q to obtain **R**, there is to be found a field $\overline{\textbf{R}}$, but which is similar with **R**. There follows that the extension of Q, obtained in the way indicated in this section, is unique up to an isomorphism.

§6. The Set **C** of Complex Numbers

The field **R** of real numbers also presents a lack of algebraic nature, that is not any algebraic equation with coefficients in **R** has solutions in **R**. For example, the equation $x^2 + 1 = 0$ does not have its roots in **R**. We proceed then to a new extension of the field of real numbers into a field in which this lack does not occur any longer.

The construction of this field, called the field of complex numbers, can be made in different ways. We present here a simple and direct way.

Let **R**$[X]$ be the ring of the polynomials over the field **R**, with the usual operations of addition and multiplication. **R**$[X]$ is a commutative ring with unity.

Proposition 6.1. $J = \{(X^2 + 1)p(X): p(X) \in \textbf{R}[X]\}$ *is an ideal of the ring* **R**$[X]$.

Indeed, for two polynomials in J, their difference belongs to J, and the product of a polynomial in J with a polynomial in **R**$[X]$ also belongs to J.

Definition 6.1. **C** = **R**$[X]/J$ *is called the set of complex numbers. From now on we will denote the real numbers by* 0, 1, a, b, c, ...

The elements of **C** have the form $p(X) + J$. The operations of addition and multiplication in **R**$[X]$ induce by factorization the operations on **C**

$$(6.1) \quad \begin{aligned} [p(X)+J]+[q(X)+J] &= [p(X)+q(X)]+J, \\ [p(X)+J] \cdot [q(X)+J] &= [p(X) \cdot q(X)]+J, \end{aligned}$$

that depend only on the complex numbers under consideration and not on their representatives.

Proposition 6.2. *Any complex number* $p(X)+J$ *can be written uniquely in reduced form* $(a+bX)+J$, *with* a, b *as real numbers.*

Proof. Let $p(X)+J$ the complex number under consideration. The polynomial $p(X)$ is written uniquely in the form $p(X) = (X^2+1) \cdot g(X)+(a+bX)$.
Then $p(X)+J = (a+bX)+J$.

Corollary 6.2. *We have:*

$$[(a+bX)+J = (a'+b'X)+J] \Rightarrow [a = a' \text{ and } b = b'].$$

It is convenient to consider the complex numbers written in the reduced form.

Proposition 6.3. *The sum and the product of complex numbers are given by :*

$$(6.2) \quad \begin{aligned} [a+bX+J]+[a_1+b_1X+J] &= (a+a_1)+(b+b_1)X+J, \\ [a+bX+J] \cdot [a_1+b_1X+J] &= (aa_1-bb_1)+(ab_1+a_1b)X+J. \end{aligned}$$

Indeed, according to (6.1), the first of these equalities is immediate. The second equality (6.1) leads to:

$$\begin{aligned} [a+bX+J] \cdot [a_1+b_1X+J] &= (a+bX)(a_1+b_1X)+J = \\ &= (aa_1-bb_1)+(ab_1+ba_1)X+bb_1(X^2+1)+J = \\ &= (aa_1-bb_1)+(ab_1+ba_1)X+J. \end{aligned}$$

Theorem 6.1. $(\mathbb{C}, +, *)$ *is a field.*

Proof. As $\mathbb{C}$ is given by the factorization of a unitary commutative ring $\mathbb{R}[X]$ by means of an ideal J of its own, $\mathbb{C}$ is a commutative ring with unit. We are still to prove that for each $(a+bX+J) \in \mathbb{C}^* = \mathbb{C} \setminus \{0\}$ there exists an inverse to multiplication and this one is from $\mathbb{C}$.

But $\dfrac{a}{a^2+b^2} - \dfrac{b}{a^2+b^2}X+J$ has the required property, q.e.d.

It is convenient to write the complex number $(a+bX)+J$ in the form $a+bi$; a is called the real part, and b the imaginary part of the complex number $a+bi$.

Theorem 6.2. *The field* $\mathbb{C}$ *of complex numbers contains a subfield isomorphic with the field of real numbers.*

Indeed, the application f: $\mathbb{R} \rightarrow \mathbb{C}$, defined by $f(a) = a+0i$ has the properties:

$$f(a+b) = f(a)+f(b), f(ab) = f(a) \cdot f(b)$$

and is injective, q.e.d.

Theorem 6.3. *There is no relation of total order on* $\mathbb{C}$ *compatible with the operations in* $\mathbb{C}$.

Proof. Let us admit by reductio ad absurdum that there exists a relation of total order $\leq$ on $\mathbb{C}$. Then we either have $0 < i$, or $i < 0$. Since $\leq$ is compatible with the operations, from $0 < i$ there result $i \cdot < i^2$ or $0 < -1$. It is absurd because the restriction of $\leq$ to the numbers of the form $a+0i$ gives the relation of total ordering on $\mathbb{R}$. If $i < 0$, then $-i < 0$ and $(-i)(-i) > 0 \Rightarrow -1 > 0$, which is also absurd, q.e.d.

Remark. It is proved in text books of algebra that $\mathbb{C}$ is a closed algebraic field. That is any algebraic equation with coefficients in $\mathbb{C}$ has its roots in $\mathbb{C}$. For example, $X^2+1 = 0$ has the roots i and $-i$.

Exercises

1. In the construction of Z, replace $\mathbb{N}$ with the cardinals of a universe U. Do Propositions 2.1, 2.2, 2.3 preserve their validity?

2. We call Dedekind cut in Q a couple $t = (T_1, T_2)$, where:

a) $T_1, \bigcup T_2 = Q$ and $T_1 \cap T_2 = \varnothing$;

b) T_1 is "inferiorly saturared", that is $x \in T_1 \wedge y < x \Rightarrow y \in T_1$, and T_2 is "superiorly saturated" that is $x \in T_2 \wedge x < y \Rightarrow y \in T_2$;

c) T_2 does not admit a prime element.

Let $T(Q)$ be the set of Dedekind cuts in Q. Point out a bijection between $T(Q)$ and $\mathbb{R}$.

3. Show that for any complex number $a + bi \neq 0$ there exist uniquely determined elements $r \in \mathbb{R}_+$ and $t \in (0, 2\pi)$ so that $a + bi = r(\cos t + i \sin t)$.

CHAPTER III

AXIOMATIC THEORIES

The theory of sets, of natural, integer, rational, real and complex numbers exposed in the preceding chapters have been get up by constructive methods. The second way of edifying the notions and the mathematical theories is the axiomatic one. In our days, this way leads to a frequently used method, with satisfying results.

§1. Deductive Systems

Definition 1.1. *A deductive system or a theory T is given by:*

1. *A system of <u>notions</u> $\{N_a\}$ and <u>relations</u> $\{R_b\}$;*

2. *The system $\{P_c\}$ of <u>properly constructed propositions</u> expressed by the notions N_a and the relations R_b;*

3. *A system of <u>deductive</u> rules $\{L_s\}$;*

4. *The subsystem of <u>true propositions</u> $\{A_d\} \subset \{P_c\}$, so that*

   a) $\{P_c\}$ is stable with respect to $\{L_s\}$ (that is any consequence of the properly constructed propositions obtained on the basis of deductive rules is a properly constructed proposition);

   b) $\{A_d\}$ is stable with respect to $\{L_s\}$.

The preceding definition of the notion of theory is a general one. For this reason it is somehow vague. We will have to precise further on the underlined terms.

The notions of *subtheory T'* is immediate: *T'* will be formulated from parts of the systems $\{N_a\}$, $\{R_b\}$ $\{P_c\}$, $\{A_d\}$, so that a) and b) may be satisfied.

Let *T* be a theory and *L* a fixed subtheory in *T* called *the logic* of the *T* theory. Then *T\L* is called *the specific part* of the theory.

We will suppose that in each theory under consideration, its logic is specified.

Definition 1.2. *We call axiomatic theory a T theory for which:*

1. *The subsystems $\{N_{a'}\} \subset \{N_a\}$, $\{R_{b'}\} \subset \{R_b\}$, which we call <u>primary-or fundamental notions and relations</u> are given, so that the other notions in $\{N_a\} \setminus \{N_{a'}\}$ and relations in $\{R_b\} \setminus \{R_{b'}\}$ may be logically deduced from the fundamental ones.*

2. *A subsystem $\{A_{d'}\} \subset \{A_d\}$ of true propositions, called <u>axioms</u> is given, so that the other true propositions from $\{A_d\} \setminus \{A_{d'}\}$ may be logically deduced from the axioms.*

The notions and relations that are not primary are called *derived* notions and relations. The true propositions in $\{A_d\}$ which are not axioms are called *theorems*.

A first classification of axiomatic theories is made by means of formalization. We understand by *formalization* of a theory T the considering of the underlined terms as just formal entities, certain symbols *a priori* given within the theory, without being defined in a certain way, making abstraction of the objective reality they come from. The formalization of an axiomatic theory can be gradually done by considering its logical and specific parts.

An axiomatic theory T is called *non-formalized* if none of the terms underlined in Definition 1.1 is formalized. This means that the primary notions and relations are determined by certain material objects we know in an intuitive descriptive manner, and the axioms are propositions whose truth in beyond doubt and expresses properties of the material objects under consideration. Within such a theory, logic is the logic of mathematical common sense achieved along man's millenary practice.

Example. Euclid's axiomatic theory of Euclidean geometry is non-formalized. The fundamental notions: point, straight line, plane are given on the basis of intuitive description, the axioms express un-doubtful proprieties of them, suggested by experience. Logic is that of common sense.

In contemporary Mathematics, the non-formalized axiomatic theories are not considered as rigorous ones.

An axiomatic theory is called *semi-formalized* if its specific part is formalized, and its logical one non-formalized. This means that primary notions and relations are just symbols initially given within the theory, and the axioms are considered as true propositions given from the beginning within the theory. The logic under consideration is the logic of mathematical common sense. Within such an axiomatic theory, in the specific part is not based on intuition any-more, the reasonings will be rigorous. This is the reason for which most of present-day mathematical theories are semi-formalized axiomatic theories. We find algebra, geometry, analysis etc. in this situation.

We call *formalized* an axiomatic theory in which the specific and logical parts are formalized. Within such a theory an utmost degree of rigour is obtained.

The pages of this book deal only with semi-formalized theories. Let us notice that formalization accepted in semi-formalized theories does not mean that primary notions and relations and the axioms given at first are pure creations of conscience. We make abstraction of their material, objective support to protect theory from impurities. In the end, the truth of a semi-formalized or formalized axiomatic theory is checked by the criterion of practice, too, as we will see in the following paragraphs.

If for an deductive theory T a selection of primary notions and relations and of axioms is effectively made, it is said that an *axiomatization* takes place, and this selection gives us a system of *axioms* for *T*.

If two axiomatizations are chosen for *T*, there are some different ways to speak about the problem of their equivalence.

§2. The Metatheory of an Axiomatic Theory

What we are interested in are the properties of an axiomatic theory, its inner structure, its relations with other theories. Such an analysis of an axiomatic theory *T*, the "*T - metatheory*", uses two points of view.

The *syntactic* point of view consists in the analysis of the constructing mode of the theory *T*. The study of *T* from this point of view is called the *syntaxis* of *T*.

The *semantic* point of view is the analysis of the interpretations and models of the axiomatic theory *T*.

We call *interpretation* of *T* in a deductive theory *T'* a representation of *T* primary notions and relations by *T'* properly constructed propositions so that each *T* properly constructed proposition becomes a *T'* proposition, properly constructed.

A *model* of *T* in *T'* is an interpretation of *T* in *T'* in which the axioms of *T* are represented on truths of *T'*.

A general way interprets any *T* primary notion N_a as an element of a set X_a for all the values of the index a in a set α. So, each *T* primary relation R_b is interpreted as a relation R_b' in the family $\{X_a\}_{a \in \alpha}$. To have a model of *T* in a *T'* theory of sets (particularly in the "naive" one) is to point out the axioms for $\{A_a\}_{a \in \alpha}$ and $\{R_b'\}_{b \in \beta}$. It is also possible to replace the family $\{X_a\}_{a \in \alpha}$ by its disjoint reunion, *X*.

Further on, let us de note by *T* a certain axiomatic theory. The problems of non-contradiction, independence, completeness and categoricity are the important problems for the metatheory of *T*.

1° The Problem of Non-Contradiction

Definition 2.1. *T is called non-contradictory if in T there is not a proposition p so that p and its negation $\neg p$, may be deduced from axioms. If T is non-contradictory it is said that the system of axioms in T is non-contradictory.*

From Definition 2.1 there results that the non-contradiction of *T* belongs to its syntaxis. The syntactic point of view in the study of the non-contradiction of T can not be applied directly. Then the following principle is adopted: *Any axiomatic non-contradictory theory admits at least a model (that is it is "consistent"), and conversely.*

There results from here that a non-contradictory axiomatic theory defines a model of its own.

To study the non-contradiction of T by the semantic method it means therefore to construct a model T' of T, but T' must itself be non-contradictory. Usually, the non-contradiction of Arithmetic is accepted, and then it serves as source of models for semi-formalized axiomatic theories.

2° The Problem of Independence

Definition 2.2. *An axiom A in T is independent from the other axioms of T if A can not be logically deduced from the remaining axioms of T. The independence of the primary notions and relations of T is analogously defined. The T system is called independent or minimal if every of its notions, relations or axioms is independent from the other ones.*

The independence of an axiom from the other ones is not always easy to be justified on syntactic way. But we can establish a useful semantic procedure to solve this problem.

Let A_n be one of the A_a axioms of an axiomatic T theory. We consider the system of axioms $\{A_a^{(n)}\}$ obtained from $\{A_a\}$ by keeping all the A_a axioms with $a \neq n$ and replacing the A_n axiom by its negation, $\neg A_n$. We denote by $T_{(n)}$ the axiomatic theory having the same primary notions and relations as T, but $\{A_a^{(n)}\}$ as system of axioms. (This operation can be achieved when A_n does not interfere in the formulation of the other axioms.)

Proposition 2.1. *The A_n axiom is independent in T if and only if the axiomatic theory $T_{(n)}$ is non-contradictory.*

Proof. If A_n is independent, we can not logically deduce it by excluding it from T, so $T_{(n)}$ and T are both non-contradictory. Conversely, as $T_{(n)}$ is non-contradictory, A_n and $\neg A_n$ are not both true, so A_n is not to be deduced from the other axioms.

Apparently, the question if the axiomatic theory T is or is not independent is not important. If it is not we may eliminate the dependent notion, relation or axiom till we get an independent one, T'. From this point of view the independence seems to be only of question of "elegance" for axiomatic theories.

In fact, to minimize an axiomatic theory may be a very difficult job, as we shall see in the following chapter.

The minimalization of an axiomatic theory can substantially contribute to the evolution of Mathematics; for example, the study of the independence of Euclidean postulate of the parallels has stimulated the quantitative and qualitative development of Geometry.

3° The Problem of Completeness and Categoricity

Definition 2.3. *An axiomatic T theory is called complete if for any properly constructed proposition p: p or* $\neg p$ *can be deduced from axioms.*

If T is complete, then the system of axioms of T is called complete. Otherwise, it is incomplete.

Two models of a theory T are called *isomorphic* if a one to one correspondence preserving their true properties can be established between them.

Definition 2.4. *An axiomatic theory T is called categorical if all its models are isomorphic with each other.*

Proposition 2.2. *Any categorical axiomatic theory is complete.*

Indeed, let T be a categorical axiomatic theory, but incomplete, and p a properly constructed proposition for which neither p, nor $\neg p$ is deducible from axioms. Let T_1 be the theory obtained by adding p to the axioms of T, and T_2 the obtained from T by adding $\neg p$ to its axioms. T_1 and T_2 are non-contradictory. If T_1' is a model of T_1 and T_2' a model of T_2, then T_1' and T_2' are isomorphic, and then T_2' is a model for T_1. Therefore, p and $\neg p$ hold in T_2, which is absurd.

We have thus put an end to the list of the most important problems in the metatheory of an axiomatic system.

We are closing this paragraph with an important notion, that of the *mathematical structure*.

We have seen that any axiomatic theory T indirectly defines a T' model of its own, and that the study of the theory T can be performed on the model T'. But T' can be made up of a system of sets and relations that satisfy certain given axioms. Consequently, the following definition can be formulated.

Definition 2.5. *We call mathematical structure a set provided with certain relations so that its elements and the given relations satisfy a system of axioms.*

The axiomatic theory of a mathematical structure is the theory of the system of axioms under consideration in which primary notions and relations are the elements of the set and given relations. The problems of metatheory are to be applied to this theory.

The notion of mathematical structure is used by the group of French mathematicians Bourbaki to classify mathematical theories, classification ranging from the simple types of mathematical theory, such as the algebraic structures, to the complex types, such as mathematical Analysis and differential Geometry. A unitary superstructure is thus obtained, based on the axiomatic method.

§3. Peano's Axiomatics of Arithmetic

The considerations in the first two sections of this chapter concerning the axiomatic theories and their metatheory will be materialized in the case of *Peano's* axiomatics of natural numbers.

In the constructive theory of natural numbers, there have been constructed the elements of the set N. Let us suppose now that they *a priori* given, namely, that we are taking for primary notions *zero*, noted with 0, and *natural number*, noted with a, b, ..., m, n, ..., x, y, ... Let us suppose as being given only a primary relation, that of "successor"; the successor of a natural number n will be denoted by n'.

We will use the symbols $=$ and $\neq$ in the meaning and with the proprieties of logical identity. With some small exceptions, we will not use the notations in the theory of sets not to confuse the theory we have in view with a mathematical structure.

The main reason for these evitation is to ensure for the formalized axiomatics P of Peano the independence with respect to a formalized axiomatic theory T for sets.

The axioms of Peano's system:

A_1. *Zero is a natural number.*

A_2. *Any natural number admits a unique successor, with is also a natural number.*

A_3. *Zero is the successor of no natural number.*

A_4. *If the successor of two natural numbers coincide, then the numbers under considerations coincide.*

A_5. *If a set of natural numbers contains zero and for each of the numbers in this set its successor belongs to the set, then the set under consideration coincides with the set of all natural numbers.*

The A_5 axiom will be called *the principle of induction* or *the axiom of induction.* It is possible to eliminate the references to a theory T of sets in the enounce of A_5 by considering a property $\mathcal{P}$ of natural numbers. As such a property $\mathcal{P}$ is not specified, A_5, is not an usual axiom but a "scheme-axiom".

To develop the theory given by Peano's axiomatic system means to obtain all the logical consequences from the primary notions and relations and from the given axioms.

Here are some immediate consequences.

Proposition 3.1. $a = b$ *implies* $a' = b'$.

The axiom A_2 is applied here.

Proposition 3.2. $a \neq b$ *implies* $a' \neq b'$.

Otherwise, $a' = b'$ and the axiom A_4 given $a = b$, which is absurd.

Remark. We can take into consideration the set N of natural numbers. 0 will be an element in N, according to the axiom A_1. Now we can conceive $"\prime"$ as an injective function defined on N with values in N.

Proposition 3.3. *For any natural number a different from zero, there exists a natural number b so that $a = b'$ and b is unique.*

Indeed, let M be the set made up of zero and those natural numbers a having the property in the first part of the enunciation. Then, 0 is in M; if a is in the M set, there is $a = b'$. But then $a' = (b')'$ has the mentioned property, therefore a' is in M. By applying the resulted axiom of induction, $M = N$. The uniqueness of b, according to the A_4 axiom, is obvious.

Theorem 3.1. *For every natural number a there exists only a function $f_a : N \rightarrow N$* so that:

1. $f_a(0) = a$;
2. $f_a(b') = [f_a(b)]'$.

Proof. Let M be set of natural numbers a for which there exists at least a function f_a with the properties 1. and 2.. Function $f_0 = 1_N$ has the proprieties 1. and 2. because $f_0(0) = 0$ and $f_0(b') = b' = [f_0(b)]'$. Therefore, zero belongs to M. If a is a certain element in M, then we consider the function $f_{a'}: N \to N$, given by $f_{a'}(b) = [f_a(b)]'$, what returns to $f_{a'} = {}' \circ f_a$. Let us show that $f_{a'}$ satisfies 1. and 2.. Indeed, $f_{a'}(0) = [f_a(0)] = a'$ and

$$f_{a'}(b') = [f_a(b')]' = ([f_a(b)]')' = [f_{a'}(b)]'$$

So a' belongs to M. By applying A_5, there results that M coincides with N. The existence of f_a for every a in N is proved.

Let us suppose by reductio ad absurdum that for a certain natural number a would exist a function h_a satisfying the conditions 1., 2. but $h_a \neq f_a$. We deduce that the set $D = \{x : h_a(x) \neq f_a(x)\}$ is non-void. But we find that $h_a(0) = a = f_a(0)$ therefore $0 \in N \setminus D$ and then:

$$n \in N \setminus D \Rightarrow h_a(x) = f_a(x) \Rightarrow h_a(x') = f_a(x') \Rightarrow x' \in N \setminus D.$$

So, according to A_5, $N \setminus D = N$, that is $D = \varnothing$, in contradiction with the hypothesis we made.

Definition 3.1. *We call addition of natural numbers the application* $+ : N \times N \to N$ *given on* $+ : (a, b) \to a + b = f_a(b)$ *for every a and b in N. Obviously:*

Theorem 3.2. *The operation of addition of natural numbers has the following properties:*
1. $a + 0 = a$, *for any a in N;*
2. $a + b' = (a + b)'$, *for any a and b in N.*

Theorem 3.3. *The addition of natural numbers has the following properties:*
1. $a + (b + c) = (a + b) + c$, *for any a, b, c in N;*
2. $a + 0 = 0 + a = a$, *for any a in N;*
3. $a + b = b + a$, *for any a, b in N;*
4. $a + 1 = 1 + a = a'$, *for any a in N; (of course $1 = 0'$).*
5. $a + b = a + c$ *implies $b = c$.*
Proof. 1. One apply induction related to c, by using theorem 3.2:

$a+(b+0) = a+b = (a+b)+0$.   Then,   if   1.   is   satisfied   by c,   then
$a+(b+c') = [a+(b+c)]' = [(a+b)+c]' = (a+b)+c'$. So, 1. is checked for all c.

2. $a+0 = a$ is true on the basis of the definition of the addition of natural numbers. To prove that $0+a = a$, we apply induction according to $a \cdot 0+0 = 0$, then, admitting $0+a = a$, it results $(0+a)' = 0+a' = a'$.

3. One prove, by induction according to n the preliminary equality $n+0' = 0'+n$. Then we prove 3. by applying the induction according to $b: a+0 = 0+a$ in agree to 2. In the hypothesis $a+b = b+a$, by using 1. and the auxiliary equality, one deduces:

$a+b' = a+(b+0) = a+(b+0') = (a+b)+0' = (b+a)+0' = b+(a+0') = b+(0'+a) =$
$= (b+0')+a = b'+a$, etc.

4. By putting $0' = 1$ in the proof of 3., it follows $a+1 = 1+a = a'$.

5. By applying the principle of induction according to a, the property immediately results.

Theorem 3.4. *For every natural number there exists a single function $g_a: \mathbb{N} \to \mathbb{N}$ having the properties:*

1. $g_a(0) = 0$;

2. $g_a(b') = g_a(b)+a$.

Proof. We apply the axiom of induction to prove the existence of the function g_a. Let M be the set of those natural numbers a for which g_a exists. The constant application $g_0: \mathbb{N} \to \{0\}$ has the properties 1. and 2. Therefore, 0 belongs to M. Let us suppose that a belongs to M, therefore that g_a with properties 1. and 2. exists. We consider $g_{a'}: \mathbb{N} \to \mathbb{N}$ given by $g_{a'}(b) = g_a(b)+b$ for any b in $\mathbb{N}$. Let us check 1. and 2. for $g_{a'}$:

$$g_{a'}(0) = g_a(0)+0 = 0+0 = 0$$
$$g_{a'}(b') = g_a(b')+b' = [g_a(b)+a]+b' = [g_a(b)+b]+[a+1] = g_{a'}(b)+a'$$

It follows that a' belongs to the set M, so M coincides with $\mathbb{N}$. The existence of g_a is ensured. To show that for every natural number a, g_a is unique, let us suppose by reductio ad absurdum that there exists at least a natural number b and two distinct functions g_b and k_b, from $\mathbb{N}$ to $\mathbb{N}$, having the properties 1. and 2. Let L be the set of natural numbers x for which $g_b(x) = k_b(x)$. From 1. $g_b(0) = 0 = k_b(0)$, so 0 is in L. If x belongs to the set L, then:

$$g_b(x') = g_b(x) + b = k_b(x) + b = k_b(x').$$

There follows that x' belongs to L, so $L = \mathbf{N}$. Therefore, $g_b = k_b$, contrary to the accepted supposition, q.e.d.

The existence and uniqueness of the function g_a enables us to give the following definition:

Definition 3.2. *We call multiplication of natural numbers the application:* $\cdot : \mathbf{N} \times \mathbf{N} \to \mathbf{N}$, *given by* $\cdot :(a, b) \to a \cdot b = g_a(b)$.

Theorem 3.5. *The operation of multiplication of natural numbers has the following properties:*
1. $a \cdot 0 = 0$, *for any a in* $\mathbf{N}$;
2. $a \cdot b' = a \cdot b + a$, *for all*
a and b in $\mathbf{N}$.
Indeed, according to Definition 3.2 and to Theorem 3.3 there are the enumerated properties.

We will also use the notation ab for $a \cdot b$.

Theorem 3.6. *The multiplication of natural numbers has the following properties:*
1. $(a+b)c = ac+bc$;
2. $a \cdot 0 = 0 \cdot a = 0$;
3. $a \cdot 1 = 1 \cdot a = a$;
4. $a \cdot b = b \cdot a$;
5. $a(bc) = a(bc)$;
6. $(ab = ac \wedge a \neq 0) \Rightarrow b = c$;
7. $ab = 0 \Rightarrow (a = 0 \vee b = 0)$.
Proof. By induction and taking into account that 2. in the theorem 3.5 is $ab' = a(b+1) = ab+a$, each of these proprieties is proved. For example, we are checking 4. If $b = 0$, the equality is checked according to 2. Let $ab = ba$. Then $ab' = ab+a = ba+a = (b+1)a = b'a$, q.e.d.

From Theorems 3.3 and 3.6 there results:

Theorem 3.7. *The triple* $(\mathbf{N}, +, \cdot)$ *is a semi-domain of integrity.*

§4. The Relation of Order on the Set N of Natural Numbers

Definition 4.1. *We say that the natural number* a *is smaller than the natural number* b *and write* $a < b$ *if there exists a non-null natural number* c *so that* $b = a + c$. *We denote with* $a \leq b$ *if* $a < b$ *or* $a = b$. *Obviously,* $\leq$ *is a binary relation for* N, *and:*

$$a \leq b \Leftrightarrow \exists c \in \mathbf{N}, \ b = a + c.$$

Theorem 4.1. *If* $a \leq b$, *there exists only a natural number* c, *so that* $b = a + c$.

Proof. The existence of c is directly ensured by the definition 4.1. If $d \in \mathbf{N}$ also exists so that $b = a + d$, there follows $a + c = a + d$, and, according to 5. in Theorem 3.3 it results $c = d$, q.e.d.

Theorem 4.2. *The set* $(\mathbf{N}, \leq)$ *is totally ordered. The structure of order is compatible with the addition and multiplication of natural numbers.*

Proof. Obviously, $a \leq a$ on the basis of Definition 4.1. If $a \leq b$ and $b \leq a$ then $a = b$; otherwise, $a < b$, $b < a$ imply, according to Theorem 4.1, $a + c = b$, $b + d = a$. These equalities lead us to $b + d + c = b$, that is $d + c = 0$, $d = 0$, $c = 0$, contrary to the hypothesis. Now, from $a \leq b$, $b \leq c$, it is easy to deduce, on the basis of Theorem 4.1, that $a \leq c$.

We shall prove now that $(\mathbf{N}, \leq)$ is *totally* ordered. For the natural number a we take the set $M_a = \{x : a \leq x \text{ or } x \leq a\}$. We therefore obliged to ascertain that $M_a = \mathbf{N}$. Obviously, $0 \leq a$, so $0 \in M_a$.

Let $x \in M_a$. In the eventuality $a \leq x$ quickly results $a \leq x'$, hence $x' = M_a$. In the other eventuality, $x \leq a$, there exists n so that $x + n = a$. We are discerning the alternative $n = 0$ or $n \neq 0$. For $n = 0$, there follow $x = a$ and $x' = a' = a + 0'$, so $a \leq x'$, which implies again $x' \in M_a$. If $n \neq 0$, according to Proposition 3.3, we find m so that $n = m'$ and deduce $a = x + m' = x' + m$, therefore $x' \leq a$ and again $x' \leq M_a$, according to the principle of induction, $M_a = \mathbf{N}$.

Then we find that $a \leq b$, $c \leq d$ imply $a + c \leq b + d$, and $a < b$, $0 \neq c$ imply $ac \leq bc$. Indeed, $a + a_1 = b$, $a_1 \neq 0$, $c + c_1 = d$ imply $a + c + (a_1 + c_1) = b + d$, $a_1 + c_1 \neq 0$. Therefore, $a + c < b + d$. Then, from $a + a_1 = b$, $a_1 \neq 0$ there are:

$$ac + a_1 c = bc, \ a_1 c \neq 0,$$

which give us $ac < bc$, q.e.d.

Theorem 4.3. *The set* $(N, \leq)$ *is well ordered.*

Proof. Let M be a certain non-void subset of N. We intend to prove that M has a prime element. Let K be the set of all natural numbers a with the property $a \leq m$ for any m in M. $N - K \neq \varnothing$, because for m in M, $m' \in K$ does not hold. Since $0 \leq m$, there results that 0 belongs to K. It is obvious that there exists a natural number m_0 in K for which its successor, m_0', is not in K. Since m_0 belongs to K, $m_0 \leq m$ for any m in M. If we had only $m_0 < m$, then $m_0' \leq m$ for any m in M and then $m_0' \in K$ contrary to the hypothesis. Therefore, $m_0 \in M$ and then m_0 is the prime element of M, q.e.d.

Remark. The preceding theorem is the immediate consequence of the theorem 4.2 in Chapter I.

The remaining results concerning the set N of natural numbers in Chapter II can be now proved again without any modifications.

§5. The Metatheoretical Analysis of the Axiomatic System of Natural Numbers

For the axiomatic system of Peano we will examine now the problems of its non-contradiction, independence and categoricity. Since the system is semi-formalized, we will deal only with its specific part.

It can be seen without any difficulties that Frege-Russell theory of natural numbers, exposed in Chapter II, is a model for Peano's axiomatics. Then the latter one is non-contradictory if Frege-Russell theory is non-contradictory. In its turn, this theory is non-contradictory if the theory of the sets used in its construction is non-contradictory. But it is more difficult to justify the non-contradiction of the theory of sets than to directly accept that Arithmetic, based on Peano's theory, is non-contradictory. Consequently, *we will admits* that Peano's axiomatics is not contradictory. This way, it becomes a basic theory and serves to the construction of certain arithmetical models for other theories. By relying on this hypothesis, we can ascertain that the Z ring and the fields Q, R, C, constructed on the basis of the set of natural numbers, are non-contradictory. For the axiomatic theories we will take under consideration we will effectively construct arithmetical models to prove their non-contradiction.

The Independence of Peano's Axiomatics

The A_1 axiom is appreciated as independent from the other ones because without it we can not formulate axioms A_3 and A_5.

To prove the independence of A_2 axiom from the other axioms in the system, let us take the set $\{0, 1, 2, 3\}$ as the set of natural numbers and the relation of succession given by the arrows:

$$0 \to 1 \to 2 \to 3.$$

Then 0 belongs to the set under consideration, therefore A_1 is checked. A_2 is not achieved on this model, because 3 has no successor. The axioms A_3, A_4, A_5 are obviously satisfied. According to the to the proposition 2.1, there results A_2 independent from the other axioms.

Let us take then the set $\{0, 1, 2\}$ as the set of natural numbers and the relation of succession given by the arrows

$$0 \to 1 \to 2 \to 0.$$

Axioms A_1, A_2, A_4, A_5 are satisfied on this model, axiom A_3 is not satisfied. So A_3 is independent from the other axioms.

Let us take $\{0, 2\}$ as set of natural numbers and the relation of succession given by $0' = 2$, $2' = 2$. All the axioms are checked on this model, except A_4 which proves its independence from the remaining axioms.

Finally, let the set $\{0, 2, 4, \dots, 1, 3, \dots\}$ be set of natural numbers and the relation of succession given by the arrows

$$0 \to 2 \to 4 \to \dots \to 1 \to 3 \to \dots.$$

The axioms $A_1 - A_4$ are checked on this model. Axiom A_5 does not hold.

Indeed, the set M of "even" numbers satisfies the hypothesis in A_5, but does not coincide with $\mathbf{N}$. We found then:

Theorem 5.2. *The axioms in Peano's axiomatic system are independent.*

The completeness of Peano's axiomatic system derives from the following

Theorem 5.3. (Dedekind). *Peano's semi-formalized Arithmetic is a categorical theory.*

Proof. Let N_1 and N_2 be two models on which all the axioms of Peano's system are achieved, and let 0_1 and 0_2 be the elements giving on zero in each of these models. We denote by an accent the succession in both the models N_1 and N_2 as interpreting the succession in Peano's system.

We will take an application $k : M_1 \to M_2$, where:

$$M_1 \subset N_1 , M_2 \subset N_2 , k\,(0_1) = 0_2 \text{ and } k\,(a') = [k\,(a)]'.$$

By complete induction in $(N_i , 0_i , ')$, we find $M_i = N_i$ for $i = 1, 2$. Then we also find that the application $h : M_2 \to M_1$, defined by $h\,(0_2) = 0_1$, $h\,(b') = [h\,(b)]'$ is the reverse of the application k and we deduce thus the existence of the isomorphism $k : (N_1 , 0_1 , ') \to (N_2 , 0_2 , ')$.

Corollary 5.1. *Peano's axiomatic system is complete.*

Remark. Semi-formalized Arithmetic can be defined as axiomatic theory of the following *mathematical structure.* Let **N** be a set in which an element noted with 0 and called the zero element has fixed itself. A function **N**: **N** $(a) = a'$, called function of succession; it is necessary to check Peano's axioms $A_1 - A_5$.

Exercises

1. For the set of homogeneous binary relation ε on a set A, the following axioms are considered:

E1 $\forall\, x \in A\ \exists\, y \in A\ x\, \varepsilon\, y$;

E2 $x\, \varepsilon\, z \wedge y\, \varepsilon\, z \Rightarrow x\, \varepsilon\, y$.

Find out that this constitutes a consistent, non-categorical and minimal system.

2. For the set of the functions $d : A^2 \to \mathbf{R}$ the following axioms are considered:

D1 $d\,(x, y) = 0 \Leftrightarrow x = y$;

D2 $d\,(x, y) = d\,(y, x)$;

D3 $d\,(x, z) \leq d\,(x, y) + d\,(y, z)$.

Show that this constitutes a consistent, non-categorical and minimal system.

3. For the set of the functions $f : \mathbf{N}^* \times \mathbf{N} \to \mathbf{N}$, show that the following axiomatic system is consistent, categorical and minimal:

F1 $f(a, 0) = 1$;
F2 $f(a, b') = a \cdot f(a, b)$.

CHAPTER IV

ALGEBRAIC BASES OF GEOMETRY

The notion of Elementary Geometry is not easy to define. Its presentation as an axiomatic theory is difficult because one needs many primary elements and relations, and axioms. In general, things simplify themselves if the set **R** of real numbers is known. Geometry may be then presented as a mathematical structure defined by a set of elements, called points, an equivalent relation on the set of the pairs of points enabling us to define the vectors and a multiplication of vectors by real numbers, subjected to a (quite small) number of axioms. To make this construction we shall have to know certain simple algebraic structures. Since we have in view non-contradictory and minimal, not necessarily complete axiomatic theories, we shall expose them in turn, under their general form.

§1. Almost Linear Spaces

Let **R** be the field of real numbers studied in Chapter II.

Definition 1.1. *We call almost linear space (real one, or over the field **R**) a non-void set M, in which an element $0 \in M$ fixed itself together with an internal operation $+ : M \times M \to M$ and with an external operation related to **R**, $* : \mathbf{R} \times M \to M$ are given, so that the following axioms may be checked:*

1. $\forall\, x, y, z \in M, x + (y + z) = (x + y) + z$;
2. $\forall\, x \in M, x + 0 = x$;
3. $\forall\, x \in M, \exists\, (-x) \in M, x + (-x) = 0$;
4. $\forall\, \alpha \in \mathbf{R}, \forall\, x, y \in M, \alpha * (x + y) = \alpha * x + \alpha * y$;
5. $\forall\, \alpha, \beta \in \mathbf{R}, \forall\, x \in M, (\alpha + \beta) * x = \alpha * x + \beta * x$;
6. $\forall\, \alpha, \beta \in \mathbf{R}, \forall\, x \in M, \alpha * (\beta * x) = (\alpha\beta) * x$.

Remark. *As one can see, within the notion of almost linear space introduced by one of the authors, [573], only a part of the axioms of linear spaces was retained.*

The elements of M will be called *vectors*, and those of **R** *scalars* or merely real numbers. From the given definition it results that the notion of almost space presents itself as a mathematical structure. We can prove:

Theorem 1.1. *The system of axioms of the notion of almost linear space is non-contradictory.*

Proof. Any real linear space is a model for the system under consideration. Especially $M = \mathbb{R}$, together with the usual operation $+$, in which $\alpha * x = \alpha x$ satisfy the axioms 1-6, q.e.d.

Theorem 1.2. *The system of axioms of the notion of almost linear space is minimal.*

Proof. Model 1. One takes $M = \{0, a, b\}$ with the operation $+$ given by

+	0 a b
0	0 a b
a	a a 0
b	b 0 b

and $\alpha * x = x$, $\forall x \in \mathbb{R}$. Axiom 1. is not checked, because we have $(a+a)+b = a+b = 0 \neq a = a+0 = a+(a+b)$. The other axioms are satisfied.

According to proposition 2.1, chapter III, it results the independence of the axiom A_1 from the other ones.

Model 2. $M = \mathbb{R}$; $x+y = y$ for any $x, y \in M$, $\alpha * x = x$.

Model 3. $M = \mathbb{R}$; $x+y = x$, $\alpha * x = x$.

Model 4. $M = \mathbb{R} \times \mathbb{R}$; $(x_1, x_2)+(y_1, y_2) = (x_1+y_1, x_2+y_2)$;

$$\alpha * (x_1, x_2) = \begin{cases} (0, \alpha x_2), & \text{if } x_2 \neq 0, \\ (\alpha x_1, 0), & \text{if } x_2 = 0. \end{cases}$$

Axiom 4. is not checked because

$$1 * [(1, 0)+(0, 1)] = 1 * (1, 1) = (0, 1) \neq (1, 1) = (1, 0)+(0, 1) =$$
$$= 1 * (1, 0)+1 * (0, 1).$$

Model 5. $M = \mathbb{R}$: $x+y$ the addition in $\mathbb{R}$, and $\alpha * x = x$, $\forall \alpha, x \in \mathbb{R}$.

Model 6. If $M = \mathbb{R} \times \mathbb{R}$; then $(x_1, x_2)+(y_1, y_2) = (x_1+y_1, x_2+y_2)$; and $\alpha * (x_1, x_2) = (\alpha x_2, \alpha x_1)$. It is clear that in this model axioms 1-5 are checked, while 6. is not.

Remark. *The system of axioms under considerations is not complete because new axioms, independent from the axioms 1-6 can be added to it.*

Propositions 1.1. *There are*

1. $x + 0 = 0 + x = x$, $\forall x \in M$;
2. $x + (-x) = (-x) + x = 0$, $\forall x \in M$.

Proposition 1.2. 1. $\alpha * (x - y) = \alpha * x - \alpha * y$;

2. $\alpha * 0 = 0$;

3. $\alpha * (-x) = -\alpha * x$.

Indeed, $\alpha * [(x - y) + y] = \alpha * x = \alpha * (x - y) + \alpha * y$, implies 1., 2. and 3. are consequences of 1.

Propositions 1.3. 1. $(\alpha - \beta) * x = \alpha * x - \beta * x$;

2. $0 * x = 0$;

3. $(-\alpha) * x = -\alpha * x$.

To prove 1. let us notice that $\alpha * x = [(\alpha - \beta) + \beta] * x = (\alpha - \beta) * x + \beta * x$, which gives us 1. The other two equalities are obtained by particularizing 1.

Theorem 1.3. *For any* α, $\beta \in \mathbb{R}$ *and* x, $y \in M$, *there is*

$$(1.1) \quad \alpha * x + \beta * y = \beta * y + \alpha * x.$$

Proof. $(\alpha + \beta) * (x + y) = \alpha * (x + y) + \beta * (x + y) = \alpha * x + \alpha * y + \beta * x + \beta * y =$
$= (\alpha + \beta) * x + (\alpha + \beta) * y = \alpha * x + \beta * x + \alpha * y + \beta * y$. But the following equality
$\alpha * x + \alpha * y + \beta * x + \beta * y = \alpha * x + \beta * x + \alpha * y + \beta * y$ implies (1.1), q.e.d.

Proposition 1.4. *If the map* $* : \mathbb{R} \times M \to M$ *is surjective, then the addition of the vectors is commutative:*

$$x + y = y + x, \forall x, y \in M.$$

Indeed, supposing $* : \mathbb{R} \times M \to M$ to be surjective, let x and y be two any elements in M. There exist then the scalars α, β and the vectors x_1, y_1 in M so that $x = \alpha * x_1$, $y = \beta * y_1$. According to (1.1), $x + y = y + x$, q.e.d.

The notion of almost linear subspace is defined in a usual manner: a non-void subset $M_1 \subset M$, which together with the restrictions of the maps + and $*$ is an almost linear space is called an almost linear subspace in M.

Theorem 1.4. *A non-void subset $M_1 \subset M$ is an almost linear subspace if and only if:*

  1. $x - y \in M_1$, $\forall x, y \in M_1$;
  2. $\alpha * x \in M_1$, $\forall \alpha \in \mathbf{R}$, $\forall x \in M$.

Indeed, condition 1. ascertains that $(M_1, +)$ is a subgroup of the group $(M, +)$. From 2. it results for $(M_1, +, *)$ that the other axioms of almost linear space are checked. The converse statement is immediate.

Definition 1.2. *We call morphism from the almost linear space $(M, +, *)$ to the almost linear space $(P, \oplus, \otimes)$ a map $f : M \to P$ which satisfies the conditions*

  1. $\forall x, y \in M, f(x + y) = f(x) \oplus f(y)$;
  2. $\forall x \in M, \forall \alpha \in \mathbf{R}, f(\alpha * x) = \alpha \otimes f(x)$.

The morphism f is called *injective* or *surjective* if the map f possesses this quality; if the map f is bijective, we say that f is an *isomorphism*.

Proposition 1.4. *If f is a morphism from the almost linear space $(M, +, *)$ to the almost linear space $(P, \oplus, \otimes)$, then:*

  1. Ker $f = \{x : x \in M$ and $f(x) = 0\}$ *is an almost linear subspace in $(M, +, *)$ (called the kernel of the morphism f)* ;
  2. Im $f = \{f(x), x \in M\}$ *is an almost linear subspace in $(P, \oplus, \otimes)$ (called the image of the morphism f).*

Theorem 1.5. *If $(M, +, *)$ is an almost linear space and $f : M \to M$ is defined by $f(x) = 1 * x$ then :*
  1. f *is a morphism of the almost linear space $(M, +, *)$ in itself*;
  2. Ker f *is an almost linear subspace in M (noted with M_0)* ;
  3. $(M_0, +)$ *is a normal divisor in the group $(M, +)$* ;
  4. Im f *is an almost linear subspace in M (noted with M_1) which satisfies the supplementary conditions*

$$
(1.2) \qquad
\begin{aligned}
&\forall\, x, y \in M_1,\ x + y = y + x, \\
&\forall\, x \in M_1,\ 1 * x = x.
\end{aligned}
$$

Proof. 1. We have $f(x + y) = 1 * (x + y) = 1 * x + 1 * y = f(x) + f(y)$; $f(\alpha * x) = 1 * (\alpha * x) = \alpha * (1 * x) = \alpha * f(x)$.

2. Is a consequence of proposition 1.4.

3. Let us notice now that $\forall\, x \in M_0$, $\forall\, y \in M$ there is $-y + x + y \in M_0$. Indeed, $f(-y + x + y) = -1 * y + 1 * x + 1 * y = -1 * y + 1 * y - 0$.

4. The first sentence derives from the proposition 1.4. Then we apply the Theorem 1.3 for $\alpha = \beta = 1$ to point out:

$$x + y = 1 * x_1 + 1 * y_1 = 1 * y_1 + 1 * x_1 = y + x.$$

Finally, $1 * x = 1 * (1 * x_1) = (1 \cdot 1) * x_1 = 1 * x_1 = x$, q.e.d.

Remark. *M_1 has a structure of almost linear space for which (1.2) are verified, so $(M_1, +, *)$ is a linear space. In the next paragraph we shall define the notion of linear space by means of independent axioms.*

Since $(M_0, +)$ is a normal divisor in M we can consider the factor set

(1.3) $\qquad M / M_0 = \{\, x + M_0 : x \in M \,\}.$

We shall put $\hat{x} = x + M_0$. In this set we define

(1.4) $\qquad \hat{x} + \hat{y} = x + y,\ \alpha * \hat{x} + \alpha * x.$

One can see that the second members do not depend on the choice of the representatives of $\hat{x}$ and $\hat{y}$, so + and * are operations in M / M_0 (the first one internal, the second one external).

There holds the following structure theorem given by R. Miron in [573]:

Theorem 1.6. 1. *The almost linear space M is the semi-direct sum of the almost linear subspaces M_0 and M_1.*

2. *M_1 is an almost linear space isomorphic with M / M_0.*

Proof. 1. Since $(M_0, +)$ is a normal divisor in $(M, +)$, we have to show only that $M_0 \cap M_1 = \{0\}$ and $M = M_0 + M_1$ hold good. Indeed, if $x \in M_0 \cap M_1$ then $1 * x = 0$, $x = 1 * x_1$.

If we multiply the second relation by the scalar 1, we get :

$$0 = 1 * x = 1 * (1 * x_1) = (1 \cdot 1) * x_1 = 1 * x_1 = x,$$

hence $x = 0$. Let now y be any element of M; we can write

$$y = (y - 1 * y) + 1 * y = y_0 + y_1 ,$$

where $y_0 = y - 1 * y$ and $y_1 = 1 * y$. But $y_0 \in M_0$ because

$$f(y_0) = 1 * y_0 = 1 * (y - 1 * y) = 1 * y - 1 * (1 * y) = 1 * y - 1 * y = 0.$$

We also have $y_1 \in M_1$ because $1 * y_1 = 1 * (1 * y) = 1 * y = y_1$. Consequently, M is the semidirect sum of M_0 and M_1.

2. The map $F : M / M_0 \to M$ given by $F(\hat{x}) = f(x) = 1 * x$ is an isomorphism.

Remarks. 1. M_1, being a linear space (over **R**), M / M_0 has the same property.

2. The semi-direct sum $M = M_0 \oplus M_1$ enable us to write uniquely any element $x \in M$ under the form $x = x_0 + x_1$, $x_0 \in M_0$ and $x_1 \in M_1$.

Corollary 1.1 (Liebeck, [571]). *If* $(M, +)$ *is a commutative group and* $(M, +, *)$ *an almost linear space, then the decomposition into a direct sum:* $M = M_0 \oplus M_1$ *holds.*

Indeed, in this case, $(M_0, +)$ and $(M_1, +)$ are normal divisors in $(M, +)$ and the first statement in the preceding theorem gives us the stated result. A construction procedure of an almost linear space, when the subspaces M_0 and M_1 are given was indicated by Professor Fr. Radó from the University in Cluj-Napoca [412]. We enounce this general result for the case when the scalars are in the field **R**.

Theorem 1.7. *Let* M_0' *be an almost linear space (over* **R**), *having the property* $1 * x = 0$, $\forall x \in M_0'$ *and* M_1' *a linear space (over* **R**). *In these conditions, there exists only one almost linear space* M *(over* **R**) *up to the isomorphisms so that its subspaces* M_0 *and* M_1 *in* $M = M_0 \oplus M_1$ *are isomorphic with* M_0' *and* M_1' *respectively.*

A last interesting result in the Geometry of almost affine spaces is given by:

Theorem 1.8. *If* $f : M \to M'$ *is a morphism of almost linear spaces (over* **R**), *then there exists only a morphism* $F : M / M_0 \to M' / M_0'$ *so that the following diagram is commutative*

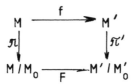

π *and* π' *being the canonical projections.*

Proof. Let be $F(\hat{x}) = \widehat{f(x)}$. If x and x' are in $\hat{x}$, then $x - x' \in M_0$. But

$1 * (x - x') = 0 \Rightarrow 1 * f(x - x') = 0 \Rightarrow f(x) - f(x') \in M_0'$, or $f(x) = f(x')$. Then $F(\hat{x})$
depends on $\hat{x}$ and not on x. To state that F is a morphism we notice:

$$F(\hat{x} + \hat{y}) = F(\widehat{x + y}) = f\widehat{(x + y)} = \widehat{f(x) + f(y)} = \widehat{f(x)} + \widehat{f(y)} = F(\hat{x}) + F(\hat{y}) \ ;$$

$$F(\alpha * \hat{x}) = F(\widehat{\alpha * x}) = f\widehat{(\alpha * x)} = \alpha * \widehat{f(x)} = \alpha * \widehat{f(x)} = \alpha * F(\hat{x}).$$

Since $\hat{x} = \pi(x)$ and $f(x) = \pi'(f(x))$, for any x in M we have
$(F \circ \pi)(x) = F(\hat{x}) = \widehat{f(x)} = (\pi' \circ f)(x)$, hence the announced diagram is commutative.
The uniqueness of the morphism F is justified by reductio ad absurdum. Let us suppose
that there are $F' \neq F$ and $F' \circ \pi = \pi' \circ f$; then $F' \circ \pi = F \circ \pi$ holds and since π is a
surjection: $F' = F$, in contradiction with the hypothesis, q.e.d.

Remark. *All the notions exposed in this paragraph can be resumed point by
point replacing the field* **R** *with any unitary ring.*

§2. Real Linear Spaces

The notion of linear space, frequently encountered, is usually defined by means
of a system of axioms, but it is not minimal. For this reason we shall modify the known
system, renouncing to the commutativity axiom of the addition of the vectors and
weakening the axiom $1 * x = x$.

Definition 2.1. *We call linear (real or over* **R***) space an almost linear space M (over* **R***), in which the axiom 7.* $\forall\ x \neq 0,\ 1 * x \neq 0$ *is verified.*

Therefore, in a linear space (over **R**) the axioms 1-6 in the definition 1.1 and the axiom 7. are verified.

Theorem 2.1. *The system of axioms 1-7 of the notion of linear space is non-contradictory.*

Indeed, for $M = \mathbf{R}$, in which + is the addition of real numbers and $\alpha * x = \alpha \cdot x$ is the product in **R**, all the axioms 1-7 are verified.

Theorem 2.2. *The system of axioms 1-7 of the notion of linear space is minimal.*

Proof. One can see easily that the models given in proving the independence of axioms 1-6 in the theorem 1.2 are also valid in the present case. By their means we prove in turn the independence of the first six theorems in the system 1-7. As for the independence of the last axiom, it is proved by the following model. Let $M = \mathbf{R}$ and + be the usual addition of real numbers. We define $\alpha * x = 0,\ \forall\ x \in \mathbf{R}$. Axioms 1-6 are checked, while 7. fails on this model. So 7. is independent from the other axioms, q.e.d.

As we have enunciated, the system of axioms 1-7 of a linear space does not contain the axiom $x + y = y + x$, and the axiom $1 * x = x$ is given under a weaker form in 7.

Theorem 2.3. *In any linear space M, over* **R***, the following properties* :

(2.1)
$$\forall x \in M,\ 1 * x = x;$$
$$\forall x,\ y \in M,\ x + y = y + x\ \text{hold.}$$

Proof. From $1 * (x - 1 * x) = 1 * x - 1 * x = 0$ and with the axiom 7. it results $x - 1 * x = 0$, so $x = 1 * x,\ \forall\ x \in M$. Then, the first equality (2.1) shows that the map $* : \mathbf{R} \times M \to M$ is surjective. According to proposition 1.4, it follows $x + y = y + x$, q.e.d.

Remark. *In a linear space M,* $1 * x = 0 \Rightarrow x = 0$. *There follows that the almost subspace* M_0 *is made up only of the vector 0, and the linear subspace* M_1 *coincides with M.*

Propositions 1.1, 1.2 and 1.3 are applied. The notion of almost linear subspace leads here to that of linear subspace. Theorem 1.4 is still true. The notions of morphism keeps its meaning for the linear spaces, as well. In this case, a morphism of linear spaces is also called *linear transformation or linear map.*

The notion of linear combination, of linear dependence and independence of the vectors and of the dimension of a linear space are known so that we shall not give them here.

Definition 2.2. *A linear (real) space M is called Euclidean linear space if it is given a map from $M \times M$ to* **R**, *called scalar product and noted with $(x, y) \rightarrow x \cdot y$, so that the axioms*:

1'. $\forall\, x, y, z \in M, x \cdot (y + z) = x \cdot y + x \cdot z$;
2'. $\forall\, x, y \in M, (\alpha * x) \cdot y = \alpha\, (x \cdot y), \forall \alpha \in \mathbf{R}$;
3'. $\forall\, x, y \in M, x \cdot y = y \cdot x$;
4'. $\forall\, x \neq 0, x^2 = x \cdot x > 0$

are verified.

The non-contradiction of the system of axioms 1-7, 1'-4' is proved by the existence of the following model:

Let be $M = \mathbf{R}^n$, with $x + y = (x^i) + (y^i) = (x^i + y^i)$. The axiom 1-7 are checked. Let now define the scalar product by

$$(x^i) \cdot (y^i) = x^1 y^1 + x^2 y^2 + \ldots + x^n y^n.$$

Axioms 1', 2', 3', 4' are easy to check. Therefore:

Theorem 2.4. *The system of axioms 1-7, 1'-4' of the notion of Euclidean linear space is non- contradictory.*

It can be proved without any major difficulties that the axioms 1'-4' are independent in the system 1-7, 1'-4', so that the above given complete system of the axioms of the Euclidean linear spaces is minimal.

Propositions 2.1. *The scalar product has the following properties*:

1. $x \cdot 0 = 0 \cdot x = 0$;
2. $x^2 = x \cdot x = 0 \Rightarrow x = 0$;
3. $(-x) \cdot y = x \cdot (-y) = -x \cdot y.$

The proofs are elementary.

Let us also notice that

$$\left(\sum_{k=1}^{n}\alpha^k x^k\right)\cdot\left(\sum_{h=1}^{p}\beta^h y^h\right) = \sum_{k=1}^{n}\sum_{h=1}^{p}\alpha^k\beta^k(x_k y_h).$$

In particular:

(2.2) $(x+y)^2 = x^2 + y^2 + 2x \cdot y.$

Definition 2.3. *We call length (or norm) of a vector x the non-negative real number noted with* $\| x \|$:

(2.3) $\| x \| = \sqrt{x^2}$.

Proposition 2.2. *There are* :
1. $\forall x \in M, \| x \|^2 = x^2$;
2. $\| x \| = 0 \Rightarrow x = 0$;
3. $\forall \alpha \in \mathbf{R}, \forall x \in M, \| \alpha * x \| = | \alpha | \cdot \| x \|$;
4. $\forall x \in M, \| - x \| = \| x \|.$

Indeed, 1. is equivalent with (2.3), and 2. is a consequence of the proposition 2.1.2. For 3. we notice $\| \alpha * x \| = \sqrt{(\alpha * x)^2} = \sqrt{\alpha^2 x^2} = | \alpha | \cdot \| x \|$. Particularizing $\alpha = -1$ we get 4.

Proposition 2.3. *For any two vectors x and y in M, the Cauchy-Schwartz inequality*

(2.4) $| x \cdot y | \le \| x \| \cdot \| y \|$

holds.

Proof. From $(\alpha * x + y)^2 \ge 0$ there is $\alpha^2 x^2 + 2\alpha \cdot x \cdot y + y^2 \ge 0$. The first member is a second degree trinomial in the variable α, and its discriminant must be negative or null. Then $(x \cdot y)^2 - x^2 y^2 \le 0$. Therefore $| x \cdot y | \le \| x \| \cdot \| y \|$, q.e.d.

Proposition 2.4. *For any two vectors , Minkowski's inequality holds* :

(2.5) $\| x + y \| \le \| x \| + \| y \|.$

Indeed,

$$\| x + y \|^2 = (x + y)^2 = x^2 + 2xy + y^2 \le \| x \|^2 + 2 \| x \| \cdot \| y \| + \| x \|^2 = (\| x \| + \| y \|)^2.$$

So $\| x + y \| \le \| x \| + \| y \|$.

Definition 2.4. *We call the angle of two non- null vectors x and y in M the real number $\varphi \in [0, \pi]$ whose cosine is given by*

(2.6) $\cos \varphi = \dfrac{x \cdot y}{\| x \| \cdot \| y \|}$.

When applying the Cauchy-Schwartz inequality, we obtain $-1 \le \dfrac{x \cdot y}{\| x \| \cdot \| y \|} < +1$,

so φ exists and is unique.

From (2.6) one deduces:

(2.7) $x \cdot y = \| x \| \cdot \| y \| \cos \varphi$.

This equality is valid when $x = 0$ or $y = 0$, too. For this reason we can give

Definition 2.5. *We shall say that the vector x is orthogonal to the vector y if*
$x = 0$ or $y = 0$, $x \ne 0$, $y \ne 0$ and $\varphi = \dfrac{\pi}{2}$. *It is written $x \perp y$.*

Proposition 2.5. 1. $x \perp y \Leftrightarrow x \cdot y = 0$;
2. $x \perp y \Rightarrow y \perp x$;
3. $x \perp y \Rightarrow \alpha * x \perp y$;
4. $\forall y \in M, 0 \perp y$;
5. $x \perp y \wedge x \perp z \Rightarrow \forall\, \alpha, \beta \in \mathbf{R}, x \perp (\alpha * y + \beta * z)$.
The proofs are easy.

A vector x is orthogonal to a subspace $M' \subset M$ if x is orthogonal to each vector in M'. When M' is generated by a system of vectors S, then x is orthogonal on M' if and only if x is orthogonal on each vector in S. This is easy to prove:

Proposition 2.6. *In an Euclidean linear space M of n dimension there exists a*

system of n vectors of 1 *length, orthogonal two by two. The vectors of such a system give a basis of M* (*called orthonormalized basis*).

§3. Real Almost Affine Spaces

To expose the notion of Euclidean space (see §5), making appeal to a simple axiomatic system satisfying the requirement of being non-contradictory and minimal, we have to present the more general concept of almost affine space. The classical construction of the affine spaces given by Weyl [544], and modified by Raşewsky [415] is not satisfactory because the axiomatic system taken into consideration by them is not minimal. We are constrained to present an alternative to this axiomatics, a distinct one which should satisfy the condition of minimality (see the papers [573-577]). The great advantage of such an axiomatics of affine spaces and then of the Euclidean ones consists in the fact that it has not to be modified at the same time with the dimension, as in the case of Hilbert's axiomatics, and this enables us to research the affine spaces and then the Euclidean infinitely dimensional ones.

Let a be a non-void set whose elements are noted with A, B, C, ... and called *points*. Any element $(A, B) \in a \times a$ is called *oriented segment* or *bipoint of origin A and extremity B*. We shall denote it by $\overline{AB} = (A, B)$.

Let ρ be an equivalent relation on $a \times a$. We denote by $T a = a \times a / \rho$ the set we call the *space of the vectors* of a or the *tangent space* of a. The elements of the quotient set $T a$ are noted with $\overrightarrow{AB}$, $\overrightarrow{CD}$, ..., x, y, ... We say then that $\overrightarrow{AB}$ is the vector determined by the oriented segment $\overrightarrow{AB}$ related to the equivalent relation ρ.

Definition 3.1. *We call almost affine* (*real or over the field* **R**) *space a non-void set* a, *together with the following data* :

1. *An equivalence relation* ρ *on the set of the oriented segments* $a \times a$ *in relation with which the vectors in* a *are defined as elements of the factor set* $T a = a \times a / \rho$;

2. *A map* $* : (\alpha, x) \in \mathbf{R} \times T a \to \alpha * x \in T a$ *so that the axioms in the groups I and II exposed below are checked.*

Group I of axioms:

$I_1.$ $\forall A \in a, \forall x \in T a, \exists B \in a : \overrightarrow{AB} = x;$

$I_2.$ $\forall A, B, A' \in a:$ $\overrightarrow{AB} = \overrightarrow{A'B} \Rightarrow A = A';$

$I_3.$ $\forall A, B, C, A', B', C' \in a:$ $\overrightarrow{AB} = \overrightarrow{A'B'} \wedge \overrightarrow{BC} = \overrightarrow{B'C'} \Rightarrow \overrightarrow{AC} = \overrightarrow{A'C'}.$

Remark. 1. *The last axiom is called by the Polish mathematician Stanislas Golab*, [46], "the Miron-Opaiţ axiom".

2. *It is clear that* $a \neq \varnothing \Rightarrow T a \neq \varnothing.$

The most important consequence of the axiom in group I is the possibility to define the operation of addition of the vectors in a. For this purpose we present some auxiliary results.

Proposition 3.1. *For any points A, A', B, B' in a*:

1. $\overrightarrow{AA} = \overrightarrow{A'A'}$;
2. $\overrightarrow{AB} = \overrightarrow{A'B'} \Rightarrow \overrightarrow{BA} = \overrightarrow{B'A'};$
3. $\overrightarrow{AB} = \overrightarrow{AB'} \Rightarrow B = B'$.

Proof. 1. By means of axiom I_1 we associate to the point A' and to the vector $\overrightarrow{AA}$ a point B' so that $\overrightarrow{A'B'} = \overrightarrow{AA}$. Let $\overrightarrow{AC}$ be arbitrary in $T a$; according to axiom I_1 there exists the point C' so that $\overrightarrow{B'C'} = \overrightarrow{AC}$. From $\overrightarrow{A'B'} = \overrightarrow{AA}$ and $\overrightarrow{B'C'} = \overrightarrow{AC}$ with axiom I_3 there follows $\overrightarrow{A'C'} = \overrightarrow{AC}$. Taking also into account that $\overrightarrow{B'C'} = \overrightarrow{AC}$, there follows $\overrightarrow{A'C'} = \overrightarrow{B'C'}$. By means of axiom I_2 it results $B' = A'$ and the equality $\overrightarrow{A'B'} = \overrightarrow{AA}$ becomes $\overrightarrow{AA} = \overrightarrow{A'A'}$, q.e.d.

2. Axiom I_1 applied to the vector $\overrightarrow{BA}$ and to the point B' ensures the existence

of a point C' so that $\overrightarrow{B'C'} = \overrightarrow{BA}$. From the premises $\overrightarrow{A'B'} = \overrightarrow{AB}$ and $\overrightarrow{B'C'} = \overrightarrow{BA}$,

with axiom I_3, we deduce $\overrightarrow{A'C'} = \overrightarrow{AA}$. We use conveniently the equality 1. and the last

equality becomes $\overrightarrow{A'C'} = \overrightarrow{C'C'}$. Axiom I_2 implies $C' = A'$, so $\overrightarrow{B'A'} = \overrightarrow{B'C'} = \overrightarrow{BA}$,

q.e.d.

3. According to 2., $\overrightarrow{AB} = \overrightarrow{AB'} \Rightarrow \overrightarrow{BA} = \overrightarrow{B'A}$; subsequently, axiom I_2 ensures

that $B = B'$, q.e.d.

According to the equality 1., the vector $\overrightarrow{AA}$ does not depend on the point A. We

shall say that $\overrightarrow{AA}$ is *the null vector* in $\mathcal{Q}$ and denote it by 0 (sometimes, in order to avoid

any possible confusions, we shall evidence its character of vector noting $\overline{0}$). The vector $\overrightarrow{BA}$

depends only on the vector $\overrightarrow{AB}$; it is called *the opposite of the vector* $\overrightarrow{AB}$ and will be

denoted by $\overrightarrow{BA} = -\overrightarrow{AB}$.

Let be $(x, y) \in T\mathcal{Q} \times T\mathcal{Q}$ and $A \in \mathcal{Q}$; then there exists only one point $B \in \mathcal{Q}$

and only one point C in $\mathcal{Q}$ so that $x = \overrightarrow{AB}$, $y = \overrightarrow{BC}$. According to the axiom I_3, the

vector $z = \overrightarrow{AC}$ does not depend on point A, but only on vectors x and y. It is called the

sum of *the vectors* x *and* y and is written $z = x + y$.

Definition 3.2. *The map* $+: (x, y) \in T\mathcal{Q} \times T\mathcal{Q} \to x + y \in T\mathcal{Q}$ *is called the*
operation of addition of the vectors of $\mathcal{Q}$. *There holds:*

(3.1) $\forall A, B, C \in \mathcal{Q} \colon \overrightarrow{AB} + \overrightarrow{BC} = \overrightarrow{AC},$

Theorem 3.1. $(T\mathcal{Q}, +)$ *is a group.*

Proof. For x, y, z in $T\mathcal{a}$ let be $x = \overrightarrow{AB}$, $y = \overrightarrow{BC}$, $z = \overrightarrow{CD}$. According to (3.1) one notices:

$$x+(y+z) = \overrightarrow{AB}+(\overrightarrow{BC}+\overrightarrow{CD}) = \overrightarrow{AB}+\overrightarrow{BD} = \overrightarrow{AD} = (\overrightarrow{AB}+\overrightarrow{BC})+\overrightarrow{CD} = (x+y)+z;$$

$$x+0 = \overrightarrow{AB}+\overrightarrow{BB} = \overrightarrow{AB} = x;$$

$$x+(-x) = \overrightarrow{AB}+(-\overrightarrow{AB}) = \overrightarrow{AB}+\overrightarrow{BA} = \overrightarrow{AA} = 0.$$

Remark. *In general, the group* $(T\mathcal{a}, +)$ *is not Abelian.*

Group II of axioms:
If the operation $+$ of addition of the vectors is given by the definition 3.2, then:

II_1. $\forall \alpha \in \mathbb{R}$; $\forall x, y \in T\mathcal{a}$: $\alpha * (x+y) = \alpha * x + \alpha * y$;

II_2. $\forall \alpha, \beta \in \mathbb{R}$, $\forall x \in T\mathcal{a}$: $(\alpha + \beta) * x = \alpha * x + \beta * x$;

II_3. $\forall \alpha, \beta \in \mathbb{R}$, $\forall x \in T\mathcal{a}$: $\alpha * (\beta * x) = (\alpha\beta) * x$.

Here closes the list of the axioms of an almost affine space.

Theorem 3.2. *If $\mathcal{a}$ is an almost affine space, then* $(T\mathcal{a}, +, *)$ *is an almost linear space.*

Indeed, Theorem 3.1 and the axioms in group II show that the tern $(T\mathcal{a}, +, *)$ is an almost linear space.

We can now expand the Geometry of the almost affine spaces, but the operation must be carefully handled, taking into account the properties of the almost linear space $T\mathcal{a}$.

Example. If $A \neq B$ are two points in $\mathcal{a}$, then we can not define the straight line determined by these points in an usual manner, as being the set of the points C in $\mathcal{a}$ with the property $\overrightarrow{AC} = \alpha * \overrightarrow{AB}$, $\alpha \in \mathbb{R}$ because we could not prove that the point B belongs to the straight line. The problem is solved as follows. The set of the vectors $\overrightarrow{AB}$, $\overrightarrow{BA}$ and

$\alpha * \overrightarrow{AB}$, $\alpha \in \mathbb{R}$ and of the finite sum of such vectors is denoted by $[\overrightarrow{AB}]$. The straight line determined by the distinct points A and B is defined as the set of the points $C \in \mathcal{A}$ for which the vector $\overrightarrow{AC}$ belongs to the set $[\overrightarrow{AB}]$.

But the geometric structure of these straight lines is not so simple. However, the problems related to the belonging of the points to a straight line, intersections of straight lines, parallelism can be solved. The notion of plane determined by three non co-linear points or by the k-plane determined by $k+1$ points which do not belong to a $k-1$ plane is analogously defined. We shall not insist upon these questions (for details the papers [572], [573], [577] can be consulted).

We shall prove now that the system of axioms under consideration is *non-contradictory and minimal*. As for completeness, it is out of question since new axioms, independent of those under consideration (such as the axiom $\forall x \neq 0: 1 * x \neq 0$), can be added to the system.

Theorem 3.3. *The axioms in group I and II are non-contradictory.*

Proof. Let be $\mathcal{A} = \mathbb{R} \times \mathbb{R}$. A homogenous binary relation ρ on $\mathcal{A} \times \mathcal{A}$ is given by:

$$[(a, b), (c, d)]\rho[(a', b'), (c', d')] \Leftrightarrow c-a = c'-a' \wedge d-b = d'-b'.$$

Obviously, ρ is an equivalence; the vector defined by the oriented segment $[(a, b), (c, d)]$ is (X^1, X^2) where $X^1 = c-a$, $X^2 = d-b$. One defines $*$ taking

$$\alpha * (X^1, X^2) = (\alpha X^1, \alpha X^2).$$

There are no difficulties to verify the axioms in group I and II. We content ourselves to draw the attention that there holds:

Theorem 3.4. *The axioms in group I and II are independent.*

§4. Real Affine Spaces

Definition 4.1. *An almost affine (real) space a is called affine (real or over the field* $\mathbf{R}$*) space if the axiom:*

III. $\forall\, x \in T\, a \setminus \{0\} : 1 * x \neq 0$ *is verified in* a.

The geometric structure of the almost linear space a is considerably particularized by adding this new axiom to the groups I and II. First of all, there is:

Theorem 4.1. *If a is an affine (real) space, then $(T\, a, +, *)$ is a linear (over* $\mathbf{R}$*) space.*

Indeed, theorem 3.2 and axiom III have as a consequence the fact that $(T\, a, +, *)$ is a linear space. We can now apply all the known properties of linear spaces to study the Geometry of a:

Definition 4.2. *We call affine Geometry the semi-formalized theory of the system made up of the groups I, II and III of axioms together with the required primary notions.*

Theorem 4.2. *Affine Geometry is non-contradictory. Its system of axioms is minimal.*

Out of the important notions in this Geometry, the first we encounter is that of affine dependence and independence of the points.

A system of points $A_0, A_1, \ldots, A_k$ is called *affinely independent* (or these points are *affinely independent*) if the vectors $\overrightarrow{A_0A_1}, \ldots, \overrightarrow{A_0A_n}$ are linearly independent. It is obvious that this property does not depend on the fixing of the point A_0 within this set. If $A_0, A_1, \ldots, A_k$ are not independent affine points, then they are called *affinely dependent*.

Example. Points A, B, C for which $\overrightarrow{AB} = \lambda \cdot \overrightarrow{AC}$ are affinely dependent.

We call straight line determined by the distinct points A and B in a the set of all points in a affinely dependent on A and B.

Propositions 4.1. *The straight line defined by the distinct points A and B in a is the set of the points $C \in a$ for which*

(4.1) $\exists\, \alpha \in \mathbf{R} : \overrightarrow{AC} = \alpha * \overrightarrow{AB}.$

Indeed, all points C affinely dependent on A and B are characterized by the equality (4.1).

(4.1) is called the vectorial equation of the straight line determined by A and B. It is equivalent to the vectorial equation

(4.1$'$) $\exists\, \alpha \in \mathbf{R} : \overrightarrow{OC} = \overrightarrow{OA} + \alpha * \overrightarrow{AB},$

in which O is an any point fixed in a.

The properties related to the belonging of the points to straight lines and to intersections of straight lines are now easy to solve by means of the vectorial equation (4.1) or (4.1$'$).

We shall say that the straight line d determined by the distinct points A, B is *parallel* to the straight line d' determined by the distinct points C, D if the vectors $\overrightarrow{AB}$, $\overrightarrow{CD}$ are linearly dependent. We shall denote this by $d \parallel d'$.

Proposition 4.2. *The relation $\parallel$ is an equivalence on the set d of the straight lines in the affine space a.*

An element of the factor set $d\, / \parallel$ (a class of parallel straight lines) is called direction of the affine space a.

Proposition 4.3. *Through a point $A_0 \in a$ there passes only one straight line of a given direction.*

Indeed, a direction is characterized by a non-null vector $m \in T\, a$. Then the straight line of equation

$$\exists \alpha \in \mathbf{R} : \overrightarrow{OC} = \overrightarrow{OA_0} + \alpha * m$$

has the required properties.

Another important notion is that of "simple ratio of three co-linear points".

Definition 4.3. *We say that the point C divides the oriented segment* $\overline{AB}$, $A \neq B$ *in the ratio* $k \in \mathbf{R} \setminus \{1\}$ *if* $\overrightarrow{AC} = k\,\overrightarrow{BC}$ *and we write* $k = (A, B \,; C)$.
(Here and in the following ones we omit the sign $*$ for the product of vectors by scalars or we replace it by $\cdot$.)

Remark. *For* $k = (A, B \,;\, C)$ *the equality* $k = 1$ *turns into* $\overrightarrow{AC} = \overrightarrow{BC}$ *which implies the impossible* $A = B$.

Proposition 4.4. 1. *Being given the distinct points* A, B *and fixed the number* $k \in \mathbf{R} \setminus \{1\}$, *there exists only one point* C *with the property* $k = (A, B \,; C)$.
2. *Being given the co-linear points* A, B, C, $A \neq B$, *there exists only one* $k \in \mathbf{R} \setminus \{1\}$ *so that* $k = (A, B \,; C)$.

Proof. For an arbitrary point O, the equality $\overrightarrow{AC} = k \cdot \overrightarrow{BC}$ is equivalent to

$$(4.2) \qquad \overrightarrow{OC} = \frac{\overrightarrow{OA} - k \cdot \overrightarrow{OB}}{1-k} \,.$$

Then: 1. If $A \neq B$ and k are given, (4.2) uniquely determines C with the required property.
2. If A, B, C are co-linear, it results uniquely defined the real number $k \neq 1$ so that $\overrightarrow{AC} = k \cdot \overrightarrow{BC}$, q.e d.

The equality $(A, B \,; C) = -1$ equivalent with $\overrightarrow{AC} = \overrightarrow{CB}$ is expressed by " C is *the middle* of the segment AB ", " C is the mid-point of AB " or by "the points A, B are *symmetric* related to C".
The notion of plane determined by three non co-linear points A, B, C is defined as the set of all points D affinely dependent on the points A, B, C. The plane defined by the non co-linear points A, B, C is characterized by the vectorial equation (for the generic point D):

$$\exists\, \alpha,\ \beta \in \mathbb{R}\colon \overrightarrow{AD} = \alpha\overrightarrow{AB} + \beta\overrightarrow{AC},$$

or by the equivalent equation:

$$\exists\, \alpha,\ \beta \in \mathbb{R}\colon \overrightarrow{OD} = \overrightarrow{OA} + \alpha\overrightarrow{AB} + \beta\overrightarrow{AC}.$$

O being fixed in $\mathcal{a}$, and $\overrightarrow{AB}$, $\overrightarrow{AC}$ vectors linearly independent.

All the properties of the plane can now be easily studied. Also, the intersection of a plane with a straight line and that of two planes is easy to be researched.

The notion of m-plane, defined by $m+1$ affinely independent points, is the immediate extension of that of straight line or plane.

Definition 4.4. *An affine (real) space is called of the dimension n if the tangent linear space* $T\mathcal{a}$ *has the dimension n.*

We use the notation $\mathcal{a}_n$ for the n-dimensional (real) affine space. We find for any point A_0 n linearly independent vectors $\overrightarrow{A_0A_i}$ ($i = 1,\ 2,\ \dots,\ n$). The points A_0, A_1, ..., A_n are affinely independent and make up a Cartesian (reference) frame in $\mathcal{a}_n$. In relation to such a frame, we can characterize the points by those n co-ordinates of theirs; the geometric notions in $\mathcal{a}_n$ are easy to be analytically expressed and the Geometry of $\mathcal{a}_n$ is algebraically edified.

Remark. In the affine Geometry exposed so far, the field $\mathbb{R}$ can be replaced by the field $\mathbb{C}$ and we shall obtain the "complex" affine Geometry.

§5. Euclidean Spaces

The Euclidean spaces are particularizations of the real affine ones. Then all the affine notions are also Euclidean notions; the Euclidean Geometry includes the real affine Geometry.

Definition 5.1. *We call Euclidean space a real affine space* $\mathcal{a}$ *for which the* *tangent space $T\mathcal{A}$ is an Euclidean linear space.*

This means that $T\mathcal{A}$ is endowed with a map from $T\mathcal{A} \times T\mathcal{A}$ with values in $\mathbb{R}$, denoted by $(x, y) \rightarrow (x \cdot y)$ and called scalar product that verifies the axioms $1' - 4'$ in Definition 2.2.

We shall denote by $\mathscr{E}$ an Euclidean space and with $T\mathscr{E}$ its tangent space.

Definition 5.2. *We call Euclidean Geometry the semi-formalized axiomatic theory of the Euclidean space $\mathscr{E}$.*

There is no difficulty in proving that the Geometry of $\mathscr{E}$ is non-contradictory and based on a minimal system of axioms.

Some important geometric notions deriving from the structure of Euclidean linear space of $\mathscr{E}$ are based on the following definition:

Definition 5.3. *We call distance between the any points A and B in $\mathscr{E}$ the non-negative real number denoted by $d(A, B)$ and given by*

$$d(A, B) = \| \overrightarrow{AB} \| .$$

Theorem 5.1. *The "distance" function $d: \mathscr{E}^2 \rightarrow \mathbb{R}$ has the following properties:*
1. $\forall A, B \in \mathscr{E} : d(A, B) \geq 0$;
2. $d(A, B) = 0 \Rightarrow A = B$;
3. $\forall A, B \in \mathscr{E} : d(A, B) = d(B, A)$;
4. $\forall A, B, C \in \mathscr{E} : d(A, B) \leq d(A, C) + d(C, B)$.

Proof. 1. From Definition 5.3 and the axiom $4'$ it results $d(A, B) \geq 0$.

2. By applying Proposition 2.2, $d(A, B) = \| \overrightarrow{AB} \| = 0 \Rightarrow \overrightarrow{AB} = 0 \Rightarrow A = B$ and conversely.

3. $d(A, B) = \| \overrightarrow{AB} \| = \| -\overrightarrow{AB} \| = \| \overrightarrow{BA} \| = d(B, A)$.

4. By using Minkowski's inequality,

$$d(A, B) = \| \overrightarrow{AB} \| = \| \overrightarrow{AC} + \overrightarrow{CB} \| \leq \| \overrightarrow{AC} \| + \| \overrightarrow{CB} \| = d(A, C) + d(C, B),$$

q.e.d.

The notions of distance and angle in the Euclidean space $\mathscr{E}$ permit to be established a lot of interesting results for the geometric notions in $\mathscr{E}$. Let A, B, C be three non co-linear points; we say that A, B, C are the vertexes of a triangle. We note with $\hat{A}$ the angle of the vectors $\overrightarrow{AB}$, $\overrightarrow{AC}$. There analogously are $\hat{B}$ and $\hat{C}$. Let then $a = d(B, C)$, $b = d(C, A)$, $c = d(A, B)$ called the lengths of the sides of the triangle ABC.

Proposition 5.1. *In any triangle ABC there holds $c < a + b$.*

The 4. point in Theorem 5.1 is applied.

Proposition 5.2. *In any triangle ABC, Pythagoras' generalized formula is true:*

$$a^2 = b^2 + c^2 - 2bc \cdot \cos A.$$

Indeed,

$$a^2 = \|\overrightarrow{BC}\|^2 = \|\overrightarrow{BA} + \overrightarrow{AC}\|^2 = \|\overrightarrow{AC} - \overrightarrow{AB}\|^2 = \|\overrightarrow{AC}\|^2 + \|\overrightarrow{AB}\|^2 - 2\overrightarrow{AC} \cdot \overrightarrow{AB} =$$
$$= b^2 + c^2 - 2bc \cdot \cos A$$

Remark. There also hold the formulae obtained by cyclicly permuting A, B, C.

Proposition 5.3. For any triangle ABC

$$c = b \cdot \cos A + a \cdot \cos B.$$

Indeed, $c^2 = \overrightarrow{AB}^2 = (\overrightarrow{AC} + \overrightarrow{CB}) \cdot \overrightarrow{AB} = bc \cdot \cos A + ac \cdot \cos B$ and one may simplify by $c \neq 0$.

Proposition 5.4. *For any triangle ABC, by denoting $p = \dfrac{1}{2}(a + b + c)$ and $k = \dfrac{2}{abc}\sqrt{p(p-a)(p-b)(p-c)}$ we have :*

$$\frac{a}{\sin A} = \frac{b}{\sin B} = \frac{c}{\sin C} = \frac{1}{k}.$$

Proof. Using the proposition 5.2 we calculate:

$$\sin^2 A = 1 - \cos^2 A = 1 - (\frac{b^2 + c^2 - a^2}{2bc})^2.$$

Hence:

$$(2bc\sin A)^2 = (2bc)^2 - (b^2 + c^2 - a^2)^2 = (2bc + b^2 + c^2 - a^2) \cdot (2bc - b^2 - c^2 + a^2) =$$
$$= [(b+c)^2 - a^2] \cdot [a^2 - (b-c)^2] = 16p(p-a)(p-b)(p-c).$$

There result $\sin A = ka$ and the enounced equalities.

Theorem 5.2. *In any triangle ABC the sum of the angles is equal to π, that is* $A + B + C = \pi$.

Proof. By noticing that $A + B + C = \pi$ is equivalent to $\cos(A + B + C) = -1$, which, in its turn, is equivalent to

$$1 + \cos A \cdot \cos B \cdot \cos C = \sin A \cdot \sin B \cdot \cos C + \sin A \cdot \cos B \cdot \sin C + \cos A \cdot \sin B \cdot \sin C.$$

There are no difficulties to verify this equality according to the propositions 5.4 and 5.2.

A triangle ABC with the property $\frac{\pi}{2} \in \{A, B, C\}$ is called rectangular.

Pythagoras' Theorem. *The triangle ABC is rectangular, having $A = \frac{\pi}{2}$, if and only if $a^2 = b^2 + c^2$.*

Proof. If $A = \frac{\pi}{2}$ in Proposition 5.2, there is $a^2 = b^2 + c^2$. Conversely, from $a^2 = b^2 + c^2$ and Pythagoras' generalized formula $a^2 = b^2 + c^2 - 2bc \cdot \cos A$ it results $-2bc \cdot \cos A = 0 \Rightarrow \cos A = 0 \Rightarrow A = \frac{\pi}{2}$, q.e.d.

An Euclidean space $\mathscr{E}$ has the dimension n if $T\mathscr{E}$ has the dimension n. It is noted with E_n.

The Geometry of E_n spaces can be analytically expanded by considering the orthonormalized Cartesian frames in E_n and by referring the points, straight lines, planes etc to them, see [109]. It has now the property of being a non-contradictory theory based on a minimal system of axioms, a complete and categorical theory. These properties will be proved in Hilbert's construction of Euclidean Geometry.

Problem

Let $(V, +, *)$ be a vectorial space over a field K, $(G, \oplus)$ an arbitrary group (with θ as neutral element) and $\sim$ a morphism of the group $(V, +)$ in the group $(\tilde{G}, \circ)$ of the automorphisms of $(G, +)$ (where $\sim : v \to \tilde{v} \in \tilde{G}$). One defines the operations

$$\uplus: (V \times G)^2 \to V \times G, \ominus: K \times (V \times G) \to V \times G$$

by $(v, g) \uplus (u, h) = (v + u, \tilde{u}(g) + h)$ and $\lambda \ominus (v, g) = (\lambda * v, \theta)$. Prove that $(V \times G, \uplus, \ominus)$ is an almost linear space over K.

CHAPTER V

THE BASES OF EUCLIDEAN GEOMETRY

In the preceding chapter we exposed the algebraic bases of Euclidean Geometry systematically using the field of real numbers we had constructed before.

But Geometry has to be edified independently of knowing the field of real numbers as a semi-formalized theory that covers the entire contents of Euclid's "Elements". Such a construction was achieved at the end of the last century by the well-known German mathematician David Hilbert in his book "Grundlagen der Geometrie" [229].

The system of axioms Hilbert enunciated is much more complicated then that used in Chapter IV. It is natural if we notice that we have to construct an axiomatic complex theory having a great number of primary notions and relations and, of course, a great number of axioms. But this also presents some important advantages:

- it does not use elements external to Geometry;
- its axioms are grouped in a natural manner;
- the expanding of the theory is done gradually and systematically;
- the common part of Euclidean and hyperbolic Geometries surprises;
- it satisfies the metatheory problems: non-contradiction, minimality, completeness and categoricity.

We shall enunciate in turn the groups of axioms and their most important consequences for the Geometry of Euclidean space.

Hilbert's axiomatic system constitutes a semi-formalized theory. Its primary notions are those of *point, straight line* and *plane*, and the primary relations are: *the incidence*, *"to be between"* and *the congruence*. The points will be denoted by A, B, C, ... the straight lines by a, b, c, ... and the planes by α, β, γ, ...; the incidence will be signalled by the symbol "belonging to" $\in$, the relation "to be between" by the succession of hyphens, $- -$, and the congruence by the sign $\equiv$. The twenty axioms are grouped in five groups; the first three groups are successively consecrated to the three fundamental relations, and the last two groups impose supplementary conditions by means of some derived notions.

We draw attention to the fact that the formulation given by Hilbert to the axioms completely avoids the set theory, and this is an advantage of scientific nature of the system since its metatheoretical analysis is not influenced by the existence or non-existence of a non-contradictory deductive theory for sets. But taking into account that the present exposition has a primary didactic purpose we shall borrow some notions, notations and formulations from the theory of sets.

The axiomatic system of Geometry does not justify either of the modalities of "illustrating" the geometric notions. But there are certain drawings that help "the intuitive following" of certain abstract reasonings. We shall provide some reasonings with figures without any "official character". To underline the conventional character of the illustrations we shall not adopt the Euclidean traditional interpretation of straight line; instead of tracing them with the rule, we shall trace them with the compasses. We shall see in Chapter VIII, §5, that, up to a point, such a convention is legitimate.

§1. Group I of Axioms

The primary relation intervening in these axioms is *the incidence*. We shall assimilate a straight line d by means of the set of points incident to the straight line and shall be able to replace the expression "point A is incident to the straight line d" by "A belongs to d". Completely analogously, we shall assimilate a plane α by means of the set of its incident points. The derived relation also seems natural: *the inclusion* benefitting from the definition: the straight line d is included into the plane α if any point incident to the straight line d is incident to the plane α.

In fact, owing to the possibility of choosing from among different formulations of the same mathematical propositions we avoid the danger of monotony and of undesirable repetitions.

We are presenting further the eight "incident" or "belonging to" axioms:

I_1. *Given two points, there is at least a straight line they are belonging to.*

I_2. *For any two distinct points there is at most a straight line incident to them.*

I_3. *At least two points belong to each straight line. There are three point so that no straight line could be incident to all these points.*

I_4. *Given three points there is at least one plane incident to them. For each plane there is at least one point belonging to it.*

I_5. *Given three points belonging to no straight line, there is at most a plane incident to them.*

I_6. *If two distinct points of a straight line belong to a plane, then any point of the straight line belongs to that plane.*

I_7. *If two planes have a point simultaneously belonging to them, then they have at least one more point having this property.*

I_8. *There are four points so that no plane may be incident to all these points.*

Of course, with the adopted notations, the axioms can be formulated more briefly; for example: I_1. $\forall A, B, \exists a: A \in a, B \in a$, etc.

The shortened form of certain expressions, derived notions whose definition is easy to imply: co-linear (points), co-planar (points), (plane or straight lines), secants etc., will be used.

Axioms I_3 and I_4 contain in fact each two axioms, but Hilbert's formulation was essentially preserved not to increase too much the number of axioms.

If we keep only axioms I_1, I_2 and the first part of axiom I_3 we obtain group I of axioms of the Geometry on the straight line.

Axioms I_1, I_2 and I_3 give the belonging axioms of the plane Geometry. The last part of axiom I_3 postulate the dimension of the plane, and axiom I_8 shows that the Geometry of the space under consideration does not reduce to the Geometry of a plane.

Here are some immediate and easy to prove consequences:

Proposition 1.1. *For any straight line a there is at least one point A non-incident to it.*

Proposition 1.2. *For any plane α there is at least one point A so that $A \notin \alpha$.*

Proposition 1.3. *For any distinct points A and B, there is only one straight line a so that $A \in a$ and $B \in a$.*

Proposition 1.4. *For any non co-linear points A, B, C, there is only one plane α incident to them.*

We shall sometimes denote the straight line a in Proposition 1.3 by (AB) and the plane α in Proposition 1.4 by (ABC) in the subsequent exposition we shall frequently use such expressions as "the straight line AB ", " the plane ABC ", considering that the expression "the straight line (AB) " constitutes a pleonasm justified only by the desire of underlining a certain aspect.

Proposition 1.5. *Two distinct straight lines have at most one common point.*

Proposition 1.6. *Two distinct planes having a common point have one and only one common straight line.*

Proposition 1.7. *Two distinct planes have at most a common straight line.*

Proposition 1.8. *A straight line a and a plane α can have the following relative positions: $a \subset \alpha$; and α have only one common point; a and α have no common points.*

Proposition 1.9. *Given a straight line a and a point A not belonging to it, there is only one plane α with the properties a ⊂α, A ∈α.*

Proposition 1.10. *For any plane there are at least three non co-linear points belonging to it.*

§2. Group II of Axioms

This group is entitled *the group of ordering axioms* and contains 4 axioms. A new primary relation is taken under consideration here, relation referring to the quality of a point of "being between" two other points, the three points being distinct two by two. Here are the axioms:

II$_1$. *If point B is between points A and C, then points A, B, C are distinct co-linear and point B is also between points C and A .*

II$_2$. *For any two distinct points A and B there is at least a point C co-linear with A and B so that point B were between points A and C.*

II$_3$. *From among three points, distinct two by two, at most one is between the other two.*

II$_4$. *For any three non co-linear points A, B, C and any straight line a in their plane, to which none of the points A, B, C belongs, if there is on a one point situated between two of the points A, B, C, then there is on a at least one more point situated between the other two of the points A, B, C.*

We use for II$_4$ the name of *Pasch's axiom*.

For the situation when B is between A and C we use the notation $A - B - C$ or $B \in |AC|$; we obviously imply $|AC|$ as a subset of the points of the straight line AC. From the axioms it results $|AA| = \varnothing$ and $|AC| = |CA|$. We call *segment* a non-ordered pair of points $\{A, B\}$; it will be noted with AB or BA, A and B being called *the extremities* of the segment, and points C with the property $A - C - B$ are internal points of the segment. It is justified to call $|AC|$: *the interior* of the segment AC. If $A \neq B$, the points of the straight line (AB) which are not extremities or internal points of the segment AB are called *external points* of the segment AB on the straight line a.

We call *triangle* a non-ordered tern of non co-linear points A, B, C. We shall denote it by ABC or $\triangle ABC$ and call points A, B, C *its vertexes* and the segments AB, BC, CA *its sides*; we shall sometimes understand by sides the straight lines determined by the vertexes. With this preparation, the reader can reformulate Pasch's axiom in a more convenient manner.

From among the consequences of the axioms in groups I and II, we retain just a few.

Theorem 2.1. *Any segment* AB, $A \neq B$, *has internal points.*

Proof. Let E be a point non co-linear with A, B and a point F chosen (according to axiom II_2) so that $A-E-F$ (fig. 1). There is a point G with the property $F-B-G$. We apply Pasch's axiom to the triangle AFB and to the straight line EG. We find that the straight line EG can not cut the side FB, so it has a common point with the interior of the segment AB, q.e.d.

Theorem 2.2. *For any co-linear and distinct two by two three points* A, B, C, *one and only one of them is situated between the other two.*

Proof. According to axiom II_3, it is enough for us to prove that at least one of the co-linear and distinct two by two points A, B, C is situated between the other two.

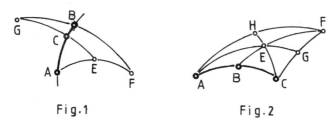

Fig.1   Fig.2

Let us admit that neither $B-A-C$, nor $B-C-A$ exist. Let us show that there is $A-B-C$. Let E be a point which does not belong to the straight line AB and F a point for which $B-E-F$ (fig. 2). Axiom II_4 applied to the triangle BFC and to the straight line AE determines a point $G \in AE$ so that $F-G-C$. By applying again Pasch's axiom to the triangle ABF and to the straight line CE we find a point $H \in CE$ so that $A-H-F$. There immediately follows $A-E-G$, and by a new applying of axiom II_4 to the triangle ACG and to the straight line FE it results $A-B-C$, q.e.d.

Theorem 2.3. *Let* ABC *be a triangle and* a *one straight line in its plane non-incident to any vertex. If* a *cuts in the inside one of the sides of the triangle, then it also cuts in the inside one and only one of the other two sides.*

Proof. According to Pasch's axiom, if a cuts in point P inside the side AB of the triangle, then it cuts inside at least one of the sides $|BC|$ and $|CA|$ (fig. 3). Let us admit that a also cuts inside $|BC|$ in a point Q and $|CA|$ in a point R. Thus, there are

(*) $A-P-B$, $B-Q-C$, $C-R-A$.

It is clear that points P, Q, R of the straight line a are distinct two by two. According to the preceding theorem, only one of these points is situated between the other two. Let Q be this one. Then $P-Q-R$. We consider the triangle APR and the straight line BC that

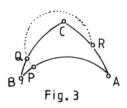

Fig. 3

cuts the side PR in the internal point Q. According to Pasch's axiom, it has to cut in the inside at least one of the remaining sides, that is $A-B-P$ or $A-C-R$. But these relations contradict the relations (*). The absurdity comes from the suppositions we made. The theorem is proved.

§3. The Orientation of the Straight Line

The axioms of order enable us to establish the existence of a relation of total order for the points of any straight line. In its turn, it will give us the possibility to show that an ordered straight line is similar to the set $(R, \leq)$. To avoid the difficulties in the usual approaches of this subject we shall present an original construction.

Lemma 3.1. *Let* A, B, C, D *be four co-linear points. If* $A-D-C$ *and* $B-D-C$ *do not hold, then* $A-D-B$ *does not hold either.*

Indeed, if E does not belong to the straight line the given points are on and F is a point with the property $C-E-F$, we can apply axiom II_4 to the triangle ACF and to the straight line DE (fig. 4). It result that there is a point H so that $A-H-F$. By also considering now $\triangle BCF$ and the straight line DE we obtain a point G so that $B-G-F$. According to the theorem 2.3 for $\triangle ABF$, the point D can not satisfy the relation $A-D-B$.

Lemma 3.2. *If for any four co-linear points the relations* $A-D-C, B-D-C$ *hold, then the relation* $A-D-B$ *does not hold.*

Indeed, let E be a point which does not belong to the straight line of the four points A, B, C, D and F a point with the property $C-E-F$ (fig. 5). Pasch's axiom and Theorem 2.3 applied to the triangles AFC, BFC and to the straight line DE shows that DE

can not cut in the inside the segments *AF* and *BF* which are sides of the triangle *ABF*. So point*D* can not be interior to the segment *AB*.

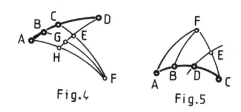

Fig.4　　　　　　　Fig.5

Let now *a* be a straight line *O* one of its points and m the set of the points of *a* except point *O*. We consider on m the following relation ∼ :

$$A \sim B \Leftrightarrow (A = B, \text{ or } O\text{-}A\text{-}B, \text{ or } O\text{-}B\text{-}A) \Leftrightarrow \text{non } A\text{-}O\text{-}B.$$

Proposition 3.1. *The relation* ∼ *is an equivalence on the set* m.

Indeed, $A \sim A$, $\forall A \in m$. From the definition of the relation ∼ , $A \sim B \Leftrightarrow B \sim A$. Let us suppose now that $A \sim B$, $B \sim C$ hold. This means that neither $A\text{-}O\text{-}B$, nor $C\text{-}O\text{-}B$ (or $O = A$, $O = B$, $O = C$) is. According to Lemma 3.1, $O\text{-}C\text{-}A$, or $A = C$, so $A \sim C$, q.e.d.

We consider the quotient set $m / \sim$. An element of this set is called an *open ray of O origin on a*.

Theorem 3.1. *Let a be a straight line and O a point of a. There exist exactly two open rays of O origin on a.*

Proof. Any of the rays of *O* origin on *a* is determined by one of its representatives which is any point $A \neq 0$ on the straight line *a*. By fixing *A* we have defined the open ray which we denote by $(O; A)$ given by

$$(O; A) = \{B \in a: O\text{-}A\text{-}B \text{ or } O\text{-}B\text{-}A \text{ or } B = A\}.$$

Let us consider now a point *C* with the property $A\text{-}O\text{-}C$. According to axiom II$_2$, such points exist. It follows $C \in a$, $C \notin (O; A)$. So $(O; C)$ is an open ray of *O* origin on the straight line *a* which has no common points with the open ray $(O; A)$. Therefore there are two open of *O* origin. Let us admit that there would exist three open rays distinct two by two on a: $(O; A)$, $(O; B)$, $(O; C)$. Then $A - O - B$, $B - O - C$ and $C - O - A$. But according to Lemma 3.2 this is impossible, q.e.d.

We shall denote $(O; A) \cup [0] = [O; A)$ and we shall call it (*closed*) *ray of O origin and passing through A*. When confusions will not be possible we shall denote the (closed) rays with h, k, ..., using also indexes, eventually.

Theorem 3.2. *Let a be a straight line and O one of its fixed points. There are two and only two rays on a with the property that their union gives us the set of the points of the straight line a and their intersection is the set $\{O\}$.*

The proof is immediate if one takes into account Theorem 3.1 and the definition of the rays $[O; A)$, $[O, B)$ with $A - O - B$.

Let us consider a straight line a and let Σ be the set of all its rays. We introduce on Σ the binary relation $\overset{\rightharpoonup}{\approx}$ defined by $h \overset{\rightharpoonup}{\approx} k \Leftrightarrow h \cap k \in \Sigma$. By using the two above lemmas we shall find:

$$h \overset{\rightharpoonup}{\approx} k \Leftrightarrow h \subset k \text{ or } k \subset h.$$

Proposition 3.2. *Relation $\overset{\rightharpoonup}{\approx}$ is an equivalence on Σ.*

The statement easily results from the definition of the relation $\overset{\rightharpoonup}{\approx}$ and from the properties of inclusion.

An equivalence class related to $\overset{\rightharpoonup}{\approx}$ in the set Σ is called an *orientation of the straight line a*. A representative of an orientation is a ray $[0; A)$ of the straight line a.

Theorem 3.3. *The straight line a has two and only two orientations.*

Indeed, according to Theorem 3.2, there are two and only two distinct representatives of O origin: $[O, A)$ and $[O, B)$ with $A - O - B$, q.e.d.

One of the two orientations of the straight line a will be called *positive*, and other one *negative*.

In conclusion, to fix an orientation on a straight line a means to specify a ray h included into a. All the other rays of the straight line a in the relation $\overset{\rightarrow}{\rightarrow}$ with h will be representatives for this orientation. If the orientation given by $h = [O, A)$ is called positive, then the negative orientation is, for example, characterized by a ray $[O, B)$ with $A - O - B$.

Definition 3.1. *We call oriented segment or bi-point an ordered pair of points* (A, B). *If* $A = B$, *the segment* (A, A) *will be called null oriented segment.*

If a is an oriented straight line and $\overline{AB}$ one of its non-null oriented segments $(A \in a, B \in a)$, then we say that $\overline{AB}$ is *positively (negatively) oriented* if the ray $[A; B)$ is positively (negatively) oriented.

All the properties known for the set of oriented segments of some straight lines can now be reproduced point by point. We only remark that for $A \neq B$ the oriented segments $\overline{AB}$ and $\overline{BA}$ have opposed orientations.

The ordering of the points of a straight line. Let a be a straight line on which we fixed an orientation called positive.

Definition 3.2. *We say that point* $A \in a$ *precedes the point* $B \in a$ *and we write* $A \prec B$ *if non-null oriented segment* $\overline{AB}$ *is positively oriented. We shall put* $A \preceq B$ *if* $A \prec B$ *or* $A = B$.

Theorem 3.4. *The relation* $\preceq$ *on the set of the points of the straight line* a *is a relation of total and dense order.*

Proof. According to the definition of the relation $\preceq$ there is $A \preceq A$; if $A \preceq B$ and $B \preceq A$ the supposition $A \neq B$ leads to absurd, so $A = B$ holds. If $A \preceq B$, $B \preceq C$, then $\overline{AB}, \overline{BC}$ are positively oriented, that is the rays $[A; B)$, $[B; C)$ are positively oriented. This implies the fact that $[A; C)$ is positively oriented, so $A \preceq C$ (the cases when $A = B$ or $B = C$ are to be separately treated).

Finally, for A, B being arbitrary, if $A \neq B$, then $\overline{AB}$ is positively or negatively oriented. Therefore exactly one of the following situations holds: $A \prec B$, $A = B$, $B \prec A$.

The property of "density" of $\preceq$ derives immediately from the theorem 2.1.

Theorem 3.5. *If $A - B - C$, then $C \prec B \prec A$ or $A \prec B \prec C$ and conversely.*

The proof resides in the observation that the situation $A - B - C$ is equivalent with "$\overline{BA}$ and $\overline{BC}$ are differently oriented". Two eventualities in the enunciation correspond to the choice of the orientation provided on (AB) by $[B \; ; A)$ and $[A \; ; B)$, respectively.

Let $\leq$ be a relation of order on the set of the points incident to a straight line a; it is said that $\leq$ is *compatible* with the relation "to be between", $? - ? - ?$, if

$$A < B \text{ and } B < C \Rightarrow A - B - C.$$

Thus it results the following theorem which summarizes the considerations in this part of the paragraph.

Theorem 3.6. *On a straight line a there exist exactly two dense and total orderings compatible with the relation "to be between" and one of these orderings is the reverse of the other one. The choice of a compatible ordering on a is equivalent with the orientation of the straight line a.*

We draw attention that the long enough expression "order relation on a straight line d compatible with the betweeness relation" in often replaced by the shortest one "sense of d ".

The notion of orientation of the straight line can be extended to the plane and space, but the problem is no more a simple one. For this reason we shall present only the notion of semi-plane.

Let α be a plane and a one of its straight lines. We denote by m the set of the points of plane α which do not belong to the straight line a. We consider on m the binary relation $\sim$ defined by: $A \sim B$ if and only if no points inside the segment AB exist on a:

$$A \sim B \Leftrightarrow \left| AB \right| \cap a = \varnothing.$$

Proposition 3.3. *The relation $\sim$ is an equivalence on m.*

Indeed, $\forall A \in m$, there is $A \sim A$. If $A \sim B$, then $| AB |$ and a have no common point, so $B \sim A$. If now $A \sim B$ and $B \sim C$, then, supposing that points A, B, C are not co-linear, the segment, $| AC |$ can not have common points with a because, according to

Pasch's axiom, it would follow that $|AB|$ or $|BC|$ had common points with a, which is absurd. If A, B, C are co-linear, by applying Lemmas 3.1 and 3.2, the conclusion is immediate, q.e.d.

An equivalence class related to $\sim$ in the set $\mathcal{m}$ is called an *open semi-plane determined by the straight line a in the plane* α. It is noted with $(a; A)$, A being a representative of the semi-plane under consideration.

Theorem 3.7. *Let α be a plane and a one straight line in α. Then there exist two and only two open semi-planes determined by the straight line a in the plane α.*

Proof. Each open semi-plane $(a; A)$ being characterized by a point A, we can fix one of them. If $C \in a$ is any point, we can find a point B so that $A - C - B$. Then $A \ne B$ and $(a; B)$ is another open semi-plane. Consequently, there are at least two open semi-planes determined by the straight line a in the plane α.

Let us admit that there would be three semi-planes $(a; A)$, $(a; B)$, (a, C) distinct two by two. Points A, B, C can be obviously supposed to be not co-linear. Then the straight line a does not pass through none of the vertexes of the triangle ABC and cuts in the inside of each of its sides, which is absurd according to Theorem 3.2, etc.

We denote by $[a, A)$ the set (a, A), to which the set of the points of the straight line a is added. We call $[a, A)$ *(closed) semi-plane determined by the straight line a in the plane α.*

Definition 3.3. *We call angle, a system of two rays $[O, A)$, $[O, B)$ with the same origin. If $[O; A) = [O, B)$, the angle is called null (or flattened) angle, and if $[O, A)$ and $[O, B)$ are distinct but belong to a straight line, the angle is called "elongated". Angles which are not null or elongated are called proper angles. The point O is called the vertex of the angles and the rays $h = [O, A)$, $k = [O, B)$ its sides. Such an angle is denoted by*
$$\sphericalangle hOk, \; \widehat{hOk}, \; \sphericalangle(h,k), \; (\widehat{h,k}) \; \sphericalangle AOB \; or \; \widehat{AOB}.$$

To the proper angle $\widehat{AOB}$ the set $\text{Int} \widehat{AOB} = ((OA); B) \cap ((OB); A)$ is associated, whose points M are called *internal* to the angle AOB and are obviously characterized by the fact that they are on that side of (OA) on which B is, and on that side of (OB) on which A is. Naturally, $\text{Int} \widehat{AOB}$ is to be read: the interior of the angle

AOB. It is easy to deduce:

Proposition 3.4. *Let be the angle AOB and the open ray* $s = (O; C)$*; the following three statements are equivalent:*

1. $C \in \text{Int}\widehat{AOB}$;

2. $s \subset \text{Int}\widehat{AOB}$;

3. *s contains a point internal to the segment AB:* $s \cap |AB| \neq \varnothing$.

Remark. Sometimes one defines angles as pairs of open rays (of the same origin); such details are not essential in the development of the theory.

The notion of *semi-space* delimited by a plane α is analogous to that of semi-plane; we shall consider it to be known.

This notion leads to the concept of dihedral angles but we shall be not interested to walk such a way.

§4. Group III of Axioms

The five axioms of this group refer to the primary relation of congruence. Of course, the name we are attributing to this relation and its notation way are arbitrary; it is desirable that neither from the name nor from the notation should result that it would be an equivalent relation not to add supplementary conditions to the axioms. The notation suggested by Hilbert for this notion, $\equiv$, has been accepted all over the world and it seems useless and difficult to be replaced. But we mention that the great Romanian mathematician Dan Barbilian (the same as the outstanding poet Ion Barbu) has proposed for this relation the symbol $\rightrightarrows$ which suggested in an inspired manner the absence within the axioms of the requirement of reflexivity of the congruence. Out of didactic reasons and with the purpose of interpreting the congruence in the set theory, we could "undo" it into two binary relations: a "congruence of the segments" and a "congruence of the angles".

Group III of Axioms

III_1. *For any non-null segment AB and any ray* $(A'; X')$ *there exists at least one point* B' *on* $(A'; X')$ *so that* $AB \equiv A'B'$.

III_2. *If* $AB \equiv CD$, *and* $A'B' \equiv CD$ *then* $AB \equiv A'B'$.

III_3. *If* $A - B - C, A' - B' - C', AB \equiv A'B', BC \equiv B'C',$ *then* $AC \equiv A'C'$.

III_4. *For any proper angle* (h, k), *any closed semi-plane* σ *delimited by the support straight line of a ray* h' *there exists in* σ *only one ray* k' *such that* $(h, k) \equiv (h', k')$. *Any proper angle is congruent with itself.*

III_5. *If* $ABC, A'B'C'$ *are two triangles for which* $AB \equiv A'B', AC \equiv A'C',$ $BAC \equiv B'A'C'$ *then,* $ABC \equiv A'B'C'$.

As one can see, the first three axioms refer to the congruence of segments, the fourth one only to the congruence of angles and last one contains both of them.

One uses to refer to the axiom III_1 as *the axiom of congruent carrying of segments* and to III_4 as the *axiom of congruent carrying of angles*. For III_3 is used the name: *the axiom of sum of segments* and III_5 is "*SAS axiom*" (*SAS* being an abbreviation for side-angle-side).

On the basis of the axioms in groups I, II and III a great number of important consequences is obtained; we shall expose just a few of them.

Proposition 4.1. *Let* $ABC, A'B'C'$ *be two triangles with the property* $AB \equiv A'B',$ $AC \equiv A'C', BAC \equiv B'A'C'$. *Then* $ABC \equiv A'B'C', ACB \equiv A'C'B'$.

The property is the immediate consequence of axiom III_5.

Theorem 4.1. *The relation* $\equiv$ *is an equivalence on the set of non-null segments.*

We shall present *the proof* of this theorem within a wider frame and intrinsic interest. Let S be set of all segments and a relation $\sim$ on S which satisfies the conditions:

E.1 $\forall s \in S, \exists s' \in S : s \sim s'$,

E.2 $s_1 \sim s_2 \wedge s_2 \sim s_3 \Rightarrow s_1 \sim s_2$.

It is easy to find out that $\sim$ is an equivalence. Indeed, for x arbitrary in S, according to E.1, there exists an y so that $x \sim y$; there follows ($x \sim y$ and $x \sim y$), and from here, according to E.2, it results $x \sim x$, that is $\sim$ is reflexive. Then we find out that $x \sim y \Rightarrow$ $\Rightarrow (y \sim y$ and $x \sim y) \Rightarrow y \sim x$, so $\sim$ is symmetrical. On the basis of the property of symmetry we can reformulate E.2 by: $(s_1 \sim s_3$ and $s_3 \sim s_2) \Rightarrow s_1 \sim s_2$, implication that expresses the transitivity of $\sim$.

One can now easily find out that $\equiv$ satisfies E.1 (as an "essentialization" of axiom III$_1$) and E.2 (which coincides with III$_2$), so $\equiv$ is an equivalence on the set of non-null segments, q.e.d.

We have thus evidenced an element of the "Aesthetics" of Hilbertean axiomatics: the "modest" requirements in the axioms lead, by means of "rational management", to major conclusions.

Theorem 4.2. *For any segment AB and any ray h of A' origin there is only one point B' on h so that $AB \equiv A'B'$.*

Proof. According to axiom III$_1$, we have to prove only the uniqueness of the point B'. Let us suppose, by reductio ad absurdum, that there is $B'' \neq B'$ so that $AB \equiv A'B'$ and $AB \equiv A'B''$. Let F be a point which does not belong to the support straight line of h. From the triangles $A'FB'$ and $A'FB''$ and from Proposition 4.1 there is $A'FB' \equiv A'FB''$ and it is obvious that $[F ; B')$, $[F ; B'')$ are in the same semi-plane determined by the straight line $A'F$. This contradicts the uniqueness of the angle $A'FB'$, in the axiom III$_4$, q.e.d.

From here, by reductio ad absurdum, there derives the implication:

$$(4.1) \quad (A - B - C \text{ and } A' - B' - C' \text{ and } AB \equiv A'B' \text{ and } AC \equiv A'C') \Rightarrow (BC \equiv B'C').$$

One uses to name it *the theorem of the difference of segments*.

In the subsequent lines, by concomitantly considering two triangles ABC and DEF, we shall understand as being also specified the bijective correspondence between their vertexes, whose graph is $\{(A, D), (B, E), (C, F)\}$ and which enables us to speak about "corresponding elements".

Definition 4.1. *Two triangles ABC, A'B'C' are called congruent if their corresponding sides are congruent.*

Proposition 4.2. *The relation of congruence of the triangles is an equivalence.* This property results from Definition 4.1 and Theorem 4.1.

Theorem 4.3. *If the triangles ABC and A′B′C′ have the property AB ≡ A′B′, AC ≡ A′C′, A = A′, then they are congruent.*

Proof. Let be, by absurd, $BC \not\equiv B'C'$. According to III$_1$, there will be D on $(B'; C')$ so that $BC \equiv B'D$. According to III$_5$, there follows $\widehat{BAC} \equiv \widehat{B'A'D}$ and from axiom III$_3$ it results that the rays $[A'; D)$ and $[A'; C')$ coincide, so $D \in (A'C')$. There follows that D coincides with C', in contradiction with the supposition we made etc.

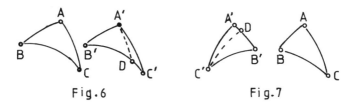

Fig.6   Fig.7

Theorem 4.4. *If the triangles ABC, A′B′C′ have the property BC ≡ B′C′, B ≡ B′, C ≡ C′, then they are congruent.*

Proof. We shall show that $BA \equiv B'A'$. Let us suppose the contrary, that is $BA \not\equiv B'A'$. Let D be on $[B'; A')$ so that $BA \equiv B'D$. From the triangles $BAC, B'DC'$ it results $\widehat{BCA} \equiv \widehat{B'C'D}$ and, according to III$_3$, $D \in (A'C')$, which implies $D = A'$ etc.

A triangle ABC with the property $AB \equiv AC$ is called isosceles. It is immediately proved:

Proposition 4.3. *If in the triangle ABC there is AB ≡ AC, then B ≡ C and C ≡ B.*

Proposition 4.4 (The difference of angles). *Let be $\widehat{BAD} \equiv \widehat{B'A'D'}, D \in \text{Int}\widehat{BAC}$ and $D' \in \text{Int } B'\widehat{A'}C'$. If $\widehat{BAC} \equiv B'\widehat{A'}C'$, then $\widehat{DAC} \equiv D'\widehat{A'}C'$ (and conversely).*

The proof considers (according to III₁) E on $(A'; B')$ so that $AB \equiv A'E$ and F on $(A'; C')$ so that $AC \equiv A'F$. Proposition 3.4 ensures the existence of a point $G \in (A'; D') \cap |EF|$. The triangle ABC and AEF are congruent (Theorem 4.3), so $\widehat{ABC} \equiv A'\widehat{EF}$, $BC \equiv EF$ and $\widehat{ACB} \equiv A'\widehat{FE}$. It results $\triangle DAB \equiv \triangle GA'E$ (Theorem 4.4), so $BD \equiv EG$. The implication (4.1) leads to $DC \equiv GF$. The enunciated conclusion appears by applying axiom III₅ to triangles CAD and $FA'G$.

The converse is easy to prove by reductio ad absurdum and deserves the title of "the sum of the angles".

One of the important consequences of the considerations made so far is given by:

Theorem 4.5. *If the triangles ABC and A′B′C′ are congruent, then their corresponding angles are congruent, too.*

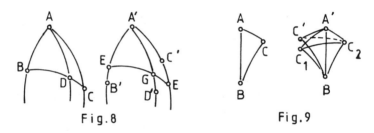

Fig.8   Fig.9

Proof. Let us suppose that $A \not\equiv A'$. We consider the angles $C_1A'B'$ and $C_2A'B$ to be congruent with the angle A and situated in different semi-planes determined by the straight line $B'A'$ in the plane of triangle $A'B'C'$, C_2 and C' being separated by $(A'B')$. Moreover, we can suppose that points C_1 and C_2 are chosen so that $AC \equiv A'C_1 \equiv A'C_2$. Triangles $A'C_2C'$ and $B'C_2C'$ being isosceles, $A'\widehat{C_2}C' \equiv A'\widehat{C'}C_2$, $B'\widehat{C_2}C' \equiv B'\widehat{C'}C_2$. One can then find out, by applying proposition 4.4 (or one of its converses), that $B'\widehat{C_2}A' \equiv B'\widehat{C'}A'$. According to theorem 4.3, $\triangle A'C_2B'$ is congruent with $_\Delta A'C'B'$. It follows that $C_2 A'B' \equiv B'A'C' \equiv B'A'C_1$, that is $A \equiv A'$, in contradiction with the adopted hypothesis.

Theorem 4.6. *The congruence of the angles is an equivalence relation.*

Proof. Axiom III$_4$ ensures reflexivity. If we suppose then that $AOB \equiv A'O'B'$, generality is not restrained if we accept $OA \equiv O'A'$ and $OB \equiv O'B'$. According Theorem to 4.3 we deduce $_\Delta AOB \equiv {_\Delta}A'O'B'$, so $AB \equiv A'B'$. Taking into account that the congruence relation of the segments is reflexive, we apply Theorem 4.5 for $_\Delta A'O'B'$ and $_\Delta AOB$; we deduce $A'O'B' \equiv AOB$ and have thus proved the symmetry of the congruence of the angles. An analogous technique of the incorporation of the congruent angles into congruent triangles enables us to transfer the property of congruence transitivity from segments to angles and complete the proof.

Remarks. The last three theorems are very useful to decide that two triangles are congruent. Using the letters S, A for sides and angles one refers to the theorems 4.3, 4.4, 4.5 by the intuitive abbreviations: *SAS* case, *ASA* case, *SSS* case, respectively.

We draw attention to the unitary character of group III of axioms, the major property of the relations between segments and angles, respectively, being gradually edified by logical implications with alternative directions.

A number of elementary properties in which segments and angles intervene can be easily enunciated and proved. The notions of angles opposed at vertex and supplementary angles are current ones.

Definition 4.2. *We call right angle a congruent angle with its supplement. The sides of a right angle are called perpendiculars.*

Theorem 4.7. *There exists right angles.*

Proof. Let a be a straight line in the plane α, $[O ; A)$ one of its rays and $[O ; B)$, $[O ; B')$, two rays for which $\widehat{AOB} \equiv \widehat{AOB'}$, $OB \equiv OB'$, $B \neq B'$. In other words, the rays $[O ; B)$ and $[O ; B')$ are in the two semi-planes defined be a in the plane α. The straight line BB' cuts a in a point D. DB is perpendicular on a. Indeed, the triangles ODB and ODB' are congruent, so $\widehat{ODB} \equiv \widehat{ODB'}$,

q.e.d.

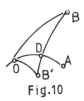

Remark. It may happen that the illustration in figure 10 did not seem convincing because our "prejudices" provide us from "seeing" $OB \equiv OB'$. But we shall find out later that we can make complete the convention of drawing the straight lines

Fig.10

with a certain convention of "appreciating the lengths" in relation with which $OB \equiv OB'$ should hold. An analogous remark can also be made referring to the angles congruences. But we repeat that, in the stage of Geometry's axiomatics, illustrations play no part and our figures, less usual, are *first of all* meant to prove this independence.

We are considering as being known the definition for the perpendicularity on a straight line on a plane and its checking criterion.

We are nor presenting the proofs of the following four propositions because they do not imply conceptual difficulties.

1. All right angles are congruent.

2. Through a point external to a straight line, in the plane detained by that point and by that straight line only one perpendicular on the straight line under consideration can be traced.

3. Each segment has a unique midpoint.

4. Each angle has a unique bisector.

The definitions of addition and substraction of segments and angles, as well as of the relation of "smaller" between segments or angles are the known ones.

Theorem 4.8. *An angle external to a triangle is bigger than each of the internal angles non-adjacent to it.*

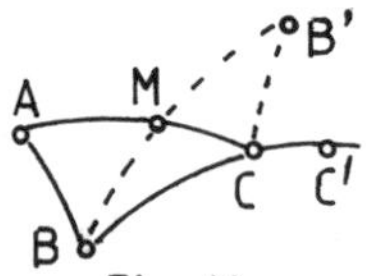

Proof. Let ABC be a triangle and C' a point with the property $B - C - C'$. We have to show that the angle $C'CA$ is bigger than the angle A and, also, that it is bigger than the angle B. We prove the statement for the angle A. Let M be the midpoint of AC and B' its symmetric of B related to M. Then B' is inside the

Fig.11

angle $C'CA$. The triangles $MB'C$ and MBA are congruent. It follows that

$A < ACB' \equiv ACC'$, q.e.d.

Of course, the notion of congruence of the triangles can extended to any figures. One can then consider bijective transformations of the space which take congruent figures into congruent figures and establish their group properties. But the problems go beyond the frame of this chapter.

§5. Group IV of Axioms

It contains two axioms, called the continuity axioms.

Axiom IV₁ (Archimedes'). *For any non-null segment AB and segment CD there exist $n \in \mathbf{N}^*$ and points C_0 , C_1 , ..., C_n on the ray $[C$; $D)$ so that $C_0 = C$, $C_{i-1} - C_i - C_{i+1}$, $C_iC_{i+1} \equiv AB$ $(i = 1, ..., n-1)$ and $D = C_{n-1}$ or $C_{n-1} - D - C_n$.*

Briefly, the axiom can also be formulated as follows: "*given the non-null segment AB and the segment CD, there exists $n \in \mathbf{N}^*$ so that $nAB > CD$*".

It is clear that if we attribute to the symbol nAB the usual significance, the two formulations are equivalent.

Axiom IV₂ (Cantor's). *For any infinite sequence of segments $\{A_nB_n\}$, $n \in \mathbf{N}^*$ of a straight line a with the property that A_iB_i is included into the inside of segment $A_{i-1}B_{i-1}$ for all $i = 2, 3, 4, ...,$ and there is not a segment situated inside all the segments in the sequence under consideration, there is a point M on a belonging to the inside of each segment in the sequence.*

Proposition 5.1. *Point M in Cantor's axiom is unique.*

Indeed, if there would exist M and M' with the property in the enunciation of Cantor's axiom and $M \neq M'$, it would follow that the segment MM' were inside all A_iB_i segments, which is absurd.

Definition 5.1. *Let a be an oriented straight line. We call Cartesian system of coordinates on a, a map $f: a \to \mathbf{R}$ having the following properties:*
a) *numbers 0 and 1 are in Ran f ;*
b) *f is monotonously increasing ;*
c) *two oriented segments $\overline{AB}$ and $\overline{CD}$ of the straight line are congruent and*

similarly oriented if and only if $f(B) - f(A) = f(D) - f(C)$.

Consequence:

Proposition 5.2. *If* $f: a \to \mathbf{R}$ *is a system of coordinates on the oriented straight line* a, *then* f *is injective.*

Let be points $O = f^{-1}(0)$ and $E = f^{-1}(0)$. The first one is called *origin* and the second one unity point for the system of coordinates $f: a \to \mathbf{R}$. There is $O \prec E$ on the oriented straight line a.

If P is a point of the straight line a, the number $x = f(P)$ is called the co-ordinate of P. If P belongs to the positive ray $[O ; E)$, then $x > 0$; if $P = 0$, then $x = 0$; if P belongs to the negative open ray, $x < 0$.

The following important result, whose proof is based on the completeness of the field of real numbers $\mathbf{R}$ holds.

Theorem 5.1. *Let* a *be an oriented straight line and* $O \prec E$ *two points fixed on* a. *Then*:

1. *There is only one Cartesian system of coordinates* $f : a \to \mathbf{R}$ *with the property* $f(O) = 0, f(E) = 1$;

2. $f: a \to \mathbf{R}$ *is a bijective map.*

Proof. 1. We shall prove this statement only on the basis of the axioms in groups I, II, III and on that of IV_1. The last part of the theorem necessarily requires that axiom IV_2 be also satisfied.

We shall construct, through the usual procedures, a Cartesian system of co-ordinates.

Let P be an any point of the straight line a ; if $P = 0$, we shall define directly $f(P) = 0$.

If $O \prec P$, we consider the sequence of points $\{P_n\}_{n \in \mathbf{N}}$ so that $P_0 = 0$, $P_0 \prec P_1 \prec P_2 \ldots$ and $P_n P_{n+1} \equiv OP$. According to Archimedes'axiom, for any index n there exists a unique natural number a_n so that

$$a_n \cdot OE \leq OP_n \leq (a_n + 1) \cdot OE.$$

We multiply this relation by a natural number $m > 0$ and notice that $m \cdot OP_n = OP_{mn}$;

we deduce the double inequality:

$$m \cdot a_n \cdot OE \le OP_{mn} < m\,(a_n + 1) \cdot OE.$$

Now there also is

$$a_{mn} \cdot OE \le OP_{mn} < m\,(a_{m \cdot n} + 1) \cdot OE\;;$$

on the basis of the choice of a_{mn} it results

$$m \cdot a_n \le a_{mn} \le m\,(a_n + 1).$$

After the division by $mn > 0$ the last inequalities give us

$$\frac{a_n}{n} \le \frac{a_{mn}}{mn} \le \frac{a_n}{n} + \frac{1}{n}\,,$$

which implies

$$0 \le \frac{a_{mn}}{mn} - \frac{a_n}{n} \le \frac{1}{n}\,.$$

By replacing n with m we immediately deduce

$$\left| \frac{a_m}{m} - \frac{a_n}{n} \right| < \frac{2}{N}\,, \quad N = 1 + \min\,(m, n).$$

The last inequality ensures that $\dfrac{a_n}{n}$ is a Cauchy sequence of rational numbers, so it will admit a limit in $\mathbf{R}$; we shall take by definition

$$f(P) = \lim_{n \to \infty} \frac{a_n}{n}\,.$$

If $P \prec 0$ holds, we consider its symmetric P' related to O and define $f(P) = -f(P')$, completing the construction of the function $f: a \to \mathbf{R}$.

To prove that f constitutes a Cartesian system of coordinates it is necessary to check the conditions a), b), c) in Definition 6.1.

Condition a) is obviously satisfied (with the specifications within conclusion 1.) because we have defined $f(O) = 0$, and if we particularize $P = E$ we shall find $a_n = n$ for any n in $\mathbf{N}^*$; it results that $f(E) = \lim_{n \to \infty} \frac{n}{n} = 1$.

We check condition b) by considering the only interesting case: $O \prec P \prec Q$. When supposing that by replacing P with Q the sequence (a_n) is replaced with (b_n), we shall deduce from $P \prec Q$:

$$a_n \cdot OE \le OP_n < OQ_n < (b_n + 1) \cdot OE,$$

so $a_n < b_n + 1$. It follows $f(P) = \lim_{n \to \infty} \frac{a_n}{n} \le \lim_{n \to \infty} \frac{b_n + 1}{n} = f(Q)$, so $f(P) \le f(Q)$. The monotony of the function f has thus been proved.

We shall present only a proving sketch of condition c). By conveniently re-noting the points it will be enough to prove that:

$$(*) \quad (AB \equiv OD \wedge A \prec B \wedge O \prec D) \Rightarrow f(B) - f(A) = f(D).$$

Let (a_n), (b_n), (d_n) be the associated sequences when P coincides with A, B and D, respectively. from $AB \equiv OD$ it follows the following inequalities for any natural number n:

$$d_n \cdot OE + a_n \cdot OE \le OD_n + OA_n = OB_n < (b_n + 1) \cdot OE,$$
$$b_n \cdot OE \le OB_n = OD_n + OA_n < (d_n + 1) \cdot OE + (a_n + 1) \cdot OE.$$

We retain from here $d_n + a_n < b_n + 1 < d_n + a_n + 3$. After having divided by n and passed to the limit, one obtains the conclusion of the implication (*).

Let g be an arbitrary system of co-ordinates on a so that $g(O) = 0$ and $g(E) = 1$. For $O < P$, $\overline{OP}$ and $\overline{P_{n-1}P_n}$ are congruent and similarly oriented; using c), we prove by complete induction: $g(P_n) = n \cdot g(P)$. It immediately follows $a_n \le g(P_n) = n \cdot g(P) < a_n + 1$. Dividing by n and passing to the limit we obtain (for $O < P$) $g(P) = f(P)$. If $P \prec O$ and P' is the symmetric of P related to O, using again c), we find out $g(P) = -g(P')$; it immediately follows in this case that $g(P) = f(P)$, as well, equality that ensures the uniqueness of the Cartesian system of co-ordinates f in the enunciated conditions.

2. To prove that the above constructed map $f: a \to \mathbf{R}$ is a bijection it is enough

to show that $\operatorname{Ran} f = \mathbb{R}$.

Let us consider that for each natural number p point E^p of the straight line a recursively defined by the conditions $E^0 = E$ and E^{p+1} is midpoint of the segment OE^p. It is easy to find out that $f(E^p) = 2^{-p}$ holds. For a natural number n we can also consider point E_n^p, where $OE_n^p = n \cdot OE^p$. It immediately results $f(E_n^p) = n \cdot 2^{-p}$.

Let u be an arbitrary real number; taking into account that Archimedes' axiom holds in $\mathbb{R}$, for any natural number p we shall determine a unique natural number z_p so that $z_p \leq u \cdot 2^p < z_p + 1$. On the basis of the provisory notation $n = z_p$, we define points $A = E^p$ and $E_p = E_{n+1}^p$.

We have thus constructed a sequence of closed segments $A_p B_p$ (for p arbitrary in $\mathbb{N}$) which is easily found out to check the hypotheses of Cantor's axiom; let M be the point inside all these segments. It follows for any p in $\mathbb{N}$:

$$f(A_p) = z_p \cdot 2^{-p} \leq u < (z_p + 1) \cdot 2^{-p} < f(B_p),$$
$$f(A_p) \leq f(M) \leq f(B_p).$$

When p tends to the infinite, it follows $f(M) = u$ and we have thus proved that $\mathbb{R}_+$ is included into $f(a)$. For any negative real number v we shall find, as above, M' so that $f(M') = -v$ and taking M as being the symmetric of M' related to O, it follows that $f(M) = v$ etc. We have thus proved that the map f is surjective and, according to those exposed above, bijective, too.

Corollary. *Any Cartesian system $f: a \to \mathbb{R}$ is a similitude.*

Remark. In the absence of axiom IV_2, $f(a)$ is a subset K of $\mathbb{R}$ which is easy to prove as being a subfield independently of the choice of the straight line a and of its points O, E. As in the above proof, we can find out the presence of all the numbers of the form $n \cdot 2^{-p}$ with n integer and p natural in K; we also deduce a lower limitation, Q_2, for the field K. A more precise lower limitation will be evidenced in §8 by considering the Pythagorean field Ω (see also [347], Chapter 19).

The applications of the last theorem are numerous and useful. From among then we remark only the notion of measure of the segments, its existence and its uniqueness.

Definition 5.2. *We call measure of the segments a map m: $S \to \mathbb{R}_+ \bigcup \{0\}$ from* *the set S of the segments to the set of non-negative real numbers satisfying the conditions:*

1. *For any null segment AA, there is $m\,(AA) = 0$;*
2. *There is a non-null segment AB for which $m\,(AB) = 1$;*
3. *If $AB \equiv CD$, then $m\,(AB) = m\,(CD)$ and conversely;*
4. *If $A - B - C$, then $m\,(AC) = m\,(AB) + m\,(BC)$.*

If we suppose that the map m exists, a non-null segment AB with the property $m\,(AB) = 1$ is called unit of measure; the number $m\,(CD)$ is called the measure of the segment CD or the length of CD measured with the unit of length AB.

Theorem 5.2. 1. *There is only a measure of the segments for which the unit of measure is a priori given.*

2. *For any real number $\alpha \in \mathbf{R}_+ \cup \{0\}$ there is a segment whose measure related to a given unit is the number .*

Proof. We shall prove 1. without making use of Cantor's axiom. Owing to condition 3. in Definition 5.2 we can admit that all the segments under considerations belong to a straight line a. We fix on a an orientation, giving a (positively oriented) ray of O origin. Let E be a point on this ray so that OE be congruent with the given unit of measure. If $f : a \to \mathbf{R}$ is the Cartesian system of co-ordinates on the oriented straight line a, uniquely determined by the condition $f\,(O) = 0$, $f\,(E) = 1$, then the map $m : S \to \mathbf{R}_+ \cup \{0\}$ defined by

$$(5.1) \qquad m\,(AB) = |\,f(B) - f(A)\,|, \ \forall\, AB \in S$$

satisfies Definition 5.2. This proves the existence of m. We prove the uniqueness of the measure m by reductio ad absurdum. Let be $m' \neq m$ with the property 1-4 and so that $m'\,(OE) = 1$. Let us consider the map $g : a \to \mathbf{R}$ given by

$$(5.2) \qquad g(P) = \begin{cases} m'(OP) & \text{if } O \prec P, \\ O & \text{if } O = P, \\ -m'(OP) & \text{if } P \prec O. \end{cases}$$

One can immediately remark the fact that g is a Cartesian system of coordinates on a and $g(O) = 0$, $g(E) = 1$. From Theorem 5.1 there is $g = f$. But then (5.1) and (5.2) with $g = f$ give $m' = m$, in contradiction with $m' \neq m$.

2. If $\alpha = 0$, then $m\,(OO) = 0$. If $\alpha > 0$, let P be the point of the straight line

with the property $f(P) = \alpha$. From (5.1) there is $m(OP) = |f(P) - f(O)| = \alpha$, q.e.d.

By applying classic methods we can determine a measure of the angles by choosing, for example, as unit of measure the right angle. We are not insisting on these problems.

Definition 5.3. *We call absolute Geometry the semi-formalized axiomatic theory made up of the notions, primary relations and groups I-IV of axioms.*

The name of absolute Geometry was given by Janos Bolyai: absolute Geometry constitutes the common part of Euclidean and hyperbolic Geometries.

§6. Group V of Axioms

This group contains only one axiom, called the axiom of the parallels.

Definition 6.1. *Two straight lines a and b are called parallels if they belong to the same plane and neither have a common point nor coincide. The parallelism of the straight lines a and b is denoted by a ∥ b, and its negation with a ∦ b.*

A first result belonging to absolute Geometry is given by:

Theorem 6.1. *In the plane determined by a straight line a and a point A not belonging to it, there is a parallel through the point A to the straight line a.*

Proof. Let α be the plane determined by a and A. In α, the perpendicular from A on a the straight line AB, situated in the plane α, is parallel to the straight line a.

In the contrary case, let D be the intersection of the straight lines a and b. In the triangle DAB, the angle DBA is congruent with the angle external to A, which is absurd according to the theorem 5.7, q.e.d.

The straight line b obtained by means of the construction in the preceding proof is called *the canonical parallel* thought the point A to the straight line a.
If there are no more parallels to a through A then the canonical parallel is briefly called *the parallel* to a through A. We can therefore notice that there is a simple procedure (given in the proof of the preceding Theorem and in Fig. 12 to determine a canonical parallel (the parallel, when the case is) through the point A to the straight line a, ($A \notin a$).
As a consequence of this theorem we also retain

Theorem 6.2. *There are distinct parallel straight lines.*

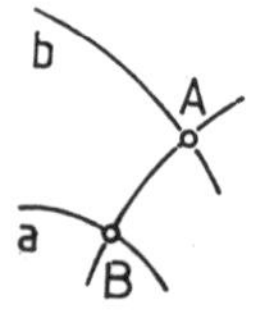

Let now *a* and *b* two distinct straight lines in a plane α and *c* a straight line sectioning *a* and *b* in distinct points. We then adopt the usual names for the angles determined by *c* with the straight lines *a* and *b*: "internal alternate", "external alternate", "correspondent" etc.

Fig. 12

Theorem 6.3. *If two straight lines in a plane cut by a secant make up congruent internal alternate angles with it, then those two straight lines are parallel.*

On the basis of Theorem 4.7, the proof of the enunciated property is immediate.

To fix the Geometry we are dealing with we shall have to discern between two alternatives. Whether in the plane α defined by the straight line *a* and point *A* (*A* ∉ *a*) there is only one parallel or there are several ones through *A* to *a*.

As we shall see, each of these statements is independent of the axioms of absolute Geometry, therefore each of them, together with the other axioms exposed so far, give an axiomatic system deserving our whole attention. In this chapter we are dealing with the first case.

Group V of Axioms

Axiom V. *Through a point A external to a straight line a (in the plane determined by A, a), there is at most a parallel to the straight line a.*

According to Theorem 6.2, there is the consequence:

Theorem 6.4. *Let A be an any point and a an any straight line which point A does not belong to. Then (in the plane determined by A, a) there is only a straight line b through point A parallel to the straight line a.*

Definition 6.2. *We call Euclidean Geometry the semi-formalized axiomatic theory of Hilbert's axiomatic system exposed in the present chapter.*

Out of the contents of Euclidean Geometry we remark certain important results.

Theorem 6.5. *If a ∥ b and c is secant to a and b, then the internal alternate angles determined are congruent.*

Proof. Let *a* ≠ *b* and *c* secant to *a* in *A* and secant to *b* in *B*. Supposing by absurd that the internal alternate angles defined by c with a and b are not congruent, we are considering through B the straight line b' in the plane, which makes with c, together with a, equal internal alternate angles. According to Theorem 6.3, it follows $b' \parallel a$. Because there also is $b \parallel a$ and b, b' pass through B and $b \neq b'$, it would follow that two distinct parallels to the straight line a can be traced through point B in the plane under consideration. Which is absurd, according to Theorem 6.4, q.e.d.

Theorem 6.6. *In any triangle ABC the sum of the angles $A + B + C$ is equal with two right angles.*

Proof. Let (AD) be the parallel through A to BC, D being in another semi-plane delimited by (AB) than C. Let also be E so that $D - A - E$. By means of Theorem 6.5 we deduce the

equalities $B \equiv \widehat{BAD}$ and $C \equiv \widehat{CAE}$ such that

$A + B + C = \widehat{DAB} + \widehat{BAC} + \widehat{CAE} = 2$ right angles, q.e.d.

Fig.13

We draw attention that figure 13 which illustrates the proof is traditionally executed, procedure used with the preceding figures having become unacceptable after we had adopted Axiom V.

Across the years there have existed many attempts at proving, within absolute Geometry, the assertion contained in Axiom V. They have failed, because, as we shall see, the system of axioms under consideration is a minimal one. However, they had a quality, that of establishing *propositions equivalent with the postulate of parallels* given by Axiom V.

We are enumerating here the most important ones:

1. *The proposition in the enunciation of theorem 6.3.*

2 (Euclid's Postulate). *If two distinct straight lines in a plane cut by a secant make up with them alternate angles and on the same side of the secant with their sum smaller than two right angles, then the two straight lines intersect each other on that side of the secant where the mentioned property holds.*

3. *The sum of the angles in any triangle is equal with two right angles.*

4. *All triangles have the same sum of their angles.*

5. *Given an acute angle AOB, the perpendicular raised on (OA) in any point M of the ray $(O; A)$ cuts (OB).*

6. *Through each point in the inside of a certain angle one can trace a straight line cutting both sides of the angle without passing through its vertex.*

7. *There is a plane quadrangle ABCD with angles $A \equiv B$ = 1 right angle for which the angles C and D are also right angles.*

(The plane quadrangles ABCD with A and B being right angles satisfying BC $\equiv$ AD are called "Sacheri quadrangles".)

8. *There is a quadrangles ABCD in plane with all the angles right. (The plane quadrangles ABCD for which $A \equiv B \equiv C$ = 1 right angle are called "Lambert quadrangles".)*

9 (Farkas-Bolyai Property). *Through any three non co-linear points a circle passes.*

10. *There exist incongruent similar triangles.*

11. *In plane, the geometrical locus of the points equally distanced from a straight line a and situated in one of the semi-planes determined by a is a straight line.*

12. *The median line of a triangle is congruent with the half of its basis.*

13. *Pythagoras' Theorem.*

14. *Two straight lines parallel with the third one are parallel with each other.*

Some of these theorems are easy to directly prove, but others need laborious proofs. We do not insist on them.

The history of parallels postulate is over two millenniums old. The problem has been solved only in the last century by N. Lobacevski in Russia, J. Bolyai in Transylvania and K.Fr. Gauss in Germany, but the general conviction as for the rightness of this solution has been delayed for another half a century. The discovery of the non-Euclidean Geometries constituted one of the decisive moments in the evolution of Mathematics as a whole. In philosophy it shook the Kantean conception on the nature of physical space.

§7. The Metatheory of Hilbert's Axiomatics

We shall metatheoreticaly analyze Hilbert's semi-formalized axiomatic system the discussing the problems of non-contradiction, minimality, completeness and categoricity, exposed in Chapter III. By accepting the non-contradiction of Arithmetic, we shall use the entire construction of the field $\mathbb{R}$ of real numbers to determine models able to prove the non-contradiction or the independence of axioms.

The Problem of Non-Contradiction

We shall construct a model, denoted by G, in which we might interpret the primary notions and relations in Euclidean Geometry and all the axioms of Hilbert's system could be checked.

We call *points* in G the ordered terns of real numbers $(x^i) = (x^1, x^2, x^3)$ and we denote them by $A = (x^i)$, $B = (y^i)$ etc.

To define the straight lines and the planes, let m be the set of the quarterns of real numbers $(a_h) = (a_1, a_2, a_3, a_4)$ for which rank $\| a_1\ a_2\ a_3 \| = 1$. For $(a_h), (b_h) \in m$ we shall put

$$(a_h) \sim (b_h) \Leftrightarrow \exists \rho \in \mathbf{R}^*, a_h = \rho b_h, h = 1, 2, 3, 4.$$

Obviously, $\sim$ is an equivalence relation. We note with $P = m / \sim$.

The elements of the set P are called *plane* in G and we denote them by $\alpha = [a_h]$, $\beta = [b_h]$, etc.

Let be $D = \{((a_h), (b_h)) \in m \times m : \text{rank} \begin{vmatrix} a_1 & a_2 & a_3 \\ b_1 & b_2 & b_3 \end{vmatrix} = 2\}$. We consider on D the relation $\approx$ given by

$$[((a_h), (b_h)) \approx ((a_h'), (b_h'))] \leftrightarrow [\exists \lambda_1, \lambda_2, \mu_1, \mu_2 \in \mathbf{R}, \begin{vmatrix} \lambda_1 & \lambda_2 \\ \mu_1 & \mu_2 \end{vmatrix} \neq 0,$$

$$\forall h = 1, 2, 3, 4; a_h' = \lambda_1 a_h + \lambda_2 b_h, b_h' = \mu_1 a_h + \mu_2 b_h].$$

It is easy to find out that $\approx$ is an equivalence on the set D. Let be $D = D / \approx$. The elements of the set D are called straight lines in G and are denote by $a = [(a_h); (b_h)]$, etc.

We define the relation of incidence for the point $A(x^i)$ and the straight line $a = [(a_h); (b_h)]$, saying that $A \in a$ holds if and only if the following equalities hold:

(7.1)
$$\begin{cases} a_1 x^1 + a_2 x^2 + a_3 x^3 + a_4 = 0, \\ b_1 x^1 + b_2 x^2 + b_3 x^3 + b_4 = 0. \end{cases}$$

As

(7.2) $\quad \text{rank} \begin{vmatrix} a_1 & a_2 & a_3 \\ b_1 & b_2 & b_3 \end{vmatrix} = 2,$

condition (7.1) does not depend on the choice of the representative $((a_h), (b_h))$ of a, but only on the straight line a. But (7.1) and (7.2), for (x^i) variables, give us the totality of the points of the straight line a; (7.1), (7.2) constitute *the equations of the straight line a*. Let us notice that they can be written under an equivalent form

(7.3) $x^i = x_0^i + tm^i$, $t \in \mathbb{R}$; $i = 1, 2, 3,$

in which m^i are not all null and are determined up to a non-null common factor, and (x_0^i) is a fixed solution of the system (7.1), (7.2).

We also define the belonging relation of a point $A = (x^i)$ to a plane $\alpha = [a_h]$ by: $A \in \alpha$ if and only if

(7.4) $a_1 x^1 + a_2 x^2 + a_3 x^3 + a_4 = 0,$

which, together with the condition

(7.5)   rank $\| a_1 \ a_2 \ a_3 \| = 1,$

shows that (7.4) does not depend on the representative $(a_h) \in m$ of the plane α, but only on the plane α.

By applying the techniques in analytical Geometry we can easily prove that the axioms in group I are checked in G.

Let $A_1 = (x_1^i)$, $A_2 = (x_2^i)$, $A_3 = (x_3^i)$ be three points of a straight line given by (7.3) for the values t_1, t_2, t_3 of t :

$$x_k^i = x_0^i + t_k m^i, \ k = 1, 2, 3; \ i = 1, 2, 3.$$

We shall say that A_2 *is situated between* A_1 and A_3 if there is $t_1 < t_2 < t_3$ or $t_1 > t_2 > t_3$. We write this as $A_1 - A_2 - A_3$.

Each axiom in group II is proved by direct calculus.

For the derived notions appearing by the consequences of this group of axioms we are presenting only a few results.

- To fix an orientation on the straight line provided by the "parametrical equations" (7.3) means to accept the amplification of "the directing parameters" m^i only by positive real factors.

   - An open ray with its origin in point $A(x_0^i)$ of the oriented straight line given by the equations (7.3) is given by fixing the sign of "the parameter" t. The ray will be closed if t is permitted to take the value 0, too.

   - For the plane $[a_h]$, a semi-space delimited by it is constituted by the set of the points $A(x^i)$ which satisfy the inequality $a_1 x^1 + a_2 x^2 + a_3 x^3 + a_4 > 0$. Choosing another representative, (b_h), for the plane under consideration, where $a_h = r \cdot b_h$, we shall obtain the same semi-space or the opposed one by replacing a_h with b_h in the above mentioned inequality, depending on $r \geq 0$ or $r < 0$.

   - A semi-plane delimited by the straight line $[(a_h);(b_h)]$ in the plane $[a_h]$ is made up of the points A situated in the plane $[a_h]$ and in a semi-space delimited by $[b_h]$.

   We denote now by $d(A_1, A_2)$ and call *distance* between the points $A_1 = (x_1^i)$ and $A_2 = (x_2^i)$ the number

$$(7.6) \quad d(A_1, A_2) = [\sum_{i=1}^{3} (x_1^i - x_2^i)^2]^{\frac{1}{2}}.$$

   We shall say that *the segment* $A_1 A_2$ *is congruent with the segment* $B_1 B_2$ and write $A_1 A_2 \equiv B_1 B_2$, if and only if $d(A_1, A_2) = d(B_1, B_2)$.

   Let then be two rays with their origin $A_0 = (x_0^i)$ given by (7.3):

$$x^i = x_0^i + t \cdot m^i, \; t \in \mathbb{R}_+ ; \; i = 1, 2, 3,$$
$$x^i = x_0^i + t \cdot n^i, \; t \in \mathbb{R}_+ ; \; i = 1, 2, 3,$$

where the factors by which m^i and n^i can be multiplied are positive. We call angle between the two rays the number $V \in [0, \pi]$ given by the solution of the equation

$$(7.7) \quad \cos V = \frac{m^1 n^1 + m^2 n^2 + m^2 n^3}{[(m^1)^2 + (m^2)^2 + (m^3)^2]^{\frac{1}{2}} [(n^1)^2 + (n^2)^2 + (n^3)^2]^{\frac{1}{2}}}.$$

   We say then that *the angles* (h, k) *and* (h', k') *are congruent* and note $\widehat{(h, k)} \equiv \widehat{(h', k')}$ if and only if the cosines of the angles are equal. A longer calculus, in fact equivalent to the edification of analytical Geometry in space shows that all the axioms in groups I-IV are checked in G.

   Consequently:

Theorem 7.1. *Hilbert's axiomatic system of Euclidean Geometry is non-contradictory.*

The Problem of Independence

As it has been seen, beginning with group II, the axioms of the system are formulated on the basis of those in the preceding groups and on their consequences. For this reason it is not possible to study the independence of the axioms as a whole.

In principle, we can approach here only the independence of a certain axiom related to the axioms in the groups that are not posterior to that axiom. From among the twenty models which would then have to be constructed to check the independence of each axiom of the remaining ones "not affected by it", we shall present here only some models which we consider as being more important.

Pasch's axiom is independent of axioms $I_1 - I_8$ and of II_1, II_2, II_3. Indeed, by keeping the model G we used above, we eliminate the point $(1, 0, 0)$. The axioms preceding II_4 are checked on. But let be $A = (2, 0, 0)$, $B = (0, 0, 0)$, $C = (0, 2, 0)$ and the straight line $a = [(0, 0, 1, 0); (1, 0, 0, 1)]$ which passes through the midpoint $M = (1, 1, 0)$ of the side of AC. The straight line a cuts neither $| AB |$ nor $| BC |$, although it is in the plane $AOC = [0, 0, 1, 0]$; II_4 is not checked, so it is independent of the axioms preceding it.

For the independence of axiom III_5 we take again the same model G. Let $A = (x^i)$, $B = (x^i + t^i)$, $C = (x^i + u^i)$ be arbitrary points; we redefine a distance δ and an angle $\sphericalangle$, replacing (7.6), (7.7) by:

$$\delta(A, B) = [(t^1 + t^2)^2 + (t^2)^2 + (t^3)^2]^{\frac{1}{2}},$$
$$\delta(A, B) \cdot \delta(A, C)\cos(\sphericalangle BAC) = t^1 u^1 + t^2 u^2 + t^3 u^3.$$

We define again in a natural way the congruence $\equiv$ by $AB \equiv CD \Leftrightarrow$ $\Leftrightarrow \delta(A, B) = \delta(C, D)$ and $\widehat{BAC} \equiv \widehat{DEF} \Leftrightarrow (\sphericalangle BAC) = (\sphericalangle DEF)$. Let G' be the interpretation constructed this way. To check the axioms in groups I and II we precede identically as with the model G, as we did when proving the non-contradiction of Hilbert's axiomatic system. Axioms $III_1 - III_4$ are checked in the model G' without any difficulties. For axiom III_5 we consider $O = (0, 0, 0)$, $A = (2, 0, 0)$ and $B = (0, 1, 0)$. For the triangles OAB and OBA we shall easily find out $OA \equiv OB$, $OB \equiv OA$ and $\widehat{AOB} \equiv \widehat{BOA}$, so the hypotheses of axiom III_5 are fulfilled. But one can find out that

$(\sphericalangle OBA) \neq (\sphericalangle OAB)$, so III$_5$ is not satisfied, proving to be independent of the axioms preceding it. Since III$_5$ does not intervene in the formulation of the axioms in the last two groups, we can even find out the total independence of this axiom in Hilbert's system.

To prove *the independence of Archimedes's axiom* in relation with the other axioms we start from the following well-known result, [523]. There are fields K totally ordered which have the field $\mathbb{R}$ of real numbers as their own sub-field, are complete (any Cauchy sequence of elements in K has its limit in K'), but are not Archimedean. Such a field has, for example, been constructed by Kolmogorov.

Let then G_K be the model constructed with the help of field K in the same way as the model G was constructed by means of $\mathbb{R}$. One can easily see that all the axioms (axiom V included) are satisfied in G_K, but that axiom IV$_1$, that of Archimedes, is not. To justify the last statement, let be the segment OA with the extremities $O = (0, 0, 0)$, $A = (\alpha, 0, 0)$, $\alpha > 0$ and the segment OB, with $B = (\beta, 0, 0)$, α and β being chosen so that $n \in \mathbb{N}^*$, for which $n\alpha > \beta$, should not exist. Then $n \in \mathbb{N}^*$ does not exist so that $nOA > AB$, because this inequality in G_K is equivalent with $n\alpha > n\beta$.

The independence of axiom IV$_2$, that of Cantor, of the other axioms is proved in the same way, considering the model $G\Omega$ in which Ω is the field constructed in the following way: Ω is the set of real numbers obtained by applying to number 1 a finite number of times the operations of addition, substraction, multiplication, division and radical extraction of the form $\sqrt{1 + a^2}$, in which a is an already constructed number. An Archimedean totally ordered field containing Q is obtained, but it is a proper sub-field of $\mathbb{R}$. But Ω is not a complete field. Consequently, on the straight lines of G_Ω (for example, on the straight line $a = [(0, 0, 1, 0); (0, 1, 0, 0)]$, which is isomorphic with Ω) Cantor's axiom does not hold.

Finally, the independence of the axiom of the parallels in Hilbert's system will be proved in Chapter VIII, showing that hyperbolic Geometry, which is obtained as a theory of the axiomatic system made up of groups I-IV in Hilbert's system and of the negation of axiom V is non-contradictory.

Thus, there is:

Theorem 7.2. *Hilbert's axiomatic system of Euclidean Geometry is a minimal one.*

The Problem of Completeness and Categoricity

Theorem 7.3. *Hilbert's axiomatics of Euclidean Geometry is complete.*

To prove it is enough, according to Proposition 2.2, Chapter III, to show:

Theorem 7.4. *The axiomatic system of Euclidean Geometry is categorical.*

Proof. Let M be a model for Hilbert's axiomatics; it follows that a system of Cartesian coordinates can be introduced into M and then we can analytically characterize the points, the straight lines and the planes of M, exactly as in the model G we used to prove the non-contradiction of Euclidean Geometry. Then the points, the straight lines, the planes and the relations in M can be regarded as points, straight lines, planes and relations in G.

We obtain a bijective correspondence between M and G which will retain the geometrical relations, since there is the same analytical Geometry both in M and in G. So M and G are isomorphic. It is written $M \xrightarrow{\sim} G$. Let M' be another model for Euclidean Geometry. There is $M' \xrightarrow{\sim} G$. It follows that $M \xrightarrow{\sim} M'$, the isomorphic relation $\xrightarrow{\sim}$ being an equivalence on the set of models. So Hilbert's system is categorical, q.e.d.

We have thus finished with the important problems of metatheory of the axiomatic system of Euclidean Geometry.

Remark. We think useful to underline the special part played by axiom IV_2 in ensuring the categoricity of the entire system of axioms. In the absence of axiom IV_2 for the arbitrary model M we shall find an isomorphic model with an analytical model G_K, where K is a sub-field of $\mathbf{R}$ satisfying only the condition of being stable in relation with the map f given by $f(x) = \sqrt{1+x^2}$. Of course, in these conditions models M, M' will no more be isomorphic if the corresponding fields K and K', respectively, are not isomorphic, either. This is a reason for which, in Hilbert's formulation, groups IV and V of axioms appear in reversed order related to the one exposed here, and the formulation of axiom IV_2 (V_2 in Hilbert's notation) aims directly at categoricity: "The set of points does not admit any effective extension in which axioms I, II, III, Archimedes' axiom and the axiom of the parallels should be satisfied". Another reason for which Hilbert places the axioms of continuity in the last group is that by denying (in a special way) the axiom of the parallels, Archimedes' and Cantor's axioms are obtained as axioms. Then, in the notation given here to axioms for the non-Euclidean Geometry we shall obtain a minimally non-contradictory axiomatic system taking groups I, II, III and axiom $\rceil$ V: "There is a point A non-incident to a straight line a so that there exist through A at least two distinct straight lines, each of then superparallel with a".

Exercices

1. Prove the proposition 1.9 and 1.10.

2. Using groups I and II of axioms, prove that the insides of triangles

constitutes a neighbourhoods basis of a Hausdorf topology of the plane.

3. Formulate a spatial analogous of the preceding exercise.

4. Let us also add tot the set U of angles the set U' of "contingent" angles; such an angle u has a ray-side, and the other one an arc of a circle tangent to this ray. The amplification by $n \in \mathbf{N}^*$ of such an angle u returns to the replacing of ray R of the arc of the circle by $\dfrac{R}{n}$. Discuss the validity of Archimedes' axiom in U' and in $U \bigcup U'$.

5. Having a model of Hilbert's axiomatics, show that there also exist bounded models.

CHAPTER VI

BIRKHOFF'S AXIOMATIC SYSTEM

As we have found out in the preceding chapter, the axiomatic system of Geometry constructed by Hilbert is a rigorous and precise one, but extremely complex. But the admiration of mathematicians for this system is not a passive one; more than one hundred axiomatizations of Geometry have appeared after Hilbert. Some of these axiomatic systems have the declared purpose of diminishing the number of primary notions and relations in order to obtain a deeper metatheoretical analysis. Other axiomatic system regroup the axioms to gradually obtain more and more specialized Geometries up to reaching Euclidean Geometry. Finally, there are axiomatizations of Geometry stating a didactic purpose: the diminishing of the number of axioms and the simplification of their enunciations for an as easy understanding as possible. Birkhoff's axiomatics belongs to this last category. We are drawing our attention to this axiomatic system because:

- the didactic of elementary Geometry is extremely important;
- from the historical point of view, this axiomatics precedes others;
- its didactic purpose is well achieved.

We do not intend to perform here a detailed analysis, but only to place Birkhoff's axiomatics as a logico-deductive theory and grasp its bilateral connections with Hilbert's "mother axiomatics".

§1. The General Framework

Birkhoff's axiomatic system supposes as being previously known *the* (naive) *theory of the sets* and *the theory of the real number*. From the scientific point of view, such suppositions may appear as hazardous, but from didactic point of view they are correct and efficient premises. Being a semi-formalized axiomatic system it will also include *the usual logic*.

The specific primary notions are: *point, straight line, plane, distance, measure*. *The point* will be regarded as elements of a set S which constitutes "the ambient space"; they will be noted with Latin capital letters eventually indexed or accentuated. *A straight line* is regarded as a set of points and will be noted with a Latin small letter; the totality of the straight lines appears as a set included into the set of the parts of S: $\mathcal{L} \subset \mathcal{P}(S)$. *The planes*, noted with small Greek letters, will make up another subset of $\mathcal{P}(S)$ which we shall note with Π. *The distance* will constitute a function $d: S \times S \to \mathbb{R}$; but we draw attention that to adopt the name "distance" for d does not suppose that we should accept *a priori* the specific properties of the distance. *The measure* will be regarded as a function

m: $\mathcal{U} \rightarrow [0, 180]$, but its domain of definition, the set $\mathcal{U}$ of "the angles" will be constructed en route, before formulating the axioms referring to m. We shall use the relation of belonging ($\in$), operations with real number ($+$, $\cdot$) or binary relations about the elements of $\mathbb{R}$ ($=$, $\leq$) and different logical symbols ($\Rightarrow$, $\wedge$, $\vee$, $\forall$, $\exists$), but we are not presenting them as primary relations because they do not belong to the specific part of this axiomatic system. We can then state that within Birkhoff's axiomatics Geometry will appear as a structure $(S, \mathcal{L}, \Pi, d, m)$.

§2. Axioms and Their Principal Consequences

We shall note the axioms with symbols B_i, where B evokes the name of the mathematician George Birkhoff and the index i permits us to number the axioms. We shall now and then interrupt the listing of axioms evidencing the derived notions, the main consequences and brief commentaries. In general, we shall use the same notations as in the preceding chapter. After having presenting each axiom in words we shall also give a formal expression.

B_1. *Through two different points one and only one straight line passes.*

$$\forall A, B \in S, A \neq B \Rightarrow \exists! a \in \mathcal{L}: A \in a \wedge B \in a.$$

(Compare with axioms I_1 and I_2; the straight line a will be denoted by (AB) or, when confusions are not possible, with AB.)

B_2. *Throughout three non co-linear points a unique plane passes.*

$$\forall A, B, C \in S: \neg(\exists a\{A, B, C\} \subset a) \Rightarrow \exists! \alpha \in \Pi: \{A, B, C\} \subset \alpha.$$

B_3. *If two distinct points are in a plane, the straight line containing them is included into that plane.*

$$\forall A, B \in S, A \neq B, \forall \alpha \in \Pi: \{A, B\} \subset \alpha \Rightarrow (AB) \subset \alpha.$$

B_4. *If two distinct planes have a non-void intersection, then their intersection is a straight line.*

$$\forall \alpha, \beta \in \Pi: \alpha \neq \beta \wedge \alpha \cap \beta \neq \varnothing \Rightarrow \alpha \cap \beta \in \mathcal{L}.$$

B$_5$. *Any straight line contains at least two distinct points. Any plane contains at least three non-collinear points. S contains at least four non-coplanar points.*

$$\forall \alpha \in \mathscr{L}: \exists A, B \in S \wedge A \neq B \wedge \{A, B\} \subset a,$$
$$\forall \alpha \in \Pi, \exists A, B, C \in \alpha \wedge \neg (\exists a \in \mathscr{L}\{A, B, C\} \subset a),$$
$$\exists A, B, C, D \in S, \forall \alpha \in \Pi \Rightarrow \{A, B, C, D\} \not\subset \alpha.$$

The axioms B$_1$ - B$_2$ constitute *the group of belonging axioms.* Although they are fewer numerically, these first five axioms make up a sub-system equivalent with the group of Hilbert's axioms of incidence.

The derived notion of *system of co-ordinates* for a straight line a is introduced: a bijective function $f: a \rightarrow \mathbb{R}$ satisfying the condition:

$$(2.1) \quad \forall P, Q \in a: |f(P) - f(Q)| = d(P, Q)$$

We shall also consider the set $SC(a)$ made up of the systems of co-ordinates for the straight line a.

B$_6$ (the axiom of the ruler). *Any straight line a admits at least a system of co-ordinates.*

$$\forall a \in \mathscr{L}, SC(a) \neq \varnothing.$$

This very strong axiom is specific for Birkhoff system; let us remember that to be sure about an analogous result in Hilbert's axiomatics (see Theorem 5.1, Chapter V) we needed groups I, II, III, IV. As a retort, we shall find out that many of Hilbert's axioms can be deduces from the axioms B$_1$ - B$_6$.

Theorem 2.1. *The "distance" function d satisfies the conditions:*
a) *for any P, Q there holds $d(P, Q) \geq 0$;*
b) $d(P, Q) = 0$ *if and only if $P = Q$;*
c) $d(P, Q) = d(Q, P)$.
Indeed, for $P \neq Q$, we can consider the straight line $(PQ) = a$ and deduce from (2.1) a), c) and $d(P, Q) \neq 0$. For $P = Q$ we shall take a arbitrary though P (see also axiom B$_5$) and from (2.1) we shall also deduce $d(P, Q) = 0$, q.e.d.

Remark. Using only axioms B$_1$ - B$_6$ we can not yet prove that d is a distance that is that it satisfies "the triangular inequality" $d(A, B) + d(B, C) \geq d(A, C)$, too. This property will later result out of geometrical considerations. For the moment, this inequality can be proved only with the supplementary hypothesis that points A, B, C are co-linear.

Theorem 2.2 ("of ruler positioning"). *Given points O, A on a straight line a, there is a system of co-ordinates h for a so that $h(0) = 0$ and $h(A) > 0$.*

Indeed, from $f \in SC(a)$ it immediately follows that function $g = -f$ and, for any real t, function $g_t = t + f$ is also in $SC(a)$. Starting from f arbitrary in $SC(a)$, we shall consider $t = -f(0)$ and find out that for $e \in [-1, +1]$ functions $h = e(t + f)$ is in $SC(a)$. In the end, we shall choose e so that $h(A) = e[f(A) - (0)]$ be positive.

The ternary relation "to be between" is introduced as a derived notion: point B is by definition situated between points A and C if A, B, C are co-linear points distinct two by two and $d(A, B) + d(B, C) = d(A, C)$. This is noted with $A - B - C$.

Theorem 2.3. *The relation "to be between" satisfies axioms* II_1, II_2, II_3 *in Hilbert's axiomatics (see* Chapter V, §2).

One can immediately check by means of a system of co-ordinates for a straight line a on which the points intervening in these axioms are chosen. Now we can re-define the inside of a segment.

Theorem 2.4. *If a is a straight line and O one of its points, the set $a - \{0\}$ is uniquely decomposed into two non-void disjoint classes ("open rays of O origin situated on a") so that points P, Q belong to distinct classes if and only if $P - O - Q$.*

Proof. Let f be a system of co-ordinates on a so that $f(0) = 0$. The set $a_1 = \{X : X \in a \wedge f(X) > 0\}$ and $a_2 = \{Y : Y \in a \wedge f(Y) < 0\}$ obviously satisfy the conditions in the enunciation. Let us consider for uniqueness an element $g \neq f$ in $SC(a)$ so that $g(0) = 0$. If, by absurd, another undoing into open rays of the set $a - \{0\}$ would correspond to the system of co-ordinates g, there existed, for instance, points P, Q in a_1 so that $g(P) \cdot g(Q) < 0$. We may denote points P, Q so that $f(P) < f(Q)$ it follows

$$(*) \quad d(O, P) + d(P, Q) = d(O, Q).$$

But from relation $g(P) \cdot g(Q) < 0$ it follows

$$(**) \quad d(P, O) + d(O, Q) = d(P, Q).$$

Subtracting the two inequalities $d(P, Q) = d(O, Q)$ is deduced, and returning into (*)

we obtain $d(O, P) = 0$, so $O = P$, which is absurd.

Remark. According to this theorem, the notions of open and closed *rays* are introduced. The orientation of the straight line a can be achieved as in the preceding chapter or an equivalence $\sim$ can be introduced on $SC(a)$, putting

$$f \sim g \Leftrightarrow f \circ g^{-1} = \text{increasing monotonous function.}$$

An orientation on a is defined as an element in $SC(a)/\sim$.

Theorem 2.5. *The system of co-ordinates h in Theorem 2.2 ("of ruler positioning") is unique.*

Indeed, it is necessary that $h(X) = d(O, X)$ should hold on the ray $(O; A)$, and $h(X) = -d(O, X)$ on the opposed ray.

Remark. For didactic purposes, the ruler axiom has in [347] a stronger formulation that includes Theorem 2.2 and 2.5, too. We also consider as obvious the fact that the derived relation of congruence of the segments can be introduced by now. In the present context, we prefer to deduce the consequences in the order in which they appear according to Hilbert's axioms, inserting here only a consequence of axioms $B_1 - B_6$ that has an outstanding importance in edifying Geometry.

Theorem 2.6. *Any non-null segment AB has a unique midpoint.*

Proof. The theorem states that there is a unique point M on the straight line AB so that $d(A, M) = d(M, B)$. Let us notice that such a point M, if it exists, can not satisfy $M - A - B$, for example, because in the system f co-ordinates which would satisfy $f(M) = 0$, $f(A) > 0$ it would follow $f(B) > 0$ and $|f(B) - f(M)| = d(B, M)$ would imply $f(M) = d(M, B) = d(M, A) = f(A)$, in contradiction with the injectiveness of f. Analogously, the hypothesis $A - B - M$ also leads to absurd. We retain then that for an eventual midpoint M, it is necessary that $A - M - B$ should hold.

We are now considering the system of co-ordinates $g: (AB) \to \mathbb{R}$ with $g(A) = 0$, $g(B) \geq 0$. The condition $d(A, M) = d(B, M)$ returns to $g(M) \equiv g(B) - g(M)$, that is $g(M) = \frac{1}{2} g(B)$. But the system of co-ordinates g under consideration is unique on the basis of the preceding theorem and according to those exposed above M exists and is unique, q.e.d.

Remark. In Chapter V, §4, we stated that Theorem 2.6 is a consequence of the axioms in groups I, II, III. Although this theorem has after that been applied in Chapter V, thus proving itself its importance, its proof was not inserted into §4. We partially do here our duty (proving only its "existence") to facilitate a comparison between typical proofs in the axiomatic systems of Birkhoff and Hilbert, respectively.

Let C be a point which is not situated on (AB). In the plane (ABC), on that side of (AB) where C is not, according to axioms III_4 and III_1, there exists a point D so that $\widehat{CAB} \equiv \widehat{ABD}$ and $AC \equiv BD$. The points C, D are separated from (AB) so there will exist a point M on (AB) inside the segment CD. We deduce

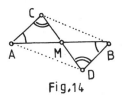

Fig.14

$\triangle ACB \equiv \triangle BDA$ (the SAS case) and so $BC \equiv AD$. It follows (the SSS case) that $\triangle ACD \equiv \triangle BCD$ and from here $\widehat{ACM} \equiv \widehat{BDM}$. Finally, the triangles ACM and BDM are congruent (ASA), so $AM \equiv BM$, q.e.d.

We go on exposing Birkhoff's axioms.

B_7 (the axiom of plane separation). *For any plane α and any straight line a included into α, the set $\alpha \setminus a$ a is decomposed into two disjoint non-void subset, H, K so that points P, Q belong to the distinct set H, K if and only if there exists on a one point X situated between P and Q.*

$$\forall \alpha \in \Pi, \ \forall a \in \mathscr{L}: \ a \subset \alpha \Rightarrow \exists H \subset \alpha \setminus a: \ H \neq \varnothing \wedge H \neq \alpha \setminus a \wedge$$
$$\wedge [\forall P \in \alpha \setminus (a \cup H): \ Q \in H \Leftrightarrow \exists X \in a: \ P - X - Q].$$

The sets H, K are of course called (open) *semi-planes* delimited by a in α.

Theorem 2.7. *On the basis on axioms* $B_1 - B_6$, B_7 *is equivalent with Pasch's axiom* II_4.

Proof. We accept B_7; let be a triangle ABC and a straight line a included into the plane α of the triangle which contains none of the vertexes A, B, C. Let H, K be the semi-planes delimited by a in α. If a contains a point inside the segment AB, the vertexes A, B are in distinct semi-planes; for instance, $A \in H$, $B \in K$. If $C \in H$, a will cut *the side BC*, and if $C \in K$, a will cut *the side AC*. The reversed implication was proved in the preceding chapter on the basis of some propositions that were deduced here as being true.

Remark. Both in Birkhoff's and Hilbert's axiomatic systems propositions II_4 and B_7 are equivalent and can be chosen as axioms. The option is a question of "elegance" and is dictated by "the concordance of conceptions". Axiom II_4 has a more geometrical formulation ànd enters better within Hilbert's system. Axiom B_7 is advantaged by the "official" adoption of the language of sets theory and is naturally taken over by Birkhoff's axiomatic system. Let us also note that B_7 is "stronger" than II_4 because the above proof is also applied when A, B, C are co-linear and ensure that a still cuts *exactly* a side (see Theorem 2.3, Chapter V).

We retain that we have so far deduced all the axioms in Hilbert's groups I and II in axioms $B_1 - B_7$. To formulate the following axioms in Birkhoff's system the derived notion of (null, proper, elongated) angle is introduced as a pair of closed rays having the same origin and being coincident, non-collinear and distinctly co-linear. Let U be the set of all angles. The notion of interior of an angle is introduced as in Hilbert's theory. Before formulating the following three axioms (referring to the measuring of angles) we also draw attention that the upper limit, 180, of the range of function m is arbitrary, expressing the option for the measure "in degrees" of angles.

B_8. $m(\widehat{AOB}) = 0$ *if and only if* $\widehat{AOB}$ *is a null angle and* $m(\widehat{AOB}) = 180$ *if and only if* $\widehat{AOB}$ *is an elongated angle.*

$$[\, m(\widehat{h,\,k}) = 0 \Leftrightarrow h = k\,] \wedge [\, m(\widehat{h,\,k}) = 180 \Leftrightarrow h \neq k \wedge \exists\, a \in \mathcal{L} \colon h \cup k \subset a)\,].$$

B_9 (the "construction" axiom of proper angles}. *For any: straight line a, closed ray h included into a, semi-plane H delimited by a and real number $u \in (0, 180)$ there is exactly a ray k so that $k \subset H \cup a$ and $m(\widehat{h,\,k}) = u$.*

We leave aside the formalized expression of this axiom because in order to offer it reasonable dimensions we would have to introduce notations for the set of closed rays and for that of the semiplanes limited by a certain straight line a; this surplus of notations would flagrantly act against the declared didactic purpose of the axiomatic system.

B_{10} (the axiom of "the addition of angles"). *If B is a point inside the angle* $\widehat{AOC}$, *then*

$$m(\widehat{AOB}) + m(\widehat{BOC}) = m(\widehat{AOC}).$$

$$B \in \text{Int}\widehat{AOC} \Rightarrow m(\widehat{AOB}) + m(\widehat{BOC}) = m(\widehat{AOC}).$$

B_{11} (the axiom of "the supplement"). *If O is between A and C, then for any point* $B \notin (AC)$, $m(\widehat{AOB}) + m(\widehat{BOC}) = 180$ *holds*.

$$A - O - C \wedge B \notin (AC) \Rightarrow m(\widehat{AOB}) + m(\widehat{BOC}) = 180.$$

We shall now introduce the derived relations of "congruence", $\equiv$, into the set S of the segments and, separately, into the set $\mathcal{U}$ of the angles, putting $AB \equiv CD \leftrightarrow d(A, B) = d(C, D)$ and $\widehat{AOB} \equiv \widehat{COD} \Rightarrow m(\widehat{AOB}) = m(\widehat{COD})$, respectively.

Theorem 2.8. *The axioms* III_1, III_2, III_3, III_4 *in Hilbert's axiomatic system are consequences of the axioms* $B_1 - B_{11}$.

The proof of this theorem is immediate. In fact, from axioms $B_1 - B_{11}$ there easily follow stronger theorems than the quoted axioms, such as: the congruences (of segments and angles, respectively) are equivalences, "the substraction" (and not only the addition) of the segments and angles. This is a great didactic advantage because theorem of large applicability are quickly and easily deduced.

At present; function d and m are not correlated; the establishing of the interdependence of these functions can be achieved in several ways. For example, a didactic version would be the direct postulation of the cosine theorem; thus, after the algebraic study of the real functions sine and cosine, the sines theorem would be also easy to deduce, the same as the fact the sum of the angles of a triangle is $180°$, the axiom of the parallels and the entire Euclidean Geometry. But we are of the opinion that such an approach of Geometry would rather annihilate than edify Geometry by subjecting it to trigonometric computations! We have presented here this imaginary version to draw attention on some more subtle requirements which had to satisfy an axiomatic theory (of Geometry) not to obturate the Aesthetics of that theory.

The carrying on version Birkhoff chose was that of Hilbert: axiom B_{12} expresses III_5 and its principal direct consequences:

B_{12} (the *SAS* axiom). *If* (A, B, C) *and* (D, E, F) *are triples of non-collinear points and* $AB \equiv DE$, $AC \equiv DF$, $\widehat{BAC} \equiv \widehat{EDF}$, *then:* $BC \equiv EF$, $\widehat{ABC} \equiv \widehat{DEF}$, $\widehat{ACB} \equiv \widehat{DFE}$.

Obviously, this axiom includes III_5 and the axiomatic system $B_1 - B_{12}$ covers the first three groups in Hilbert's axiomatics. But it is easy to find out that, owing to the "ruler axiom", the axioms of continuity are also consequences of the axioms $B_1 - B_{12}$. We can then conclude:

Theorem 2.9. *Axioms* $B_1 - B_{12}$ *have the contents of absolute Geometry in Hilbert's axiomatic system as consequence.*

Only after formulating the *SAS* axiom we can find out that the function $d: S^2 \to \mathbb{R}_+ \cup \{0\}$ is a distance (it satisfies the triangular inequality, too). We remind that we have proved Theorem 4.3 of the isosceles triangle and Theorem 4.7 of the external angle in the preceding chapter. The following two theorems have the same proofs both in Hilbert's and Birkhoff's systems.

Theorem 2.10. *In a triangle ABC, a bigger angle is opposed to the longer side, and conversely.*

Proof. Let us suppose $AB < AC$, hence (Fig. 15) there is D so that $A - D - C$ and $AB \equiv AD$. Using the above reactualized Theorems, we deduce $C < \widehat{BDA} = \widehat{ABD} < B$. The converse is proved by reductio ad absurdum.

Theorem 2.11. *In a triangle, a side is shorter than the sum of the other two.*
Proof. We consider the triangle ABC and a point D so that $C - A - D$ and $AB \equiv AD$ (Fig. 16). It results

$$\widehat{ADB} \equiv \widehat{ABD} < \widehat{DBC},$$

Fig.15

Fig 16

so $d(B, C) < d(A, B) + d(A, C)$, q.e.d.

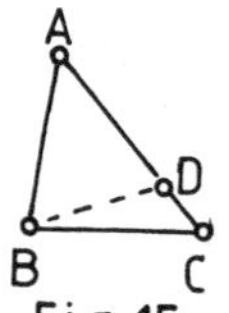

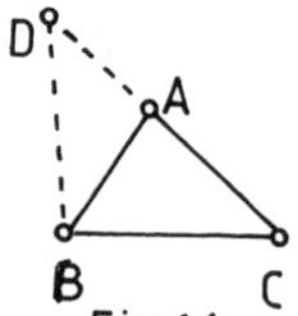

Remark. We deduced the triangular inequality $d(A, B) + d(B, C) \geq d(A, C)$ for the function d; taking also into account Theorem 2.1 we can now state that the name of distance for function d is correct.

The last axiom of Birkhoff's system, B_{13}, coincides with axiom V "of the parallels" in Hilbert's system:

B_{13}. *Given a straight line a and a point A which does not belong to it, in the plane determined by a and A there is at most a straight line b containing the point A and parallel with the straight line a.*

With this, the list of axioms in Birkhoff's axiomatic system comes to an end.

§3. Elements of Metatheory

The contents of the preceding paragraph was directed towards the deduction of the following proposition:

Theorem 3.1. *Birkhoff's axiomatic system implies Hilbert's axiomatic system.*

This theorem is more important for the metatheoretical analysis of Hilbert's system (since it provides it with a model) and less important for Birkhoff's system. Therefore, the directing of the deductions in the preceding paragraph proved to be useful as concerns the edifying of Geometry on the basis of Birkhoff's axiomatics, but not for the metatheoretical analysis. It is imperiously necessary for us to prove here a converse of Theorem 3.1.

Theorem 3.2. *Hilbert's axiomatic system implies Birkhoff's axiomatic system.*

Proof. As we had remarked after having formulated axiom B_5, the axioms in Hilbert's group I imply the axioms $B_1 - B_5$. Let us consider now as being fixed two points U, V in "Hilbert's space"; according to Theorem 6.2 in the preceding Chapter, there is a measure function we shall denote by d, where $d: S^2 \to \mathbb{R}_+ \cup \{0\}$ and $d(U, V) = 1$. To simplify the writing we shall note AB instead of $d(A, B)$. Theorem 5.1, Chapter V states that axiom B_6, of the ruler, is a consequence of Hilbert's axiom. By means of Theorem 2.6 we find now that axiom B_7 (of the plane separation) is a consequence of Hilbert's axioms.

Since in §6 of the preceding Chapter we did not effectively proceed to introducing the measure of angles we shall adopt here a more technical way, a less elegant but faster one. We shall construct a function $m: \mathcal{U} \to [0, 180]$ putting

$$(3.1) \quad m(\widehat{AOB}) = \frac{180}{\pi} \arccos \frac{OA^2 + OB^2 - AB^2}{2 \cdot OA \cdot OB}.$$

It is necessary to assure ourselves about the consistence of this definition that is about the independence of the value of $m(\widehat{AOB})$ of the choice of points A, B on the sides $(O; A)$ and $(O; B)$, respectively. By just replacing A with C we can for example suppose that $O - A - C$ holds (Fig. 17) and the proving of the equality

$$\frac{OA^2 + OB^2 - AB^2}{2 \cdot OA \cdot OB} = \frac{OC^2 + OB^2 - BC^2}{2 \cdot OC \cdot OB}$$

is necessary; this equality is equivalent with Stewart's relation:

$$OB^2 \cdot AC + BC^2 \cdot OA = AB^2 \cdot OC + OA \cdot OC \cdot AC.$$

Of course, we suppose the way of proving this equality in Euclidean Geometry as being known; on this occasion, we draw attention that Stewart's relation has an intrinsic importance and it is not a peripheric element of Geometry!

When checking axiom B_8 we shall find out:

$$m(\widehat{AOB}) = 0 \Leftrightarrow \frac{OA^2 + OB^2 - AB^2}{2 \cdot OA \cdot OB} = 1 \Leftrightarrow (OA - OB)^2 = AB^2 \Leftrightarrow [O; A) = [O; B).$$

Analogously, $m(\widehat{AOB}) = 180$ returns to $OA + OB = AB$, so to $A - O - B$, that is to fact that AOB is an elongated angle.

Axiom B_9 (of angles construction) expresses just axiom III_4 (of congruent conduct of angles).

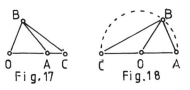

Fig.17   Fig.18

The calculuses in checking axiom B_{10} do not present special difficulties, but they are long and lack in geometrical interest.

To check axiom B_{11}, we shall place ourselves in the particular case (which does not restrain generality) $OA \equiv OB \equiv OC$. On the basis of the trigonometric property $\cos(\pi - u) = -\cos u$, the proving of the conclusion $m(\widehat{AOB}) + m(\widehat{BOC}) = 180$ returns to

$$\frac{OA^2 + OB^2 - AB^2}{2 \cdot OA \cdot OB} = -\frac{OB^2 + OC^2 - BC^2}{2 \cdot OB \cdot OC},$$

which, in its turn, is equivalent with $AC^2 = AB^2 + BC^2$, therefore with Pythagoras' theorem.

Axiom B_{12} expresses axiom III_5 and Theorem 4.3, Chapter V, and axiom B_{13} coincides with axiom V, so Theorem 3.2 is completely proved. In conclusion, we can state that Birkhoff's axiomatic system admits a model in Hilbert's axiomatic theory, therefore:

Theorem 3.3. *Birkhoff's axiomatic system is non-contradictory.*

But we draw attention that Birkhoff's and Hilbert's axiomatic system are not equivalent because their primary notions and relations in the two systems do not correspond. On the basis of Birkhoff's axiomatic system we easily obtain the primary notions and relations from Hilbert: $\Delta = \mathcal{L}$; $\Pi = \Pi$; the congruence of the segments is introduced by $AB \equiv CD \Leftrightarrow d(A, B) = d(C, D)$, and the congruence of the angles by

$(\widehat{h, k}) \equiv (\widehat{h', k'}) \Leftrightarrow m(h, k) = m(h', k')$. But with the primary notions and relations

from Hilbert we can obtain only $\mathcal{L}$, Π and m. To determine d, it is necessary to arbitrarily choose a standard OE, after which we can define $d(A, B)$ as the unique non-negative real·number x determined by $AB = x \cdot OE$. The change of the standard OE returns to the replacing of the function $d: S^2 \to \mathbb{R}$ by an analogous function $d': S^2 \to \mathbb{R}$, where $d'(A, B) = k \cdot d(A, B)$, k being a positive real number. Therefore, starting from Hilbert's axiomatic theory, we can not determine Birkhoff structure $(S, \mathcal{L}, \Pi, d, m)$, but only a more general mathematical structure $(S, \mathcal{L}, \Pi, \delta, m)$, where $\delta = \{k \cdot d: k \in \mathbb{R}_+\}$.

The change $d \to \delta$ has profound significations. One of them has already been evidenced: the impossibility of a canonical choice of a standard for measuring the lengths within Euclidean Geometry. (Such a choice is possible in non-Euclidean Geometries.) Another significance, with a practical character, is given by the possibility of the co-existence of some distinct standards to measure the lengths, such as: the meter, the yard, the millimetre, the light year, the parsec etc. Two other significances will be evidenced later.

Taking into account that we have already found out that Hilbert's axiomatic system is categorical and complete, we shall now analyze what metatheoretical implications can the choice of an element $d \in \delta$ have, choice that determines Birkhoff's Geometry. But we have seen that the choice of the distance d in the δ family returns to the fixing (in Hilbert's Geometry) of a standard OE. We can replace such a standard by another one, $O'E'$, and we do not restrain generality supposing that $OE \parallel O'E'$. In the case that presents interest, OE and $O'E'$ are not congruent and there is a point U common for the straight lines OO' and EE' (it also supposes that E and E' are on the same side of the straight line OO'). Then there will exist a homothety F of centre U and ratio $r = \dfrac{UO'}{UO} = \dfrac{O'E'}{OE}$ which turns OE into $O'E'$. This homothety F does not affect either of the primary notions with Hilbert and, therefore, constitutes a automorphism of Hilbert's model. The homothety F does not preserve the distance d, but replaces it with $d' = \dfrac{1}{r} d$ and thus constitutes an isomorphism of Birkhoff's *distinct* models $(S, \mathfrak{L}, \sqcap, d, m)$, and $(S, \mathfrak{L}, \sqcap, d', m)$. But we are interested in the that these *distinct* models are *isomorphic* and therefore:

Theorem 3.4. *Birkhoff's axiomatic system is categorical and complete.*

Now we draw attention that the replacing $d \to \delta$ has also a certain group significance. The group of geometric transformations preserving the Birkhoff structure is made up of isometries, and the group of geometric transformations preserving the Hilbert structure is made up of *similitudes* (compositions of isometries and homotheties).

We point out that we have presented in this chapter a *didactic* version of Birkhoff's axiomatics; this version contains more axioms than the *scientific* one and this supplementation can lead to the damaging of the minimality of the system. But we underline that minimality may be minor quality as compared with accessibility.

For certain axioms we can not even raise the question of their being "totally" independent, that is of those axioms that contributed to the forming of some notions intervening in the axiom under analysis, as well.

But we can remark the total independence of axiom B_{13}, that of the parallels, absolute Geometry also containing Euclidean Geometry in which the negation of B_{13} is valid.

Axiom B_{12} (SAS) is also independent of the others, as one can find out by replacing the distance d of a model with another distance d' associated to a fixed straight line a :

$$d'(A, B) = \begin{cases} d(A, B) & \text{if } \{A, B\} \not\subset a, \\ 2d(A, B) & \text{if } \{A, B\} \subset a. \end{cases}$$

Triangles such as ABC and DEF which satisfy $d(A, B) = d(D, F)$, $d(A, C) = d(D, F)$ and $m(\widehat{BAC}) = m(\widehat{DEF})$ may not satisfy the axiom (B_{12}) for the new distance d' if, for example, $\{A, B\} \subset a$ and $\{D, E\} \not\subset a$. This modification does not alter the validity of the other axioms.

Problem

The major disadvantage of Birkhoff's axioms, the consideration of an un-geometrical distance was eliminated in [579]. There appear only two primary notions: *point* and *ratio of two segments*. We present here the axioms (noted by $V_1, \ldots, V_6$), the enounces of theorems (noted by $t_1, \ldots, t_{16}$) and the derived notions (introduced by definitions $d_1, \ldots, d_{19}$) and we propose to the reader to imagine the proofs.

V_1 (Axiom of ratio). *For any ordered pair of segments* AB *and* $CD \neq 0$ *a real non-negative number* $\frac{AB}{CD}$ *is associated so that*

a) $\dfrac{AB}{CD} = 0$ *iff* $AB = 0$ *(that means* $A = B$*)*.

b) $C \neq D \wedge E \neq F \Rightarrow \dfrac{AB}{CD} \cdot \dfrac{CD}{EF} = \dfrac{AB}{EF}$.

t_1. *If* $AB \neq 0$, *then* $\dfrac{AB}{AB} = 1$.

t_2. *If for* $i = 1, \ldots, n$, A_iB_i *are arbitrary segments and* λ_i *arbitrary real numbers, for any* $CD \neq 0$ *there exist an unique* $\gamma \in \{=, <, >\}$ *so that*

(1) $$\sum_{i=1}^{n} \lambda_i \frac{A_iB_i}{CD} \gamma 0$$

and the symbol γ *is independent of* CD.

Convention: As in (1) the denominator CD is not revelant, we agree to present (1) under the more economical expression

(2) $\sum_{i=1}^{n} \lambda_i A_i B_i \, \gamma \, 0$.

d_1. *If for AB, CD one of the type (2) relation AB = CD, AB < CD, AB > CD
we say: AB is equal, smaller or greater than CD. Obviously $AB \gamma CD \Leftrightarrow \frac{AB}{CD} \gamma 1$.*

d_2. *If for three distinct points A, B, C we have*

(3) $AB + BC = AC$

*we say: B is between A and C and we write $A - B - C$. The set of all points B between A
and C is "the interior of the segment AC".*

t_3. *If A, B, C are three distinct points, no more than one is between the other
two.*

d_3. *Three points A, B, C are co-linear if either at least two of them coincide or
one of them is between the others. If A, B, C are not co-linear we say that $\{A, B, C\}$
is a triangle; in the usual manner we define its vertexes and its sides (as interior of
segments).*

d_4. *Straight line is the set of all points co-linear with two given distinct points.*
(If the given points are A, B we denote such a straight line by $|A, B|$ but we can not
say yet if it is unique.)

t_4. *For any points A, B there exists at least a straight line belong at least two
points.* (We can not say that there are more than two points on a straight line.)

d_5. *If $A - B - C$ we say that A and C are on distinct sides of B and we say that B
and C are on the same side of A.*

d_6. *For n points A_i, $i = 1, 2, \ldots, n$ we say that they are coherent indexed if $A_i - A_j - A_k$
holds iff $i < j < k$ or $i > j > k$. For $n = 1$ and $n = 2$ we conventionally agree the
coherence of any indexation.*

d_7. *We call figure any set of points. Straight lines are figures and benefit of
notations with small latin letters a, b, c, …; other figures are denoted by $\mathscr{F}$, G, … .*

d_8. *Two figures $\mathscr{F}$ and $\mathscr{F}'$ are similar if there exists a function S: $\mathscr{F} \to \mathscr{F}'$ and a real positive number k so that for any $A \neq B$ in $\mathscr{F}$ and for $A' = S(A)$, $B' = S(B)$, $\frac{AB}{A'B'} = k$.* (*S* is named *similitude* and it is obviously a bijection; *k* is *the ratio of the similitude S*).

d_9. *Two figures $\mathscr{F}$ and $\mathscr{F}'$ are equal if they are similar and the ratio of at least one similitude from $\mathscr{F}$ to $\mathscr{F}'$ is* 1. (For such a similitude the name of *isometry* is more precise than that of similitude.)

V_2 (Moore's axiom). *Any subset of no more than four points of a straight line admits at least a coherent indexation.*

t_5. *Any three points of a straight line are co-linear.*

t_6. *Throw distinct points A, B passes a unique straight line, $|A, B|$.*

t_7. *Two distinct straight lines have in common either one point or none.* (Respectively we say: they are secant or parallel.)

V_3 (Hilbert's axiom). *Let d be a straight line and A one of its points. Let UV be a non-null segment and r a positive number. Then, there exist exactly two points B, B' so that $\frac{AB}{UV} = \frac{AB'}{UV} = r$; B and B' are on distinct sides of A.*

t_8. From any straight line d, there exists at least a bijective correspondence f to the set $\mathbb{R}$ of real numbers, to that

$$M - N - P \Rightarrow f(M) < f(N) < f(P) \lor f(M) > f(N) > f(P).$$

t_9. *The previous correspondence f induces an order $\leq$ on d similar with the usual one on $\mathbb{R}$; for such ordered set d the Dedekind's principle is valid.*

t_{10}. *The previous order $\leq$ of d is total, dense and without extremal elements.*

t_{11}. *Hilbert axiom III$_1$ holds.*

t_{12}. *Hilbert axioms III$_2$ and III$_3$ hold.*

Remark. Above considered axioms edify the geometry on a straight line if we would add now-as a new axiom- that all points belong to a straight line. The following axiom will assures the bidimensionality.

V4 (Pash's axiom). *There exists at least a triangle. If a straight line does not cut two sides of a triangle it does not cut nor the third one.*

t_{13}. *If a straight line d does not contains none of the vertexes of the triangle ABC and cuts one of its side, it should cut one side more.*

t_{14}. *A straight line d separates the points which not belong to it in two disjoint, non-void ... "semiplanes".*

t_{15}. *No straight line cuts all sides of a triangle.*

d_{10} introduce in the usual manner the notion of angle. *The equality of angles derives from d_9.*

V_5 (Kerekjarto's axiom). *Let ABC be a triangle and AB = DE. There exist exactly two points F, F', on distinct sides of DE such that $\triangle ABC = \triangle DEF = \triangle DEF'$. If two triangles are equal, then their corespondent angles are equal.*

t_{16}. *The axiom III_5 and III_4 of Hilbert's system hold.*

V_6 (Euclid's axiom). *Throw a point A not belonging to the straight line d no more than one parallel to d passes.*

Obviously the axiomatic system $V_1 - V_6$ is equivalent with Hilbert's one; therefore it is non-contradictory and categorical.

CHAPTER VII

GEOMETRICAL TRANSFORMATIONS

§1. Generalities

Given the sets m, n, we call (geometrical) transformation of m into n a map $T: m \to n$; the presence of the qualificative in bracket is justified only when certain "geometrical structures" are indicated or implied on the two given sets and T satisfies certain conditions referring to these structures. When there is one more transformation $T': m' \to n'$, in the hypothesis that $n \subset m'$, we can consider the transformation $T' \circ T: m \to n'$ called *composition of transformations* T and T' defined for any $M \in m$ by $(T' \circ T)(M) = T'(T(M))$. The partial operation of compounding the (geometrical) transformations is associative, that is transformations $T'' \circ (T' \circ T)$ and $(T'' \circ T') \circ T$ exist simultaneously and coincide. (The coincidence of two transformations returns to the coincidence of domains, ranges and correspondences.)

For the map T considered above, in the supplementary hypothesis $n \subset m$, composition $T \circ T$ more often noted with T^2 can be defined. The more general symbol T^n is defined for any non-null natural number n, recoursively, by $T^1 = T$ and $T^{n+1} = T \circ T^n$. If n is the smallest non-null natural number for which $T^n = T$ holds, it is said that transformation T is *involutive of n order*; when the order is not specified we shall take $n = 2$.

For any set m the identical transformation $1_m: m \to m$ is evidenced, where for any M in m: $1_m(M) = M$; when confusions are not possible we shall note I instead of 1_m.

Given the transformation $T: m \to n$ we shall say about a transformation U that it is *the inverse* (or *the reverse*) of T if $T \circ U = I_n$ and $U \circ T = I_m$; in this case we shall use the notation $U = T^{-1}$.

Theorem 1.1. *Transformation T admits an inverse when and only when it is bijective; if a transformation T admits an inverse then the inverse is unique.*

Theorem 1.2. *The set of bijective transformations of a set m in itself constitutes together with the compounding operation a group (which we shall note with B_m).*

An element $F \in m$ satisfying $T(F) = F$ is said to be a fixed element of transformation T.

Let be a transformation $T : m \to n$ and a subset u of m; the set $\{T(M) : M \in u\}$ is denoted by $\check{T}(u)$) and called *image through T of the set u.* Let $\mathcal{P}(m)$ be the power set of m. It is easy to find out that we can conceive $\check{T}$ as transformation from $\mathcal{P}(m)$ to $\mathcal{P}(n)$; $\check{T}$ is called the "globalized" of T. Usually, the symbol ˇ which appeared in the notation is omitted; we shall also do so when this omission does not generate confusions.

§2. Isometries

One calls *distance* on the set m a map $d : m^2 \to \mathbf{R}$ satisfying the conditions:
1. $d(A, B) = 0 \Leftrightarrow A = B$;
2. $d(A, B) = d(B, A)$;
3. $\forall A, B, C \in m : d(A, B) + d(B, C) \geq d(A, C)$.

One calls *metric space* a couple (m, d), where d is a distance for m . Within the metric space (m, d) one calls *sphere* of centre O and ray r the geometrical locus of points M satisfying $d(O, M) = r$. In the particular case when m is an Euclidean plane, instead of the word sphere that of *circle* is used.

We call isometry of (m, d) a surjective map $T : m \to m$ which *preserves the distance*, that is :

$$\forall A, B \in m : d(T(A), T(B)) = d(A, B).$$

Theorem 2.1. *Any isometry is injective; the isometries of a metric space make up a group related to the compounding operation.*

Remark. The condition of surjectiveness included here into the definition of isometries simplifies essentially the proof.

Theorem 2.2. *An isometry T of an Euclidean plane preserves the ternary relations of "to be between" and "to be co-linear": the image through T of a straight line, a ray, a segment, a semi-plane, a circle, an arc of a circle is a set characterized by the same name.*

Proof. The condition $A - B - C$, equivalent with $B \in |AC|$, is expressed here by $d(A, B) + d(B, C) = d(A, C)$ and $d(A, B) \cdot d(B, C) \cdot d(A, C) \neq 0$ It results that the isometry T preserve the relation "to be between" and $T(|AC|) = |T(A)T(C)|$. If $A - O - B$, then $X \in [O; A) \Leftrightarrow d(O, X) + d(B, O) = d(B, X)$, so T applies rays on rays. Let s be the semi-plane limited by d which does not contain A. We note $T(A) = A'$, $T(d) = d'$ and let M' be the image through T of an arbitrary point M. $M \in s \Leftrightarrow |AM| \cap d \neq \varnothing \Leftrightarrow M' \in s'$, where we denoted by s' the semi-plane delimited by d' which does not contain A'. If C is a circle of centre O and ray R, then $T(C)$ is the circle of centre $O' = T(O)$ and the same ray R. An arc of circle appears intersecting a circle with a semi-plane.

Theorem 2.3. *An isometry T preserves the angles, that is from $T(A) = A'$, $T(B) = B'$, $T(C) = C'$ it follows $\widehat{A'B'C'} \equiv \widehat{ABC}$.*

It is found out that the triangles ABC and $A'B'C'$ are congruent, therefore they have congruent corresponding angles.

Theorem 2.4. *If an isometry T admits the distinct points A, B as fixed points, then any point M of the straight line AB is a fixed point; an isometry T of the plane π in itself which admits three non co-linear fixed points A, B, C is the identical transformation.*

The proof is immediate. Let be $M \in (AB)$ and let us suppose, for example, that $M - A - B$ holds, so $d(M, A) + d(A, B) = d(M, B)$. For $M' = T(M)$ it will result $d(M', A) = d(M, A)$ and $d(M', B) = d(M, B)$, conditions that ensure $M = M'$.

In the hypothesis $T(A) = A$, $T(B) = B$, $T(C) = C$, for N arbitrary in π, we can choose one of the vertexes of the triangle ABC, A, for example, so that (AN) may cut (BC) in a point M. As M is on (BC), it will result, according to the preceding phase, $T(M) = M$ and then, for N on (AM), it will result $T(N) = N$, q.e.d.

We invite the reader to formulate and prove similar results related to isometries of the three-dimensional Euclidean space.

§3. Symmetries

Let be a plane π and a straight line d in π; one calls *symmetry* (of the plane π) *related to* d a transformation $S: \pi \to \pi$ which associates to the point $M \notin d$ the point M' characterized by the condition that d be the midperpendicular of the segment MM' and preserve points $M \in d$; the straight line d is called *the axis of the symmetry*.

Theorem 3.1. *A symmetry related to a straight line d is an isometry.*

Proof. Let A' and B' be the images of points A, B and A_1, B_1 the orthogonal projections of A, B on d. We shall suppose that we are in the generic case $A_1 \neq B_1$. We shall successively find out $\triangle BB_1A_1 \equiv \triangle B'B_1A_1$, $A_1B \equiv A_1B'$, $\triangle ABA_1 \equiv \triangle A'B'A$, $AB \equiv A'B'$. The case $A_1 = B_1$ is to be dealt with separately. Surjectiveness is easy to prove.

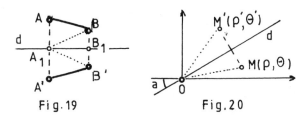

Fig.19   Fig.20

We shall insert here some reasons referring to analytical expressions of plane symmetry related to a straight line; these considerations may be omitted at the first reading.

Let us consider a polar frame having a ray Ox with $O \in d$ as a polar axis. We shall also note with a the acute angle formed by d with Ox. Let $M(\rho, \theta)$ and $M'(\rho', \theta')$ be its image through, the symmetry S. We shall easily find out $OM \equiv OM'$ so $\rho = \rho'$ and $\frac{\theta+\theta'}{2} = a$, therefore $\theta' = 2a - \theta$. We conclude that in the polar frame under consideration the symmetry S is expressed by $\rho' = \rho$, $\theta' = 2a - \theta$.

Let Oy be a ray perpendicular on Ox; we shall make explicit the effect of symmetry S upon the Cartesian co-ordinates in the frame we have obtained. We shall easily deduce: $M(x, y)$, where $x = \rho\cos\theta$ and $y = \rho\sin\theta$. We get $x' = \rho'\cos\theta' = \rho\cos(2a-\theta) = \cos 2a(\rho\cos\theta) + \sin 2a(\rho\sin\theta) = x\cos 2a + y\sin 2a$ and $y' = \rho'\sin\theta' = \rho\sin(2a-\theta) = \sin 2a(\rho\cos\theta) - \cos 2a(\rho\sin\theta) = x\sin 2a - y\cos 2a$.

So $S: M(x, y) \rightarrow M'(x', y')$, where $x' = x \cdot \cos 2a + y \cdot \sin 2a$ and $y' = x \cdot \sin 2a - y \cdot \cos 2a$.

Theorem 3.2. *Any symmetry* S *related to a straight line satisfies* $S^{-1} = S$ *(so* $S^2 = I$, *as well). If* I *is the identical transformation, the set* $\{S, I\}$ *constitutes a group related to the composition.*

Theorem 3.3. *A point* X *of the plane is made invariant by a symmetry* S *related to a straight line if and only if it is situated on the axis of the symmetry.*

Theorem 3.4. *A straight line is made invariant to a symmetry* S *of axis* d *if and only if* $a \perp d$ *or* $a = d$.

Theorem 3.5. *A circle* $C(C, r)$ *is made invariant by a symmetry* S *of axis* d *if and only if* $C \in d$. *(It is found out that* $S(C)$ *is a circle having the centre in* $C' = S(C)$ *and the same ray* r).

Theorem 3.6. *For any congruent triangles* ABC *and* $A'B'C'$ *in a plane* π *there is a unique isometry* $T: \pi \rightarrow \pi$ *so that* $T(A) = A'$, $T(B) = B'$ *and* $T(C) = C'$. *Any isometry of the plane is the product of at most three symmetries related to conveniently chosen straight lines.*

Proof. Let us suppose that $A \neq A'$ and let d_1 be the mediatrix of the segment AA'. We note with S_1 the symmetry related to d_1 and let be $S_1(B) = B_1$, $S_1(C) = C_1$. We find out easily: $\triangle ABC \equiv \triangle A'B_1C_1 \equiv \triangle A'B'C'$. It results $A'B_1 \equiv A'B'$. Let us suppose $B_1 \neq B'$ and let d_2 be midperpendicular of the segment B_1B'. It is obvious that $A' \in d_2$, so the symmetry S_2 of axis d_2 admits A' as a fixed point. Let be $S_2(C_1) = C_2$. It results $\triangle ABC \equiv \triangle AB_1C_1 \equiv \triangle A'B'C_2 \equiv \triangle A'B'C'$ and therefore $C_2A' \equiv C'A'$, $C_2B' \equiv C'B'$. According to these congruences, C_2 coincides with C' (and $T = S_2 \circ S_1$ is an isometry that satisfies the enunciated conditions) or C_2 is symmetrical with C' related to $A'B'$. Let S_3 be in this case the symmetry related to $(A'B')$; the transformation $T = S_3 \circ S_2 \circ S_1$ is an isometry that applies A, B, C in A', B', C'.

If, besides T an isometry U so that $U(A) = A'$, $U(B) = B'$, $U(C) = C'$ also existed, the transformation $U^{-1} \circ T$ would be an isometry having the co-linear points A, B, C fixed. According to Theorem 4, §2, $U^{-1} \circ T = I$, so $T = U$.

For any arbitrary isometry T of the plane π and for an arbitrary triangle ABC we shall find points $A' = T\ (A)$, $B' = T(B)$, $C' = T(C)$,   and shall construct the decomposition of T into at most three axial symmetries as in the first phase of the proof.

Let E_3 be the three-dimensional space and a straight line $d \subset E_3$; one calls (spatial) symmetry of axis d a transformation $S : E_3 \to E_3$ that preserves the points of d and associates to M, which is not situated on d, the symmetrical point M' related to d in the plane containing M and d. Theorems 3.1, 3.2 and 3.3 preserve their enunciations (and proofs).

Theorem 3.7. *Let S be a spatial symmetry of axis d. A straight line   distinct from   is made invariant by S if and only if $a \perp d$ and $a \cap d \neq \varnothing$. A plane is made invariant by   if and only if it is perpendicular on .*

Theorem 3.8. *A sphere $S(C,\ r)$ is made invariant by a spatial symmetry S related to an axis d if and only if $C \in d$.*

One calls *symmetry of centre O* (of a plane π through O or of a space) the transformation that preserves the point O and applies $M \neq O$ in M' so that O may be the midpoint of MM'. One calls symmetry related to a plane π the transformation $S : E_3 \to E_3$ that preserves the points in the plane π and transforms a point M, which is not situated in π, into the point M' characterized by the property that π is the mediating plane of MM'. We consider the reformulation of the theorems in this paragraph (and their proof) referring to symmetries related to points or planes to be a useful and accessible exercise.

§4. Vectors

This paragraph has the immediate purpose of preparing a vocabulary for the next paragraph but also the more general one of presenting new connections between the preceding chapters.

Chapter IV showed the way in which elementary Geometry can be edified on the basis of the notion of vectorial space; a presentation of the way in which one can obtain the notion of vector as derived notion related to an axiomatic system of Geometry is necessary.

First of all, we shall present the "classic" modality of introducing the notion of vectors as a class of "equipollence" in the set of oriented segments, where two segments are equipollent if they "have the same size, direction and sense". Of course, from the intuitive point of view, this modality is simple, quick and accessible. But we appreciate that in order to rigorously construct the notion of vector this way, we need to overcome certain substantial technical difficulties.

Let us consider within the Euclidean Geometry the set $\mathcal{L}$ of the straight lines. The relation of parallelism, $\|$, constitutes a binary relation on $\mathcal{L}$ that proves to be an equivalence. We do not present here the proof of this proposition which is well done in the handbooks, but we draw attention that this stage also contains appreciable technical difficulties.

Let us consider the factor set $\Delta = \mathcal{L} / \|$; an element in Δ is an equivalence class $[d]$ and is called *direction*; instead of the expression $a \in [d]$, "a has the direction $[d]$" or, more frequently, "a and d have the same direction" are preferred.

Within an axiomatic theory of Euclidean Geometry the concept of *orientation* of a straight line is specified; we presented for Hilbert's axiomatic theory (Chapter V, §3) a modality of defining which we appreciate to be original; the introducing of this concept for Birkhoff's axiomatics is sketched in the remark below Theorem 2.4.

We draw attention here that the two defining modalities do not coincide because an equivalence class in the set of the rays situated on a straight line a can not be the same thing as an equivalence class in the set of Cartesian systems of that straight line; we think it is suggestive to say that the modalities indicated are "synonymous" as they make up "words" having the same meaning out of "distinct letters". We can also say that we have successively described the same objective reality in different "languages", but we also know that there is a "dictionary" enabling us to translate "word by word" an expression into the other one. (A system of co-ordinates $f: a \to \mathbb{R}$ bijectively corresponding to the ray $f^{-1}(\mathbb{R}_+ \cup \{0\})$ thanks to Theorem 2.5. in the preceding chapter.) We would have not insisted upon this "detail" if it had not expressed a typical situation in the edification of Mathematics and if we had not been pleased to combat the "mean" tendency of looking in mathematical theories for "coincidences" instead of "resonances".

Further on, as for the concept of orientation, we shall prefer the significance furnished by the first definition which we appreciate to have a more pronounced geometrical character.

Let d be a straight line and Ω_d the set of all the orientations of the straight lines which have the direction $[d]$. An element ω in Ω_d is a family of rays situated on the same straight line we shall denote by $\sigma(\omega)$ and call *support of* ω. Obviously,

(4.1) $[\omega \in \Omega_d \wedge h \in \omega] \Leftrightarrow [k \in \omega \Leftrightarrow (h \subset k \vee k \subset h)]$.

We introduce a binary relation $\uparrow$ on Ω_d saying that $\omega \uparrow \omega'$ holds if: $\sigma(\omega) = \sigma(\omega')$ and $\omega = \omega'$, or $\sigma(\omega) \neq \sigma(\omega')$ and that $[O; A)$ exists in ω and

$[O'; A')$ exists in ω' so that A and A' be in the same semi-plane delimited by the straight line OO'.

The symbol ↑ will be read: the orientation ω *is coherent* with the orientation ω'. To give more precision to the definition of the relation of coherence of the orientations we shall offer two formulae in which the notation $(OO'; A)$ introduced in §3, Chapter V, for the (open) semi-plane delimited by the straight line OO' and containing point A will also appear:

(4.2) $\sigma(\omega) = \sigma(\omega') \Rightarrow [\omega \uparrow \omega' \Leftrightarrow \omega = \omega']$.

(4.3) $\sigma(\omega) \neq \sigma(\omega') \Leftrightarrow [\omega \uparrow \omega' \Leftrightarrow \exists [O; A) \in \omega \exists [O'; A') \in \omega':$
            $(OO'; A) = (OO'; A')]$.

Lemma 4.1. *If $\omega \uparrow \omega'$ and $\sigma(\omega) \neq \sigma(\omega')$, then for any rays $[U; B)$ in ω and $[U'; B')$ in ω', points B, B' are in the same semi-plane delimited by the straight line UU', that is*

(4.4) $\forall [U; B) \in \omega, \forall [U'; B') \in \omega': (UU'; B) = (UU'; B')$.

Proof. We take the statement $\omega \uparrow \omega'$ according to (4.3) by the existence of the rays $[O; A)$, $[O'; A')$ included into the same semi-plane delimited by (OO'). We replace in a first stage only one of the rays: $[O'; A')$ with $[U'; B')$. As these rays define the same orientation they will be linked by an inclusion relation. We are fixing the ideas by supposing that $[U', B') \subset [O'; A')$. It follows $O' - U' - B'$ and $U' \in \text{Int} \widehat{AOO'}$. By means of Proposition 3.4 in Chapter V we deduce that the straight line OU' cuts the segment AO' in a point M. As this straight line also cuts the side $O'B'$ in U', it cannot cut the side AB' anymore (it would contradict Theorem 2.3, Chapter V), so A, B' are in the same semi-plane delimited by (OU'). The proof ends by showing that, owing to symmetry, the replacing of the ray in ω does not affect the validity of the condition in relations (4.3) and (4.4), either.

Theorem 4.1. 1) *For any straight line d, ↑ is an equivalence in the set Ω_d of the orientations of the straight lines having the same direction d.*

2) *The quotient set $\Omega_d / \uparrow$ contains exactly two elements (called senses for the*

direction [d]).

3) *For any equivalence classe s in* $\Omega_d / \uparrow$ *and any straight line a of direction* [d] *there exists in s exactly on orientation* ω *which has a as support.*

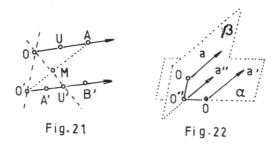

Fig. 21   Fig. 22

Proof. 1) By means of (4.1) it is easy to find out that $\uparrow$ is reflexive and symmetrical. As for transitivity, we assume $\omega \uparrow \omega'$ and $\omega' \uparrow \omega''$. Let a, a', a'' be the support straight lines of the orientations ω, ω', ω'' . If the straight lines a, a' a'' are not co-planar, we consider points O, O', O'' on them and let γ be the plane containing these points. We consider then the rays $[O; A) \in \omega$, $[O'; A') \in \omega'$, $[O''; A'') \in \omega''$, If H is the semi-space delimited by γ which contains A, it immediately follows that A and A' are in H. We restrain our attention to the plane β which contains a and a'' ; in γ, the straight line OO'' delimits a semi-plane containing A and A'' , so $\omega \uparrow \omega''$ holds. The same conclusion is also valid when a, a', a'' are co-planar, and we can now choose points O, O', O'' (according to Lemma 4.1) as co-linear in order to simplify the reasoning. The transitivity of the relation $\uparrow$, which is therefore an equivalence, has thus been proved.

Formula (4.2) immediately leads to conclusion 3). Then we find out that each of the opposed orientations, ω and $-\omega$, of the straight line d represents a class s and $-s$ in $\Omega_d / \uparrow$; if a class s' different from s and $-s$ also existed, this would have a representative ω' having the support d which could not coincide either with ω or with $-\omega$, which is absurd.

We shall put an end to these considerations by introducing a notation which will be useful later. If A, B are distinct points, we can evidence a sense $\underrightarrow{AB}$ (called "the sense from A to B"), namely the element in the quotient set $\Omega_d / \uparrow$ admitting the ordering ω

of the straight line (AB) characterized by $[A; B) \in \omega$ as a representative. Let now S be the set of the points and $B = S \times S$ the set of bi-points. A binary relation $\sim$, called "equipollence", is introduced into B by:

$$(4.5) \quad (A, B) \sim (C, D) \Leftrightarrow (A = B \wedge C = D) \vee (A \neq B \wedge AB \equiv CD \wedge AB \parallel CD \wedge \overrightarrow{AB} \uparrow \overrightarrow{CD}).$$

According to the above ones, it immediately follows:

Theorem 4.3. $\sim$ *is an equivalence on the set* B.

The equivalence class of the bi-point (A, B) related to the equipollence relation $\sim$ is called *vector* and is noted with $\overrightarrow{AB}$. Sometimes in an obviously improper way, it is said that A is the "origin" and B the "extremity" of the vector $\overrightarrow{AB}$.

Remark. We have not made so far explicit referring to the "dimension" of the ambient Euclidean geometrical space; we have implied that S would be the set of the points in the three-dimensional Euclidean space E_3 only by referring to the proof of the transitivity of the relation of parallelism and by analyzing the case of the non-coplanar straight lines in the proof of Lemma 4.1. Of course, if we want to restrain the construction to a plane or to a straight line, a part of the technical difficulties disappears. From the physical point of view, the notion of "sliding vector" which can be obtained by particularizing S as the set of the points of a straight line is useful (in this case the vector can be characterized by "size" and "sense" because direction is unique for all the vectors). We do not think that there are reasons to introduce the mathematical notion of "tied vector" which would coincide with that of bi-point.

We appreciate as being advantageous from scientific and didactic points of view the introducing of the equipollence relation (and of the notion of vector) by the construction we are sketching below.

Let $g: S \times S \to S$, the function defined by: $g(A, B)$ be the midpoint of the segment AB; (in particular $g(A, A) = A$). Let us introduce now the equipollence $\sim$ as binary relation for the set B of the bi-points by:

$$(4.6) \quad (A, B) \sim (C, D) \Leftrightarrow g(A, D) = g(C, B).$$

The quotient set $V = B / \sim$ is called *vectorial space*. The name is easy to be justified by finding out that properties 1-7 in Chapter IV, §1 and 2 are checked for the thus introduced notion of vector. For this purpose it is also necessary to define the vector $\alpha \cdot \overrightarrow{AB}$, where α is an arbitrary real number.

- If $B = A$ or $\alpha = 0$, by definition $\alpha \cdot \overrightarrow{AB} = \overrightarrow{AA}$.

- If $B \neq A$ and $\alpha \neq 0$, $\alpha \cdot \overrightarrow{AB} = \overrightarrow{AC}$ holds, point C being on the straight line (AB), satisfying $d(A, C) = |\alpha| \cdot d(A, B)$ and being uniquely determined by the condition $C \in [A; B) \Leftrightarrow \alpha > 0$.

We draw attention only that the property of distributivity of the product with scalars to the sum of the vectors constitutes the condensed expression of the fundamental theorem of similitude.

Moreover, by defining the scalar product $\cdot$ on V by $\overrightarrow{AB} \cdot \overrightarrow{AC} = d(A, B) \cdot d(A, C) \cdot \cos BAC$, we find out that conditions $1' - 4'$ in Chapter IV, §2 are satisfied, so the structure of Euclidean (vectorial) space can be naturally introduced on V.

§5. Translations

Let A, A' be points of the Euclidean space E_3. One calls *translation of vector AA'* the transformation $T: E_3 \rightarrow E_3$ defined by $T(M) = M' \Leftrightarrow \overrightarrow{AA'} = \overrightarrow{MM'}$. In the particular case $A = A'$ it is obvious that the translation of vector $\overrightarrow{AA'}$ is the identical transformation of E_3.

We shall also use the same term of translation of vector $\overrightarrow{AA'}$ for restrictions of the above defined transformation T to proper subsets S of E_3 made invariant by T; here are some examples of such sets:

1. a semi-space bounded by a plane parallel with (AA');
2. a plane parallel with (AA');
3. sets obtained by intersecting sets of the characterized type from among which we can quote insides of dihedral angles or semi-planes delimited by straight lines parallel with (AA').

Theorem 5.1. *Any translation is an isometry; the inverse of a translation is a translation; the composition of two translations of vectors $\vec{u}, \vec{v}$, is the translation of vector $\vec{u} + \vec{v}$.*

Theorem 5.2. *A translation T of non-null vector $\vec{u}$ does not have fixed points, preserves the straight lines parallel with $\vec{u}$, planes α parallel with $\vec{u}$ and does not preserve circles or spheres.*

Theorem 5.3. *The composition $S_2 \circ S_1$ of two symmetries related to the parallel axes d_1, d_2 is a translation of vector $\vec{u}$ perpendicular on d_1, having a size equal to the double of the distance between d_1 and d_2 and the sense from d_1 to d_2. Any translation is decomposed into product of symmetries.*

Proof. At the beginning, we examine the case when S is a plane. Let us consider $OO_1 \perp d_1$, $O \in d_1$ and an axis Ox oriented from O to $O_1 \in d_2$. Let also O_2 be symmetrical to O related to O_1. Let us suppose $O_1 (a, O)$. Point $A(x, y)$ passes through symmetry S_1 into $A'(x', y')$, where, obviously, $x' = -x', y' = -y'$. Point $A'(x', y')$ passes through symmetry S_2 into $A''(x'', y'')$, where $x'' = 2a - x' = 2a + x$ and $y'' = y' = y$. Transformation $T = S_2 \circ S_1$ leads $A(x, y)$ into $A''(x + 2a, y)$, so it is a translation of vector $\overrightarrow{AA''} = \overrightarrow{OO_2}$.

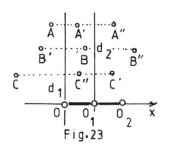

Fig.23

If S coincides with E_3, we shall consider for the arbitrary point A the plane α, incident with A and perpendicular on d_1. Let U, U_1 be the points in which d_1, d_2 pricks α. With the above same notations it is easy to find out that UU_1 is a median line in the triangle $A'AA''$ and therefore $\overrightarrow{AA''} = 2 \cdot \overrightarrow{UU_1}$. The vector $\overrightarrow{UU_1}$ does not depend on point A so the conclusion of the theorem is proved.

A simple analytical expression of transformation $T = S_2 \circ S_1$ is obtained taking d_1 as axis Oz (so d_1 has equations $x = y = 0$) and d_2 included into the plane xOz (d_2 of equations $x - a = 0 = y$). Then, for A (x, y, z), point T (A) has the co-ordinates

$(x + 2a, y, z)$.

In a more general meaning, translation T of vector $a \cdot \vec{i} + b \cdot \vec{j} + c \cdot \vec{k}$ transforms $M(x, y, z)$ into $M'(x', y', z')$, where $x' = x + a, y' = y + b, z' = z + c$.

§6. Rotations

To define the title notion that of *oriented angle* is necessary. For the reader who appreciates that he correctly intuits this notion, the reading of the following intended lines may be postponed.

Let π be an Euclidean fixed plane and O a point in π. We consider the set L of the rays of origin O in π. We remind that a (non-oriented) angle of vertex O is a subset $h, k = \{h, k\}$ of L made up of two elements. Let U_0 be the set of these angles. We also remind that we introduced in the preceding chapter the measure in degrees of the angles admitting a restriction $m : U_0 \to [0, 180]$.

We call *oriented angle of vertex* O an element $u = (u_1, u_2)$ of the set $U_0 = L \times L$. The binary relation $\approx$ of congruence of the oriented angles is introduced on the set U_0 by the condition $(u_1, u_2) \approx (v_1, v_2)$ if there exists a symmetry S related to a straight line (which passes through O and is in π) so that $S(u_1) = v_2$ and $S(u_1) = v_1$. It is easy to prove:

Theorem 6.1. *For any* u_1, u_2, v_1 *in* L *there exists a unique element* v_2 *in* L *so that* $(u_1, u_2) \approx (v_1, v_2)$.

Let us suppose as being fixed an element $c = (a, b)$ in U_0 which satisfies the condition that a, b is a proper angle. Let H be the semi-plane delimited by the support straight line of a containing $b - \{0\}$. For any oriented angle $u = (u_1, u_2)$ we shall determine, according to Theorem 6.1, a unique ray u^* so that $u \approx (a, u^*)$.

We shall now define a function $\mu : U_0 \to (-\pi, +\pi)$ which will constitute "*the measure in radians of the oriented angles*", specifying the value $\hat{u} = \mu(u_1, u_2)$ in all the possible eventualities.

- if $u^* = a$, then $\hat{u} = 0$;

- if u^* is the ray opposed to a, then $\hat{u} = \pi$;

- if u^* is in H, then $\hat{u} = \dfrac{\pi}{180} m(a, \overset{\frown}{u^*})$;

- if u^* is not included into $a \bigcup H$, then $\hat{u} = -\dfrac{\pi}{180} m(a, \overset{\frown}{u^*})$.

Now it is easy to remark

$$(u_1, u_2) \approx (v_1, v_2) \Leftrightarrow \mu(u_1, u_2) = \mu(v_1, v_2)$$

which ensures us that $\approx$ is an equivalence on U_0.

If C is a point in π, $C \neq 0$, the translation of vector $\overrightarrow{OC}$ gives us a bijection of U_0 in the set U_c of oriented angles having their vertex in C. One accepts that this bijection preserves the measure of oriented angles. The choice of the above oriented angle c is achieved by the well known *convention* of the "trigonometric sense"; within this convention an intuitive concept of "anti-clockwise rotation" which does not affect the rigorous definition of the concept of rotation is used.

We appreciate as being particularly important the operation of *addition of oriented angles* which will benefit from the notation $\hat{+}$; by definition, $(u_1, u_2) \hat{+} (v_1, v_2) = (u_1, w)$, where the ray w is uniquely determined by the condition $(v_1, v_2) \approx (u_2, w)$, having a geometrical significance easy to intuit.

An algebraic operation $\hat{+}: (-\pi, \pi]^2 \to (-\pi, \pi]$ is defined by

$$a \hat{+} b = a + b + 2k\pi \wedge k \in \mathbf{Z}.$$

It is easy to deduce the following *property of additivity* of μ:

$$\forall u, v \in U_0, \; \mu(u \hat{+} v) = \mu(u) \hat{+} \mu(v).$$

Let a point O and a measure $\hat{u}$ of oriented angle be in the plane π; one calls *rotation of centre O and oriented angle $\hat{u}$* a transformation R of π in itself (sometimes noted with $R_{0, \hat{u}}$ to precise the defining elements) which associates to the arbitrary point M be point $M' = R(M)$ characterized by the conditions: 1. $OM' \equiv OM$ and 2. $\mu(\overset{\frown}{MOM'}) = \hat{u}$ (one must notice that, for $M = 0$, it results $M' = 0$ and condition 2. is not necessary anylonger, but it has no significance, either). In the case when $\hat{u} = 0$, rotation R coincides in the identical transformation; for $\hat{u} = \pi$, R is the symmetry related to the point O. If $\hat{u} \neq 0$ and $\hat{u} \neq \pi$ we say that $R_{O,\hat{u}}$ is a *proper rotation*.

Theorem 6.2. *Any rotation of the plane π is an isometry;* $(R_{O,\hat{u}})^{-1} = R_{O,-\hat{u}};$ $R^{-1} = R$ *if and only if R is an identical transformation or symmetry related to a point;* $R_{O,\hat{u}} \circ R_{O,\hat{v}} = R_{O,\hat{u}+\hat{v}}.$

To as expeditively as possible prove this theorem, we shall consider a polar frame of axis Ox. Point $M'(\rho, \theta \hat{+} \hat{u})$ through $R_{O,\hat{u}}$ corresponds to the point $M(\rho, \theta)$. The establishing of the truthfulness of the statements in the enunciation within this polar frame is immediate (Fig. 24).

In the Cartesian frame (Ox, Oy) associated to the polar frame under consideration, the co-ordinates (x, y) of M are given by $x = \rho \cdot \cos\theta$, $y = \rho \cdot \sin\theta$. For the co-ordinates (x', y') of M' we immediately obtain:

$$x' = x \cdot \cos\hat{u} - y \cdot \sin\hat{u}, \quad y' = x \cdot \sin\hat{u} + y \cdot \cos\hat{u}.$$

Theorem 6.3. *A proper rotation $R = R_{O,\hat{u}}$ admits a single fixed point (O), does not admit fixed straight lines and makes invariant the circle (C, r) if and only if $C = 0$.* (We precise that for an arbitrary straight line d, the image straight line $R(d) = d'$ forms with d angles among which at least one is equal to $|\hat{u}|$.)

Theorem 6.4. *The composition of two symmetries S_1, S_2 of the plane π related to axes d_1, d_2 secant in O is a rotation of centre O; any rotation R can be written as a product of two symmetries of the plane.*

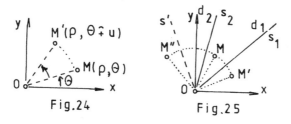

Fig.24   Fig.25

Proof. Let Ox be an arbitrary ray taken as polar axis. Let a be the angle of the straight lines d_1, d_2 and let us consider that the axis Ox forms with the straight lines d_1, d_2 the oriented angles α and $a \hat{+} \alpha$, respectively. Let us consider the point M of polar co-ordinates ρ and θ. It immediately follows that $M' = S_1(M)$ has polar co-ordinates ρ and $\theta' = 2\alpha - \theta$; point $M'' = S_2(M')$ will have polar co-ordinates ρ and $\theta'' = 2(a \hat{+} \alpha) - \theta' = 2a \hat{+} \theta$. We conclude that the transformation $S_2 \circ S_1$ is a rotation of centre O and angle $2a$. The usefulness of considering angles $a . \theta$, α in the proof as being oriented is obvious; we considered $\hat{-}$ as operation opposed to addition $\hat{+}$ and multiplication as "repeated addition".

Given the rotation R of centre O we shall consider an arbitrary ray s_1 of origin O; let $R(s_1) = s'$ and s_2 be the ray bisecting $\sphericalangle(s_1, s')$. Let d_1, d_2 be the support straight lines of s_1, s_2; by noting the symmetries of axes d_1, d_2 with S_1, S_2 it is easy to find out that $R = S_2 \circ S_1$ holds.

Theorem 6.5. *If T is a translation of non-null vector and R and R' are proper rotations of the plane π, then $T \circ R$ and $R \circ T$ are distinct rotations and $R \circ R'$ is a rotation or a translation.*

Proof. The idea of making explicit the transformation $T \circ R$ is that of determining symmetries S_1, S_2, S_3 related to the axes d_1, d_2, d_3 so that $R = S_2 \circ S_1$, $T = S_3 \circ S_2$ and therefore $T \circ R = S_3 \circ S_1$. Let be $R = R_{O,a}$ and $T(O) = O'$. The case $O' = O$ is excluded. Let s_2 be a ray of origin O perpendicular on $s = R^{-1}(s_2)$. We consider (Fig. 26) the bisector s_1 of the angle $(s, \overset{\wedge}{s_2})$. We shall note with d_1, d_2 the support straight lines of s_1, s_2 and with d_3 the midperpendicular of the segment OO'. The conditions enunciated when we exposed the idea in the preamble of the proof are satisfied. Therefore, $T \circ R$ is the rotation of centre $U \in d_3 \cap d_1$ and oriented angle $\hat{a}$ (the same as that of the rotation R). The situation $d_1 \parallel d_3$ is equivalent with $d_1 = d_2$ and requires that R be the identical transformation.

In order to make $R \circ T$ explicit, we shall notice that $(R \circ T)^{-1} = T^{-1} \circ R^{-1}$ and we have thus returned to the initial problem. We shall find out that $R \circ T$ is a rotation $R_{V,a}$, where $V = T^{-1}(U)$, point U being the one described above. We find out that:
$$R \circ T = T \circ R \Leftrightarrow U = V \Leftrightarrow T = I.$$

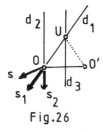

Fig.26

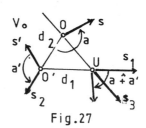

Fig.27

In order to make $R \circ R'$ explicit let us suppose $R = R_{O,a}$ and $R' = R_{O',a'}$. If $O = O'$, then $R \circ R' = R_{O,a \hat{+} a'}$, according to Theorem 6.2. Let us suppose $O \neq O'$ and let s_2 be a ray of origin O' situated on $d_2 = (OO')$. We determine s' so that $R'(s') = s_2$ and let d_1 be the support straight line of the bisector of the angle formed by s_2 and s'. We also consider the support straight line d_3 of the bisector of the angle formed by $[O; O')$ and $s = R([O; O'))$. We also consider the symmetries S_i related to the straight lines d_i (for $i = 1, 2, 3$). It results $R' = S_2 \circ S_1$, $R = S_3 \circ S_2$ and so $R \circ R' = S_3 \circ S_1$. $R \circ R'$ is a translation if and only if $d_3 \parallel d_1$, that is $\hat{a} \hat{+} \hat{a}' = 0$; in this case, the translation applies O in $O'' = R(O)$. If $\hat{a} \hat{+} \hat{a}' \neq 0$, let be $U \in d_1 \cap d_3$; $R \circ R'$ is a rotation of centre U and angle $a \hat{+} a'$.

Remark. In figure 27 there also appears the centre V of the rotation $R' \circ R$, symmetric with U related to (OO'); the planned conditions make impossible the equality $R \circ R' = R' \circ R$.

One calls *rotation related to an axis d* an isometry R of E_3 which has the straight line d or all the points of the space as set of fixed points.

Theorem 6.6. *The restriction of a rotation R related to an axis d to a plane π perpendicular on d is a rotation of the plane π of angle a independent of the choice of π. Each rotation of a plane extends to a rotation related to an axis.*

Proof. Let O be the common point of d and π. For M arbitrary in π, $OM \perp d$ holds; if $M' = R(M)$, it follows $OM' \perp d$ and so $M' \in \pi$. Therefore, $R(\pi) \subset \pi$. If N is arbitrary in π, R being surjective, a point N_1 will exist so that $R(N_1) = N$ and N_1 will obviously be in π. We have thus proved that $R(\pi) = \pi$. Let R_π be the restriction of R to π. Obviously, R_π is an isometry of π preserving the point O, so it is a rotation $R_{O,a}$ of the plane π.

Let α be a semi-plane delimited by the straight line d. Its image through the isometry R is a semi-plane $\beta = R(\alpha)$ delimited by the same straight line d. The dihedral angle $\widehat{\alpha, \beta}$ is sectioned by a plane π perpendicular on d according to a plane angle whose size and orientation do not depend on the choice of π.

For a rotation $R_{O,a}$ of a plane π we shall define a transformation $T: E_3 \to E_3$ specifying the image $M' = T(M)$ of a generic point $M \in E_3$. We consider the orthogonal projection N of M in the plane π and the point $N' = R_{O,a}(N)$; M' is defined by the condition that $\overrightarrow{NM} = \overrightarrow{N'M'}$. Obviously, for $M \in \pi$ we take $T(M) = R_{O,a}(M)$. It is easy to find out that T is an isometry preserving all the points of the straight line d perpendicular in O on the plane π. If T admits one more fixed point $A \notin d$, then $R_{O,a}$ also admits the orthogonal projection B of A is π as a fixed point, so $\hat{a} = O$ etc.

Theorem 6.7. *Any rotation R related to a straight line d can be presented in an infinity of ways as composition of two symmetries related to planes by the straight line d.*

Indeed, the restriction of R to a plane $\pi \perp d$ is a plane rotation $R_{O,a}$ which, according to Theorem 6.3., is expressed as composition of symmetries related to the straight lines b, c secant in O: $R_{O,a} = S_c \circ S_b$. Let β, γ be the planes through d, containing b and c, respectively; it is easy to find out that $R = S_\gamma \circ S_\beta$, where S_β, S_γ are symmetries related to the "index" planes.

§7. Homotheties

Let be a point O of a plane π and a non-null real number k; we call *homothety of centre O and ratio k* the transformation H of the plane π in itself (or of the space E_3 in itself) which associates the point $M' = H(M)$ defined by $\overrightarrow{OM'} = k \cdot \overrightarrow{OM}$ to each point M.

For $k > 0$, we say that H is a *direct* homothety; for $k < 0$, *indirect* homothety. (The name is of Greek origin: homo = the same, thesis = position.) We shall also use the notation $H(0, k)$ to precise the centre O and the ratio k. For $k = -1$, the homothety is obviously an identical transformation, and for   we obtain the symmetry related to a point.

Theorem 7.1. *For any homothety H of ratio k, from $H(M) = M'$ and $H(N) = N'$, $(MN) \parallel (M'N')$ and $M'N' = k \cdot MN$ result.*

Theorem 7.2. *Any homothety $H(0, k)$ is bijective; its inverse has the same centre and ratio equal to k^{-1}.*

Theorem 7.3. *If we note with a stress the images through homothety $H = H(0, k)$ of the points, the following properties for H hold :*

1. *it transforms a straight line (AB) into a parallel straight line $(A'B')$;*
2. *transforms the ray $(A ; B)$ into the ray $(A' ; B')$;*
3. *transforms the "segment" $|AB|$ into the "segment" $|A'B'|$;*
4. *transforms the semi-plane $(d ; A)$ into the semi-plane $(d' ; A')$, where $d' = H(d)$;*
5. *transforms the circle $C(C, r)$ into the circle $C'(C', |k| \cdot r)$;*
6. *transforms an arc of circle ABC into the arc of circle A'B'C '.*

Theorem 7.4. *If C' is the image of a circle C in a homothety H of centre O, then any straight line d passing through O simultaneously cuts C and C': the angles under these circles are cut are congruent.*

Proof. If there is M common to C and d, then $M' = H(M)$ will obviously be common to C' and d (Fig. 28). In this hypothesis, triangles OMC and $OM'C'$ are similar, so $CM \parallel C'M'$. It immediately results that tangents t, t' traced into M, M' to the circles C and C', respectively, are parallel. The angles formed by d with C, C' are by definition equal to the angles formed by d with t, t' and these are congruent as correspondent angles.

Theorem 7.5. *In a Cartesian frame within which the centre C of the homothety $H(C, k)$ admits co-ordinates (a, b), point $M(x, y)$ is transformed into $M'(x', y')$, where $x' = kx + (1 - k)a$, $y' = ky + (1 - k)b$.*

Theorem 7.6. *Let H be the homothety of centre O and ratio $k \neq 1$, and T the translation of vector $\vec{u}$. The equalities $T \circ H = H(O', k)$ and $H \circ T = H(O'', k)$, where $(k-1) \cdot \overrightarrow{O''O} = k \cdot u$, hold.*

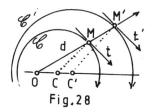

Fig.28

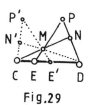

Fig.29

Theorem 7.7. *If $h \cdot k \neq 1$, $H(D, h) \circ H(C, k) = H(E, kh)$, where $h(k-1)\overrightarrow{EC} = (1-h)\overrightarrow{ED}$, holds. If $hk = 1$, $H(D, h) \circ H(C, k)$, is a translation of vector $(1-h)\overrightarrow{CD}$.*

The proof will for the moment analyze the case $C = D$. Let M be arbitrary and $N = H(C, k)(M)$, $P = H(D, h)(N)$. It follows $\overrightarrow{CM'} = k \cdot \overrightarrow{CM}$, $\overrightarrow{CP} = h \cdot \overrightarrow{CM}$, so $\overrightarrow{CP} = (k \cdot h) \cdot \overrightarrow{CM}$, that is $P = [H(D, h) \circ H(C, k)](M) = H(C, kh)(M)$, and the equality of transformation in the enunciation is checked (for $hk \neq 1$) when $E = C$. If $hk = 1$, $H(C, kh)$ is an identical transformation and thus also coincides with the translation of null vector. (This particular case, $C = D$, will be reformulated in the corollary 1 of this theorem.)

For $C \neq D$ (Fig. 29), we take M which is not situated on (CD) and points N, P defined as above. With Thales' theorem we easily find out $MP \parallel CD \leftrightarrow hk = 1$. In the hypothesis $hk \neq 1$ a point E common to the straight lines CD and PM will thus exist. With Menelaos' relation for the triangle CND and the transversal (PM) we find the characterization in the enunciation of point E which proves also its independence of the chosen point M. The same relation of Menelaos applied to the triangle PMN and to the transversal (CD) ensures $\overrightarrow{EP} = hk \cdot \overrightarrow{EM}$, that is $P = H(E, hk)(M)$ etc.

Finally, if $C \neq D$, but $hk = 1$, by the fundamental theorem of similitude, we find $\overrightarrow{MP} = (1 - \frac{1}{k}) \cdot \overrightarrow{CD} = (1 - h) \cdot \overrightarrow{CD}$, and the second part of the conclusion of the theorem follows.

Corollary 1. *For any point* C, *the homotheties of centre* C *form (in relation with composition) a group isomorphic with the multiplicative group of non-null real numbers.*

Corollary 2. *If* H, K *are different homotheties from the identical transformation,* $H \circ K = K \circ H$ *holds if and only if the centres of the homothetiaes* H *and* K *coincide.*

Proof. In Figure 29 there also appears the centre E' of homothety $H(C, k) \circ H(D, h)$ in the hypothesis that $C \neq D$ and $hk \neq 1$. It is easy to find out that $E \neq E'$ on the basis of the characterization in the enunciation of the theorem of point E and on that of the analogous characterization for E': $k(h-1)\overrightarrow{E'D} = (1-k)\overrightarrow{E'C}$. Indeed,

$E = E'$ would return to $\frac{k(h-1)}{1-k} = \frac{1-h}{h(k-1)}$, that is $h \cdot k = 1$, which is contrary to the hypothesis.

For $C \neq D$ and $h \cdot k = 1$, the two compositions of homotheties are translation of vectors $(1-h)\overrightarrow{CD}$ and $(1-k)\overrightarrow{DC}$, respectively, which would coincide if and only if $h + k = 2$, which would imply $h = k = 1$, so H and K would be identical transformations.

Corollary 3. *For any homothety* H, K *transformation* $H^{-1} \circ K^{-1} \circ H \circ K$ *is a translation.*

Proof. The transformations $T_1 = H \circ K$ and $T_2 = H^{-1} \circ K^{-1} = (K \circ H)^{-1}$ are made explicit as above, and compositions $T_2 \circ T_1$ proves in the end to be a translation.

Theorem 7.8. *If* S *is a symmetry (related to a straight line* d, *or to a plane* α, *or to a point* U) *and* H *is the homothety of centre* O *and ratio* k, *having the same domain and co-domain as* S, *then* $S \circ H \circ S$ *is a homothety of centre* $O' = S(O)$ *and ratio* k.

The proof is immediate, For any arbitrary point A, let be $B = S(A)$,

$C = H(B)$, $D = S(C)$. From $C = H(B)$ it follows $\overrightarrow{OC} = k \cdot \overrightarrow{OB}$. Through the

symmetry S, this relation becomes $\overrightarrow{O'D} = k \cdot \overrightarrow{O'A}$, that is $D = H'(A)$, where

$H' = H(O; k)$. Therefore, $(S \circ H \circ S)(A) = H'(A)$ etc.

Corollary. *If T is an isometry and H a homothety of the Euclidean three-dimensional plane or space, there exists a homothety H' (having the same ratio as H) so that $T \circ H = H' \circ T$.*

Proof. If T and H are transformations of the Euclidean plane, we can express, according to Theorem 3.6, the isometry T as composition of symmetries related to straight lines, $T = S_3 \circ S_2 \circ S_1$. According to the theorem, we successively find homotheties H_i so that

$$T \circ H = S_3 \circ S_2 \circ S_1 \circ H = S_3 \circ S_2 \circ H_1 \circ S_1 =$$
$$= S_3 \circ H_2 \circ S_2 \circ S_1 = H_3 \circ S_3 \circ S_2 \circ S_1 = H_3 \circ T.$$

If T and H are transformations of E_3, T is presented under the form of a composition of at most four symmetries related to planes, and the proof is analogous.

For $n = 2$ or $n = 3$ we shall use the notation G_n for the set of transformations $U = T \circ H$, where T is an isometry of the Euclidean space E_n and H a homothety of the same space.

Theorem 7.9. *For $n = 2$ and $n = 3$, $(G_n, 0)$ is a group.*

Proof. Let $U_1 = T_1 \circ H_1$ and $U_2 = T_2 \circ H_2$ be elements in G_n. According to the corollary of Theorem 7.8, we find out:

$U_1 \circ U_2 = T_1 \circ (H_1 \circ T_2) \circ H_2 = T_1 \circ (T_2 \circ H') \circ H_2 = (T_1 \circ T_2) \circ (H' \circ H_2)$.
According to Theorem 7.7, there is the eventuality that $H' \circ H_2$ be a homothety K or a translation T_3. In both cases we shall deduce $U_1 \circ U_2 = T \circ H$: in the first case $T = T_1 \circ T_2$ and $H = K$, and in the second one $T = T_1 \circ T_2 \circ T_3$ and K is the identical transformation. The identical transformation I is expressed by $I = I_1 \circ I_2$, where I_1 is the identical transformation regarded as isometry and I_2 is the identical transformation regarded as homothety. We also find out $(T \circ H)^{-1} = (H' \circ T)^{-1} = T^{-1} \circ (H')^{-1}$ (the

notations being the ones in the above corollary) and so the elements in G_n have inverses in G_n, q.e.d.

Remarks. Group G_n is called *the group of similitudes*. All the primary notions in Hilbert's axiomatic system (point, straight line, plane, incidence, "to be between", congruence) are preserved by the group G_n. For Birkhoff's axiomatic system, the above sentence is no more true, distance being preserved only by the sub-group of the isometries strictly contained into G_n.

§8. Inversions

Let $C(P, r)$ be a circle situated in the plane π; one calls *inversion related to* C the transformation $T : \pi - \{P\} \to \pi - \{P\}$ which applies M in $M' = T(M)$ characterized by $\overline{PM} \cdot \overline{PM'} = r^2$ (by also implying the co-linearity of points P, M, M'). It is also said that T is *the pole* P *and power* r^2 inversion, phrase summarized by the notation $T = T(P, r^2)$.

Theorem 8.1. *An inversion related to the circle* C *is involutive; its fixed points make up the circle* C.

Theorem 8.2. *If* M', N' *are the images of points* M, N *in the inversion* $T = T(P, r^2)$ *and* $P \not\subset (MN)$, *then* $\triangle PMN \sim \triangle PN'M'$ *and*

$$M'N' = r^2 \cdot \frac{MN}{PM \cdot PN}.$$

Points M, N, M', N' *are situated on a circle* O *orthogonal to the circle* $C(P, r)$.

Proof. From $PM \cdot PM' = PN \cdot PN'$ it results $\dfrac{PM}{PN'} = \dfrac{PN}{PM'}$ and then $\triangle PMN \sim \triangle PN'M'$. From the proportionality of the sides it results: $M'N' = \dfrac{MN \cdot PM'}{PN} =$

$= r^2 \cdot \dfrac{MN}{PM \cdot PN}$. Let $O(O, R)$ be the circle through M, N and N' we deduce that $M' \in O$ from the power of point P or by summing the opposed angles. Let (PS) be a tangent to O and $S \in O$; $PS^2 = \overrightarrow{PM} \cdot \overrightarrow{PN} = r^2$, that is $S \in O$ and $PS \perp SO$, q.e.d.

Theorem 8.3. *Let be the inversion* $T = T(P, r^2)$ *and a straight line* d. *If*

$P \in d$, the set $d - \{P\}$ is made invariant by T. If $P \notin d$, the circle $D(D, DP)$ exists so that $T(d) = D - \{P\}$; any ray $[P; X)$ forms equal angles with d and D.

Proof. In the case $P \in d$, from $M \in d - \{P\}$ it easily results that $M' = T(M)$ is on d; we also use the obvious equality $T^{-1} = T$. For the case $P \notin d$, let A be the projection of P on d and $A' = T(A)$. Let $M \neq A$ be situated on d and $M' = T(M)$.

According to the preceding theorem, $\triangle PAM \sim \triangle PM'A$, so $\widehat{PM'A'} \equiv \widehat{PAM} = 90°$, so M' is situated on the circle D of $|PA'|$ diameter. For $X \in D - \{P\}$, we consider the point N common to the straight lines d and PX (obviously non-parallel). It is easy to prove that point X coincides with the image N' of N through T. Therefore, $T(d) = D - \{P\}$.

Considering the tangent $M'Y$ to D, we shall easily find out:

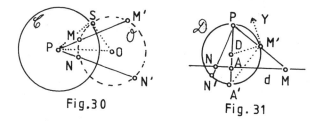

Fig.30   Fig. 31

$\widehat{YM'P} = 90° - \widehat{PM'D} = 90° - \widehat{M'PA} = \widehat{PMA}$ and the last statement in the enunciation is proved.

Remark. The centre D of the circle D has appeared as the midpoint of the segment PA' (with $A' = T(A)$), where $A \in d$, $PA \perp d$. Considering also P_1 symmetric to P related to d we can easily prove that $D = T(P_1)$.

Theorem 8.4. *Let T be the inversion related to the circle $C(P, r)$ and a circle $O(O, R)$. If $P \in O$, then $T(O - \{P\})$ is a straight line. If $P \notin O$, then $T(O)$ is a circle $O'(O_1, R')$, where, denoting by $O(P)$ the power of point P related to the circle O:*

$$\overline{PO_1} = \frac{r^2}{O(P)} \cdot \overline{PO} \text{ and } R' = \frac{r^2 R}{|O(P)|}$$

hold; a ray $(P; X)$ *cuts the circles* O *and* O' *under congruent angles.*

Proof. The case $P \in O$ constitutes a reformulation of the conclusion in the preceding theorem. We shall analyze the case $P \notin O$. Let A, B be the points of O situated on the straight line PO. We consider the images A', B' of A, B through T. Let be M which is not situated on (AB) and $M' = T(M)$. Using Theorem 8.2, we shall deduce: $\triangle PMA \sim \triangle PA'M'$, $\triangle PMB \sim \triangle PB'M'$. In the case sketched in Figure 32, the rays $(P; A)$ and $(P; B)$ coincide, so $(P; A')$ will also coincide with $(P; B')$. It follows this way: $\widehat{B'M'A'} = \widehat{PM'A'} - \widehat{PM'B'} = \widehat{PAM} - \widehat{PBM} = \widehat{AMB}$. We easily find out

$M \in O \leftrightarrow \widehat{AMB} = 90° \leftrightarrow \widehat{B'M'A'} = 90° \leftrightarrow M' \in O'$, where, obviously, O' is the circle of $A'B'$ diameter. The evidenced equivalence is also valid when $M \in (AB)$. We have thus proved that $T(O) = O'$. The characterizations in the enunciation for O' are obtained as follows:

$$\overline{PO_1} = \frac{1}{2}(\overline{PA'} + \overline{PB'}) = \frac{1}{2}r^2(\overline{PA}^{-1} + \overline{PB}^{-1}) = r^2 \cdot \frac{\overline{PA} + \overline{PB}}{2} \cdot (\overline{PA} \cdot \overline{PB})^{-1} = \frac{r^2 \cdot \overline{PO}}{O(P)};$$

$$R' = \frac{1}{2}|\overline{PA'} - \overline{PB'}| = \frac{1}{2}r^2|\overline{PA}^{-1} - \overline{PB}^{-1}| = r^2|\frac{\overline{PB} - \overline{PA}}{2}| \cdot |\overline{PA} \cdot \overline{PB}|^{-1} = \frac{r^2 R}{|O(P)|}.$$

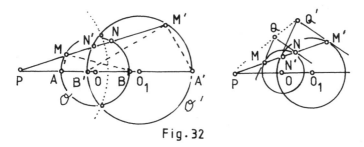

Fig. 32

Let us also consider now the homothety H of centre P and ratio equal with $\frac{r^2}{O(P)}$. By this homothety, the circle O is taken to the circle O'. The enunciated conclusion is an immediate consequence of Theorem 4, §6. However, we draw attention that by homothety H point M is not taken to $M' = T(M)$, but to the other point N' in which PM cuts O'; but this change of point (M' instead of N') does not influence the value of the angle under which the straight line PM cuts the circle O'.

Remark. The inversion T related to the circle C preserves a circle O when and only when: $O = C$ (all the points of O being fixed) or when O is orthogonal to C.

Theorem 8.5. *An inversion T preserves the angles under which straight lines or circles are cut.*

The proof is based on the last statements in Theorem 8.3 and 8.4. Let M be common to the lines d and e (d being, in Figure 33, a straight line and e a circle) and lines d', e', images through T of d, e; then the straight line PM forms with d', e' congruent angles, but "reversely" situated related to those formed by (PM) with d, e. Theorem 8.5 admits a more general formulation we are omitting here.

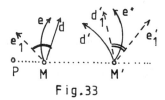

Fig.33

In the Euclidean space E_3, to a sphere $S(P, r)$ one associates the inversion T, the bijective transformation of the set $E_3 - \{P\}$ defined by:

$$T(M) = M' \Leftrightarrow \overline{PM} \cdot \overline{PM'} = r^2$$

(the last equality also implying the co-linearity of points P, M, M'). By analogy with Theorem 8.3-8.5 one can prove:

Theorem 8.6. *The inversion T related to the sphere $S(P, r)$:*

1. *preserves the plane containing the pole P;*

2. *transform the non-incident with P planes into sphere through P from which point P is excluded, and conversely;*

3. *transforms the sphere $\Omega(O, R)$ that does not contain P into a sphere $\Omega'(O_1, R')$, where*

$$\overline{PO_1} = \frac{r^2}{\Omega(P)} \cdot \overline{PO} \text{ and } R' = \frac{r^2 \cdot R}{|\Omega(P)|};$$

the spheres Ω *and* Ω' *coincide if and only if* Ω *is orthogonal with S or coincides with S;*
4. *the inversion preserves the angles under which planes and spheres are cut.*

Theorem 8.7. *Let be circles (spheres)* $C(C, r)$, $O(O, R)$ *and* T, J *the inversions related to these circles (spheres). Let* T' *be the transformation defined this way:*

- *if* $O \in C'$, T' *is the symmetry related to the straight line (plane)* $J(C - \{O\})$;

- *if* $O \notin C'$, T' *is the inversion related to the circle (sphere)* $J(C)$. *In these conditions, the transformation* $J \circ T \circ J$ *is a restriction of the transformation* T'.

Proof. We shall consider the set of points $D = J(C - \{O\})$ which is obviously made invariant by transformation $J \circ T \circ J$.

If $O \in C$, D is a perpendicular plane on the straight line OC. A straight line d perpendicular on D (which does not contain O) becomes through J in an orthogonal circle on D which is made invariant by T and restored by J into d. So $J \circ T \circ J$ preserves the perpendicular straight lines on D (OC included). A circle (sphere) a orthogonal circle (sphere) to D, with is preserved by T and returned by J into the same circle (sphere) a. Therefore, $J \circ T \circ J$ preserves the orthogonal circles (spheres) to plane D. We deduce that, for points M on which transformation $J \circ T \circ J$ is defined, $(J \circ T \circ J)(M)$ is symmetrical to M related to D etc.

The case $O \notin C$ is analogously analyzed by specifying that D is a circle (sphere) and that an inversion overtakes the role of the symmetry related to D. The theorem concludes only $J \circ T \circ J \subset T'$ because the composition $J \circ T \circ J$ is not defined in O and in $J(C)$ [in $J \circ T(O)$, T' is not defined, either].

Problems

To point out the importance of the transformation in the "Elementary Geometry" we appreciate that it is opportune to give some problems with usual enunciations (without transformations) but whose solutions with transformations are very conclusive.

1. Let $C(O, R)$ be a circle and A, B two fixed points on C so that $O \notin AB$. Find the locus of the orthocentre H of triangles ABM when M runs over C.

2. Let a, b, c be parallel straight lines, b between a and c, at the distances m, n from a and c. Find the length ℓ of the side of the equilateral triangle ABC, where $A \in a$, $B \in b$, $C \in c$.

3. Let ABC be a triangle, $C(O, R)$ its circumcircle, δ a given direction. Let A', B', C' be on C so that AA', BB', CC' have the direction δ. Let H, H_a, H_b, H_c, H_a', H_b', H_c' be the orthocentres of the triangles ABC, $A'BC$, $AB'C$, ABC', $AB'C'$, $A'BC'$, $A'B'C$. Prove that the points H, H_a, H_b, H_c, H_a', H_b', H_c' are co-linear.

4. Let $A(A, a)$, $B(B, b)$, $C(C, c)$ be circles, any two of them externally tangent. We consider circles $X(X, x)$, $Y(Y, y)$ both tangent to A, B and C. (One of these circles may degenerate into a straight line and its radius becomes infinite.) Prove

$$\frac{1}{x} + \frac{1}{y} = 4 \cdot \sqrt{\frac{a+b+c}{abc}}.$$

5. Let $ABCD$ be an inscribed quadrilateral. Deduce the following two Ptolomeus' equalities

$$AC \cdot BD = AB \cdot CD + AD \cdot BC,$$

$$AC \cdot (AD \cdot CD + AB \cdot CB) = BD \cdot (BA \cdot DA + BC \cdot DC).$$

CHAPTER VIII

THE ERLANGEAN PROGRAM

Towards the middle of the last century, the accumulations of geometric results have permitted delimitations of separate disciplines with specific problems and methods: the projective Geometry, non-Euclidean Geometries, affine Geometry, differential Geometry. A unitary interpretation of Geometry was drawn up by Felix Klein at the University in Erlangen in 1882. This unification enabled the quantitative and qualitative development of Geometry and is quoted under the name of *"Erlangen Program"*.

§1. Klein Spaces

Definition 1.1. *We call concrete Klein space a couple* $(\mathcal{M}, G)$, *where* $\mathcal{M}$ *is a set (whose elements are called points), and G is a group transformations of the set* $\mathcal{M}$ *(an element T in G is a bijective map T:* $\mathcal{M} \to \mathcal{M}$ *called motion).*

An equivalence ε defined by:

$$F \varepsilon F' \Leftrightarrow \exists T \in G, \ \check{T}(F) = F'$$

is introduced on the set $\mathcal{P}(\mathcal{M})$ of the subset of $\mathcal{M}$. The elements $F \subset \mathcal{M}$ are called *concrete figures* of the Klein space. The equivalence class $[F]$ of such a figure F constitutes an *abstract figure* or a *configuration. The study of the properties of the configurations in* $\mathcal{P}(\mathcal{M}) / \varepsilon$ constitutes the Geometry of a concrete space $(\mathcal{M}, G)$. In other words, the Geometry of $(\mathcal{M}, G)$ is *the study of the properties of figures* $[F]$ *invariants of the transformations of group G.*

Remark. This way of defining Geometries is one of the best ones, but it also has serious disadvantages. Certain domains of Geometry, such as algebraic Geometry, do not get into this scheme. On the other hand, non-geometric theories can easily get into the above definition: if $\mathcal{M}$ is a group (H, e) and the transformations in G are the interior self-morphisms $T_a \colon H \to H$, where $T_a(h) = a \circ h \circ a^{-1}$, the study of Klein Space (H, G) is the *algebraic* study of the group H. But the importance of Erlangen program consists first of all in that it enables a "hierarchization" of Geometries. We shall presents this "hierarchization" in the modern language of the theory of categories.

Let us consider a category K whose objects are concrete Klein spaces, (m, G). A morphism from (m, G) to (m', G') is a couple (m, γ), where: m is a map of m in m', and γ is a homomorphism of group G in G', so that the enclosed diagram may be commutative. (The vertical arrows express the actions of the groups as transformations $p: (X, T) \to T(X)$ etc.) The composition of morphisms is achieved on components:

$$(m', \gamma) \circ (m, \gamma) = (m' \circ m, \gamma' \circ \gamma).$$

$$
\begin{array}{ccc}
m \times G & \xrightarrow{\ m \times \gamma\ } & m' \times G' \\
{\scriptstyle p}\downarrow & & \downarrow{\scriptstyle p'} \\
m & \xrightarrow{\quad m \quad} & m'
\end{array}
$$

An isomorphism f of the category K is a couple $f = (b, i)$, where b is a bijection and i an isomorphism of groups (which ensures the commutativity of a diagram).

Thus, an equivalence $\sim$ is defined on the class of the objets of K, where $(m, G) \sim (m', G')$ if a K-isomorphism f exists between these objects. The class of element (m, G) related to the equivalence $\sim$ is denoted by $[m, G]$ and called *abstract Klein space*. The Geometry of two concrete Klein spaces (m, G) and (m', G') including themselves into the same abstract Klein space is identical since we can "translate" by the bijection b any property of the former into a property of the latter one, and these properties are simultaneously made invariant by groups G and G', respectively.

In particular, we can consider a group G' "associated to G", where $G' = \{b \circ T \circ b^{-1} : T \in G\}$ for any concrete Klein space (m, G) and for any bijection b of m, the composition in G' being the composition of maps. Of course, there also exists a K-isomorphism $f = (b, \hat{b})$, where $\hat{b}: G \to G'$ and $\hat{b}(T) = b \circ T \circ b^{-1}$. According to this example, Kleinean Geometries on the same set m are not modified *if we replace the group G by an associated group*.

Let $[K]$ be the "skeleton" of category K; the objects of $[K]$ are abstract Klein spaces, the morphism of $[K]$ identify themselves as K-morphisms between representatives. We obtain a subcategory G of $[K]$ keeping as morphisms only the $[K]$-monomorphisms $f = (b, i)$, where b is a bijection. The existence of a G-morphism

$f = (b, i)$ from (m, G) to (m', G') enables us to identify m with m'. It also results that i is an injective map and we do not restrain the generality by accepting that i is an inclusion. It is said under these conditions that the Geometry $[m, G]$ is *subordinated* to the Geometry $[m', G']$. All true propositions in $[m', G']$ are also true in $[m, G]$, but in general, not conversely; a part of the notions and relations in $[m, G]$ does not constitute notions or relations in $[m', G']$ because they are not made invariant by the larger group G'. Under these conditions, the category G constitutes a unitary image Geometries, its morphisms marking subordination relations between (Kleinean) diverse Geometries.

An efficient procedure of evidencing certain Geometries $[m, H]$ subordinated to a Geometry $[m, G]$ consists in the choice of a subset $a \subset m$; the transformations T in G will be retained in H if and only if they satisfy $T(a) = a$. Under these conditions, a is called *the absolute* of Geometry $[m, H]$.

We shall present in the coming paragraphs some Kleinean Geometries, evidencing their subordination relations. We intend to illustrate on examples the study methods of these Geometries in order to gradually surprise the degrees of generality of these methods.

§2. Plane Affine Geometry

One takes the set R^2 for m, One calls affine transformation of R^2 a map $T: R^2 \to R^2$ which associates the element (x', y') to the element (x, y) in R^2, where:

(2.1) $\quad x' = a_{11} \cdot x + a_{12} \cdot y + a_{10}$ and $y' = a_{21} \cdot x + a_{22} \cdot y + a_{20}$,

a_{ij} being the real numbers specified so that the matrix

$$M_T = \begin{bmatrix} a_{11} & a_{12} \\ a_{21} & a_{22} \end{bmatrix}$$

may be non-singular.

Let A_2 be the set of the affine transformations of R^2.

Theorem 2.1. *The set A_2 forms a group related to the composition operation.*

The proof is simple; to find out that by means of the composition of affine transformations T and S an affine transformation $S \circ T$ is obtained we shall notice $M_{S \circ T} = M_S \cdot M_T$ and it will follow $\det M_{S \circ T} = (\det M_S) \cdot (\det M_T) \neq 0$. The identical transformation I is obtained by particularizing $a_{11} = a_{22} = 1$ and $a_{12} = a_{21} = a_{10} = a_{20} = 0$. The condition $\det M_T \neq 0$ enables us to solve the system of equations (1.1) related to (x, y) and to evidence the reverse T^{-1} of transformation T, its corresponding matrix being $M_{T^{-1}} = (M_T)^{-1}$.

Consequently, we will consider *the Plane Affine Geometry* as study of the couple (R^2, A_2).

We call point an element in R^2. For a point (x_i, y_i), when no confusions are possible, we shall also use the notation A_i.

Given two distinct points A_1, A_2 there will exist no affine invariant associated to them; we convince ourselves of this finding out that for any other two distinct points B_1, B_2 we can determine an affine transformation T so that $T(A_1) = B_1$, $T(A_2) = B_2$.

Let us now consider three distinct points A_1, A_2, A_3. When trying to determine an affine transformation T that should apply then in points B_1, B_2, B_3, respectively, we are constrained to take into consideration the determinant:

$$(2.2) \quad S(A_1, A_2, A_3) = Det \begin{pmatrix} x_1 & y_1 & 1 \\ x_2 & y_2 & 1 \\ x_3 & y_3 & 1 \end{pmatrix}.$$

We may determine a unique "transformation" T if $S(A_1, A_2, A_3) \neq 0$ and its matrix will be non-singular if and only if $S(B_1, B_2, B_3) \neq 0$, since the equality:

$$(2.3) \quad S(B_1, B_2, B_3) = \det(M_T) \cdot S(A_1, A_2, A_3).$$

We retain from here that $S(A_1, A_2, A_3)$ *does not constitute an affine invariant*, but condition $S(A_1, A_2, A_3) = 0$ *has a geometric character* and represents the "co-linearity" condition of points A_1, A_2, A_3. We draw attention that the name of "co-linearity" is arbitrary in principle, but it reflects an intuitive property manifesting itself when we are representing in a usual manner the couples (x, y) by points of the plane. We shall now introduce the concept of straight line.

Let us consider a maximal subset d of $\mathbb{R}^2$ such that

$$\{A_1, A_2, A_3\} \subset d \Rightarrow S(A_1, A_2, A_3) = 0.$$

This kind of special elements $d \in \mathcal{P}(\mathbb{R}^2)$ will be named straight line. The family $\mathcal{L}$ of straight lines is conceived as a subset of $\mathcal{P}(\mathbb{R}^2)$.

Theorem 2.2. *For the set of points R^2 and for the set $\mathcal{L}$ of the straight lines, if we interpret the incidence by belonging, the axioms of incidence of the plane Geometry I_1, I_2, I_3 are satisfied.*

We shall now find out that, in the hypothesis $A_1 \neq A_2$, the condition $S(A_1, A_2, A_3) = 0$ returns to the existence of a real number t so that

$$(2.4) \quad x_3 = t \cdot x_1 + (1-t) \cdot x_2, \; y_3 = t \cdot y_1 + (1-t) \cdot y_2.$$

Now we get a maximal set $d = (A_1 A_2) \in \mathcal{L}$ keeping fixed values for x_1, y_1, x_2, y_2 and letting t free in $\mathbb{R}$.

As we do not know if the real parameter t has a geometric significance, since we do not *a priori* know what formulae (2.4) become after having applied the generic affine transformation (2.1), we replaced points A_i by $A_i' = T(A_i)$ by $i = 1, 2, 3$. Let be $A_i'(x_i', y_i')$. But it is easy to find out the equalities:

$$(2.4') \quad x_3' = t \cdot x_1' + (1-t)x_2'; \; y_3' = t \cdot y_1' + (1-t) \cdot y_2'.$$

After having established these equalities, we find out that the affine transformation T does not change the real parameter t, so this parameter is an *invariant* of A_2. The name of this invariant is again arbitrary, but the geometric intuition recommends to call it *simple ratio* (of three co-linear points) and note $t = (A_3, A_1; A_2)$ (it is easy to see on a traditional drawing that, if we had the right to measure lengths, we should find $|t| = d(A_2, A_3):d(A_2, A_1)$ and the name of "ratio" is thus justified). A simple calculus leads to:

$$(2.5) \quad S(A_1, A_2, A_3) = 0 \wedge A_1 \neq A_2 \Rightarrow (A_3, A_1; A_2) = \frac{x_3 - x_2}{x_1 - x_2} = \frac{y_3 - y_2}{y_1 - y_2}.$$

It is possible to develop affine Geometry in the direction planned by Hilbert's axiomatic system by introducing the ternary relation of betwenness by:

(2.6) $A_1 - A_2 - A_3 \Leftrightarrow A_1 \neq A_3 \wedge S(A_1, A_2, A_3) = 0 \wedge (A_3, A_1 ; A_2) < 0.$

Theorem 2.3. *Within affine Geometry* $(R^2, \mathcal{L}, ? - ? - ?)$, *the axioms* II_1, II_2, II_3 *in Hilbert's system are verified.*

The proof is simple to be done. To verify axiom II_3 it is useful to point out the following formulae:

(2.7) $$\begin{cases} (A_1, A_2 ; A_3) + (A_1, A_3 ; A_2) = 1, \\ (A_1, A_2 ; A_3) \cdot (A_2, A_1 ; A_3) = 1. \end{cases}$$

Remark. For distinct points A_1, A_2 the condition $A \in |A_1A_2|$ which returns to $A_1 - A - A_2$ is expressed by $(A, A_1 ; A_2) \in (0, 1)$.

We shall find out later that Pasch's axiom is also satisfied. But let us notice that in order that point $A(x, y)$ may belong to the straight line A_1A_2, according to (2.2), it is necessary that there should hold:

$$a \cdot x + b \cdot y + c = 0, \text{ where } a = y_1 - y_2, b = x_2 - x_1,$$
$$c = x_1 y_2 - y_1 x_2 \text{ and } a^2 + b^2 \neq 0.$$

The equality $a \cdot x + b \cdot y + c = 0$ represents *the equation of the straight line* A_1A_2. By replacing one or both points A_i by other points of the straight line, the coefficients a, b, c are amplified by the same non-null factor. The condition $a^2 + b^2 \neq 0$ is essential; it expresses $A_1 \neq A_2$. So, a certain straight line d is given by a triplet (a, b, c) of real numbers determined up to a factor of proportionality provided that numbers a and b should not be simultaneously null.

Theorem 2.4. *Given a straight line d of equation $ax + by + c = 0$, each of the inequalities $ax + by + c \geq 0$ and $ax + by + c < 0$ defines a semi-plane delimited by the straight line .*

Proof. Let $M(m, n)$ and $P(p, c)$ be points non-situated on d. An arbitrary point $Z(u, v)$ of the straight line (MP) is given by an arbitrary value of the parameter t by $u = tm + (1 - t)p$, $v = tn + (1 - t)q$. This point Z is on d if and only if

$$\frac{t}{t-1} = \frac{ap + bq + c}{am + bn + c}.$$

By means of formulae (2.7) one can notice that $\frac{t}{t-1} = (P, M; Z)$ holds and therefore Z will exist on d so that $P-Z-M$ if and only if the real numbers $ap+bq+c$ and $am+bn+c$ have contrary signs. Therefore, the condition that P, Q should be in the same semi-plane coincides with the definition of the equivalence $\sim$ formulated in Chapter V, §3, q.e.d.

Theorem 2.5. *Within the plane affine Geometry axiom* II_4 *of Pasch holds.*

Proof. In the preceding theorem we found out that axiom B_7 (of plane separation) in Birkhoff's system holds; and Theorem 2.7 in Chapter VI pointed out the equivalence between B_7 and II_4.

The condition that the straight lines of equations $ax+by+c = 0$ and $a'x+b'y+c' = 0$ should by parallel is known: $\frac{a}{a'} = \frac{b}{b'}$.

The validity of Hilbert's axiom V is immediately checked.

Remark. The contents of this paragraph largely reproduces the ideas in the construction of the analytical model G (Chapter V, §7). For variation and expeditivity, *the affine* plane was studied here, but no conceptual difficulties appear in extending this study to *n-dimensional affine spaces*. To simplify things, the construction started from the field R of real numbers, but it can not be resumed without essential modifications for any ordered field K. With these extensions it is easy to achieve the first steps in the study of *the n-dimensional affine Geometry over a field K*, that is the study of Kleinean Geometry (K^n, A_n), where A_n contains the transformations $T: K^n \to K^n$ which apply the point $A(x^i)$ in $\overline{A}(\overline{x}^i)$, where there exists $a_j^i \in K$ (for $i = 1, 2, \ldots, n$ and $j = 0, 1, \ldots, n$) so that $\overline{x}^i = \sum_{j=1}^{n} a_j^i x^j + a_0^i$ for $i = 1, 2, \ldots, n$ and which have the matrix $M_T = (a_j^i)$ (with $i, j = 1, 2, \ldots, n$) non-singular.

We hope the exposition in this paragraph to have correctly suggested the effective way to obtain the specific notions by the analysis of the action of the chosen group over the simple figures. We shall now present the way of obtaining certain Geometries subordinated to real plane affine Geometry.

By formula (2.3) we deduced the law of transformation of $S(A_1, A_2, A_3)$. If we retain only the transformations T in A_2 for which it is achieved the condition:

$$(2.8) \quad \det M_T \in \{-1, +1\},$$

we obviously obtain a subgroup $\mathscr{E}_2$ of A_2. In Kleinean Geometry, $(R^2, \mathscr{E}_2)$, we shall also have an invariant associated to an arbitrary triplet of points (A_1, A_2, A_3) called "the area of the triangle" and given by

$$(2.9) \quad |A_1A_2A_3| = \frac{1}{2} \cdot |S(A_1, A_2, A_3)| \,.$$

(The name ignores the possibility of the degeneration of "triangle" $A_1A_2A_3$.) For Kleinean Geometry $(R^2, \mathscr{E}_2)$ the name of (plane, real) *equi-affine* Geometry is used. The equi-affine Geometry is obviously subordinated to the affine one. A further specialized Geometry $(R^2, \mathscr{E}_2^+)$ is obtained by replacing the condition (2.8) by:

$$(2.8') \quad \det M_T = 1 \,.$$

This time $\operatorname{sign} S(A_1, A_2, A_3)$ is also invariant and we have the possibility to decide whether a triangle $A_1A_2A_3$ is positively or negatively oriented.

Let us consider point $(0, 0) \in R^2$ as individualized and retain only the transformations T in A_2 which preserve this point; in formulae (2.1) it is then imposed to take $a_{10} = a_{20} = 0$. According to the general theory in §1, we chose an absolute $a = (0, 0)$ and therefore we obtain what is called (plane, real) *centro-affine* Geometry $(\mathbb{R}^2, C_2)$, obviously subordinated to the affine one.

By choosing a as absolute in (R^2, A_2) the set of the points of the straight line d which has the equation $x = 0$, we particularize in (2.1) $a_{12} = a_{10} = 0$ and offer the condition $\det M_T \neq 0$ the simpler form $a_{11} \neq 0 \neq a_{22}$. This way, we obtain the (plane, real) *axial* Geometry subordinated to the affine one.

Of course, for the centro-affine and axial Geometries one can evidence specific invariants, but we do not deal with them here.

§3. The Real Projective Plane

Let n be the set $\mathbb{R}^3$ from which point $(0, 0, 0)$ is excluded. Let us consider on n the binary relation γ defined by:

$$(a, b, c)\gamma(a', b', c') \Leftrightarrow \exists \gamma \in \mathbb{R}^* \quad a' = ka, \ b' = kb, \ c' = kc \,.$$

It is easy to find out that γ is an equivalence on n; the quotient set n / γ will be denoted by P_2 and the equivalence class of the element (a, b, c) by $[a, b, c]$. We shall refer to the operation of re-choosing a representative for $[a, b, c]$ by the term of "normalization".

We are initiating the construction of a Klein space (m, G) by taking $m = P_2$. As the element of P_2 are considered to be *points*, let us agree on a certain mechanism of notation: point A will be given by $[a^1, a^2, a^3]$, point B by $[b^1, b^2, b^3]$ etc. If the capital letter designing the point is (lowerly) indexed we shall also use the same index for the components of the tern representing that point; for instance, M_i will be given by $[m_i^1, m_i^2, m_i^3]$.

We associate to a non-singular matrix of order 3

$$T = \begin{pmatrix} t_1^1 & t_1^2 & t_1^3 \\ t_2^1 & t_2^2 & t_2^3 \\ t_3^1 & t_3^2 & t_3^3 \end{pmatrix}$$

a projective transformation $T : P_2 \to P_2$ defined by $Y = T(X)$, where

$$(3.1) \quad y^j = \sum_{i=1}^{3} t_i^j x^i, \, j = 1, 2, 3.$$

We draw attention that we are using the same letter both for the transformation and for the matrix; we shall take care that any eventuality of confusion should be eliminated by the context. It is obvious that by amplifying a non-singular matrix T of order 3 by a non-null real factor k we do not affect the projective transformation T associated to it, but only the normalization of the representative (y^1, y^2, y^3) of Y.

The set of projective transformations T obviously makes up a group (related to the operation of composition); we shall denote this group by G.

The group G acts transitively on the set P_2 and on the family of the subsets of P_2 made up by exactly two points each. Therefore, we shall not be able to associate notions or invariant values to figures with at most two points.

We are researching the problem of determining a transformation T which should apply given points A, B, C in prescribed points A_1, B_1, C_1; like in the preceding paragraph, there appears the necessity of considering the determinant:

$$S(A, B, C) = Det \begin{pmatrix} a^1 & a^2 & a^3 \\ b^1 & b^2 & b^3 \\ c^1 & c^2 & c^3 \end{pmatrix}.$$

By amplifying by a non-null real factor k the representative (a^1, a^2, a^3) of point A, we amplify by the same factor the set $S(A, B, C)$, so S does not constitute even a map from $(P_2)^3$ in $\mathbb{R}$ as it would imply the notation. However, the condition $S(A, B, C) = 0$ is independent of representatives and is also made invariant by the projective transformations, that is *it has a geometrical character*. Instead of the quite confusing notation $S(A, B, C) = 0$ we shall prefer the formulation: "*A, B, C* are co-linear points". By means of this relation of co-linearity *the notion of straight line* is introduced: the straight line (AB) determined by the distinct points A and B in P_2 is the set of points C in P_2 co-linear with A, B.

Theorem 3.1. *In plane projective Geometry the following "axioms of incidence" are satisfied:*

P_0: *For any two distinct points A, B there exists exactly one straight line (AB) containing them.*

P_1: *For any two distinct straight lines a, b there exists exactly one point $a \frown b$ belonging to them.*

P_2: *There exist three non co-linear points.*

Proof. By expressing the condition $S(A, B, X) = 0$ we shall find out that a straight line d in (P_2, G) is given by an element $d = [D_1, D_2, D_3]$ in P_2 so that point X belongs to this straight line if and only there holds:

(3.2) $D_1 x^1 + D_2 x^2 + D_3 x^3 = 0$.

We draw attention upon the mechanism of notation: a small letter of the Latin alphabet marks the straight line and the same capital letter is used for the components of the representative tern. To remove any possible confusions, the numbering indexes in the tern are placed in the lower part. Let us also notice that the straight line d is not identified by the element d. The images through T of the points of the straight line d make up a straight line c and for the associated tern $\tilde{c} = (C_1, C_2, C_3)$ there holds the equality:

$(3.1')$ $D_i = \sum_{j=1}^{3} t_i^j C_j$ for $i = 1, 2, 3,$

which proves that, in general, does not hold $T(\tilde{d}) = \tilde{c}$, equality which would justify the identification $d = \tilde{d}$.

To check P_0 we consider:

(3.3) $D_1 = Det \begin{bmatrix} a^2 & a^3 \\ b^2 & b^3 \end{bmatrix}$, $D_2 = Det \begin{bmatrix} a^3 & a^1 \\ b^3 & b^1 \end{bmatrix}$, $D_3 = Det \begin{bmatrix} a^1 & a^2 \\ b^1 & b^2 \end{bmatrix}$.

Before checking P_1 it is necessary to notice that the sign $\frown$ used in the enunciation does not coincide with the sign $\cap$ of the intersection since by intersecting two sets of points we obtain *a set* eventually made up of a point. So it is stated that $a \cap b = \{a \frown b\}$ would hold. We shall read here the sign $\frown$ as "intersected with", too, but referring to the *geometrical* operation of intersecting, not to that in the theory of sets. We check P_1 by showing that for $a \neq b$, the matrix $\begin{bmatrix} A_1 & A_2 & A_3 \\ B_1 & B_2 & B_3 \end{bmatrix}$ has the rank two and therefore the linear and homogenous system:

$$A_1 x^1 + A_2 x^2 + A_3 x^3 = 0, \; B_1 x^1 + B_2 x^2 + B_3 x^3 = 0$$

has a unique solution up to a factor of proportionality, that is only a point $C \in P_2$, where

$(3.3')$ $c_1 = Det \begin{bmatrix} A_2 & A_3 \\ B_2 & B_3 \end{bmatrix}$, $c_2 = Det \begin{bmatrix} A_3 & A_1 \\ B_3 & B_1 \end{bmatrix}$, $c_3 = Det \begin{bmatrix} A_1 & A_2 \\ B_1 & B_2 \end{bmatrix}$.

Finally, three obviously non co-linear points are: $[1, 0, 0]$, $[0, 1, 0]$ and $[0, 0, 1]$.

We hope that the way of drawing up the proof of the theorem has suggested the existence of a "duality" between points and straight lines in plane projective Geometry. This impression will turn into certainty in what follows.

Let us re-consider the operation associating the straight line (AB) to two distinct points A, B, and, to evidence that it is an operation, we shall re-note here (AB) with $A \smile B$. We shall read the sign $\smile$ as "united with" and we shall refer to the respective operation by the term of *unity*. The sign ⌣ is similar to that of the union, ∪, but benefits from distinct names and the implied operations have clearly differentiated results.

Definition 3.1. *Let P be a proposition referring to points and straight lines in plane projective Geometry and to the operations* ⌣ , ⌣ *of unity and intersection. One calls the dual of proposition P the proposition P′ which is obtained from P substituting by each other the words point-straight line and the operations* ⌣ , ⌣ .

Theorem 3.2 (The principle of duality). *Within plane projective Geometry, for any correctly constructed proposition P, if P′ is the dual proposition of P, then P and P′ are simultaneously true.*

Proof. Let $\widetilde{P_2}$ be the set associated to P_2; there are thus two disjoint sets between which there exists a bijective correspondence $f: P_2 \to \widetilde{P_2}$. We shall keep calling the elements of P_2 points and those in $\widetilde{P_2}$ straight lines. A group G of transformations is given on the set P_2 so that we can endow $\widetilde{P_2}$ with the associated group $\widetilde{G}$ made up of the transformations $\widetilde{T} = f \circ T \circ f^{-1}$ where $T \in G$. We shall consider the map $F: G \to \widetilde{G}$, where $F(T)$ has the adjoint of the matrix T as matrix. It is easy to find out that F constitutes an isomorphism of groups of transformations. The couple (f, F) constitutes a morphism in the category of concrete Klein spaces from (P_2, G) to $(\widetilde{P_2}, \widetilde{G})$ which is an isomorphism. By this morphism, proposition $P′$ in the second model corresponds to proposition P in the first one according to definition 3.1. So P and its dual $P′$ are simultaneously true.

Remark. Plane projective Geometry can also be constructed axiomatically. Propositions P_0, P_1, P_3 constitute its first axioms. For such an axiomatic construction Theorem 3.2 benefits from another type of proof, that is: it is proved that the duals $P_i′$ of axioms P_i are true propositions and then that axioms P_i are consequences of the dual system of axioms $P_i′$. Here, it is to be currently noticed that P_0 and P_1 are dual with each other and that P_2 is equivalent to proposition $P_3′$: *There exist three non-secant straight lines.* As new axioms are enunciated, this equivalent property is followed. Of course, if besides the primary notions (point, straight line, belongs to) new notions appear, either primary or derived, the extending of Definition 3.1 is necessary to "translate" directly the new notions, as well.

Now we get back to the condition $S(A, B, C) = 0$ and find out that, in the hypothesis $A \neq B \neq C$, it is equivalent with the existence of a real number r so that the matrix

$$\begin{bmatrix} a^1 - rb^1 & a^2 - rb^2 & a^3 - rb^3 \\ c^1 & c^2 & c^3 \end{bmatrix}$$

should have the rank 1. If we fix the proportionality factor for each of the terns (a^n), (b^n), then the real number r is uniquely determined and there holds:

$$r = \frac{Det \begin{bmatrix} a^1 & a^2 \\ c^1 & c^2 \end{bmatrix}}{Det \begin{bmatrix} b^1 & b^2 \\ c^1 & c^2 \end{bmatrix}} = \frac{Det \begin{bmatrix} a^2 & a^3 \\ c^2 & c^3 \end{bmatrix}}{Det \begin{bmatrix} b^2 & b^3 \\ c^2 & c^3 \end{bmatrix}} = \frac{Det \begin{bmatrix} a^3 & a^1 \\ c^3 & c^1 \end{bmatrix}}{Det \begin{bmatrix} b^3 & b^1 \\ c^3 & c^1 \end{bmatrix}} .$$

By replacing the point C by another point $D \neq B$, we shall replace the real number r by another one, s. It is easy to find out that the real number

$$\frac{r}{s} = \frac{Det \begin{bmatrix} a^i & a^j \\ c^i & c^j \end{bmatrix}}{Det \begin{bmatrix} b^i & b^j \\ c^i & c^j \end{bmatrix}} : \frac{Det \begin{bmatrix} a^i & a^j \\ d^i & d^j \end{bmatrix}}{Det \begin{bmatrix} b^i & b^j \\ c^i & c^j \end{bmatrix}}$$

does not depend anylonger either on the normalization chosen for terns (a^n), (b^n) or on the pair of indexes $\{i, j\} \subset \{1, 2, 3\}$. We shall use for $\frac{r}{s}$ the notation $(A, B; C, D)$ and the name of cross ratio of (distinct) co-linear points A, B, C, D taken in this order. A not quite easy algebraic calculus ensures:

Theorem 3.3. *If (A, B, C) and (A_1, B_1, C_1) are terns of distinct co-linear points, then:*

1. *There exists an infinity of projective transformations T that makes these terns correspond to each other.*

2. *The restriction to the straight line $A \smile B$ of all these transformations is a*

bijective map t which has the straight line $A_1 \smile B_1$ as co-domain and, for an arbitrary point D situated on $A \smile B = \{A, B, C\}$, $t(D) = D_1$ holds if and only if $(A, B; C, D) = (A_1, B_1; C_1, D_1)$.

From this theorem we deduce:
- the projective group acts transitively upon the triplets of co-linear points, therefore no invariants exist for three co-linear points;
- the cross ratio of four co-linear points is invariant.

The cross ratio of four concurrent straight lines$(a, b; c, d) = (\bar{a}, \bar{b}; \bar{c}, \bar{d})$ is also naturally introduced (here, four co-linear points appear in the second member). The two notions of cross ratio are connected by means of:

Theorem 3.4. *In plane projective Geometry there holds property:*

P_3: *If A, B, C, D are four distinct co-linear points and O is a point non-situated on the straight line $A \smile B$, then, for the straight lines $a = O \smile A$, $b = O \smile B$, $c = O \smile C$, $d = O \smile D$ there holds: $(a, b; c, d) = (A, B; C, D)$.*

We also avoid the algebraic calculuses necessary to check this theorem. We only draw attention upon the fact that Proposition P_3 is usually adopted as an axiom for edifying plane projective Geometry as axiomatic theory. It is interesting that the dual of proposition P_3 is a mere reformulation of its own. Then we say that: P_3 is a self-dual proposition.

Further on, we shall mention two Kleinean Geometries subordinated to the projective Geometry.

Let us take as absolute $\mathcal{A}$ the straight line of equation $x^3 = 0$. The projective transformation T given by (3.1) will preserve this absolute if and only if $t_3^1 = t_3^2 = 0$. The condition that the matrix T should be non-singular imposes $t_3^3 \neq 0$; taking into account that we are allowed to amplify the whole matrix by a non-null real factor, we can suppose, without restricting generality, that $t_3^3 = 1$ holds. Now we make explicit transformation T by saying that $Y = T(X)$ if

(3.4) $y^1 = t_1^1 x^1 + t_2^1 x^2 + t_3^1 x^3$, $y^2 = t_1^2 x^1 + t_2^2 x^2 + t_3^2 x^3$, $y^3 = x^3$,

$$(3.5) \quad Det \begin{pmatrix} t_1^1 & t_2^1 \\ t_1^2 & t_2^2 \end{pmatrix} \neq 0.$$

(Under these conditions, the representative (y^1, y^2, y^3) of Y is normalized according to the representative chosen for X.)

Let us denote by A the sub-group of G made up only of the projective transformations T which preserves the absolute $\mathcal{a}$. We shall provisorily re-note $P_2 - \mathcal{a}$ with m and analyze the Kleinean Geometry (m, A). (We are here implying that we have noticed that any projective transformation T preserving $\mathcal{a}$ has a restriction which is a bijection of m and we identify this restriction with the initial transformation.) For the generic point $M = [m^1, m^2, m^3]$ in m we can uniquely choose a representative $(m^1, m^2, 1)$. This way we achieve a bijection $f: m \to \mathbb{R}^2$, where $f(M) = (m^1, m^2)$. On the basis of this bijection we identify m by $\mathbb{R}^2$ and readjust the notation mechanism by: $X = (x^1, x^2)$, $Y = (y^1, y^2)$ etc. Thus, (3.4) becomes:

$$(3.4') \quad y^1 = t_1^1 x^1 + t_2^1 x^2 + t_3^1; \; y^2 = t_1^2 x^1 + t_2^2 x^2 + t_3^2,$$

the (3.5) condition being preserved. It is easy now to find out that the Geometry obtained this way is the (plane, real) affine Geometry studied in the preceding paragraph.

Another Kleinean Geometry subordinated to (P_2, G) is obtained by choosing as absolute $\mathcal{a}$ "the oval conic" of equation $(x^1)^2 + (x^2)^2 - (x^3)^2 = 0$. Since no points with $x^3 = 0$ can exist on this conic, we can imagine it as a circle of equation $(x^1)^2 + (x^2)^2 - 1 = 0$. The projective transformation T given by (3.1) will preserve this absolute if and only if a real number ρ exists so that:

$$(3.6) \quad \begin{pmatrix} t_1^1 & t_1^2 & -t_1^3 \\ t_2^1 & t_2^2 & -t_2^3 \\ t_3^1 & t_3^2 & -t_3^3 \end{pmatrix} \cdot \begin{pmatrix} t_1^1 & t_2^1 & t_3^1 \\ t_1^2 & t_2^2 & t_3^2 \\ t_1^3 & t_2^3 & t_3^3 \end{pmatrix} = \rho \cdot \begin{pmatrix} 1 & 0 & 0 \\ 0 & 1 & 0 \\ 0 & 0 & 1 \end{pmatrix}.$$

We shall note with H' the subgroup G made up of projective transformations T satisfying condition (3.6).

We also consider the set C of the "interior" points of a. These points $X = [x^1, x^2, x^3]$ can be algebraically characterized by the condition $(x^1)^2 + (x^2)^2 - (x^3)^2 < 0$ or *geometric* by the condition that *any* straight line d through X should have exactly two points common with a.

The geometric conditions return to: For any $(d_1, d_2, d_3) \neq (0, 0, 0)$ so that $d_1 x^1 + d_2 x^2 + d_3 x^3 = 0$, the system $d_1 u^1 + d_2 u^2 + d_3 u^3 = 0$, $(u^1)^2 + (u^2)^2 - (u^3)^2 = 0$ should admit two non-proportional real solutions. By eliminating the unknown u^3, we express this condition by the trinomial of second degree

$$[(d_3)^2 - (d_1)^2](u^1)^2 - 2d_1 d_2 u^1 u^2 + [(d_3)^2 - (d_2)^2](u^2)^2 = 0$$

should have distinct real solutions. By the calculus of the discriminant, the condition returns to $(d_1)^2 + (d_2)^2 - (d_3)^2 > 0$. We have so summarized the geometric condition to:

$$(d_1, d_2, d_3) \neq (0, 0, 0) \wedge d_1 x^1 + d_2 x^2 + d_3 x^3 = 0 \Rightarrow (d_1)^2 + (d_2)^2 - (d_3)^2 > 0.$$

By amplifying the inequality by $(x^3)^2$ and relating $d_3 x^3$ from the equality, we deduce that for any real d_1, d_2 there should hold:

$$[(x^3)^2 - (x^1)^2](d_1)^2 - 2x^1 x^2 d_1 d_2 + [(x^3)^2 - (x^2)^2](d_2)^2 > 0.$$

A new calculus of trinomial discriminant finally offers to the analyzed condition the presumed algebraic form.

Theorem 3.5. *A projective transformation T in H' preserves the quality of a point X in P_2 of being interior to a that is $T(C) \subset C$.*

Proof. Let be $T(X) = Y$. Any straight line d through Y is the image through T of a straight line c through X. If X is in C, the straight line c will intersect the absolute a in exactly two points L_1, L_2. The image through T of these points constitute the intersection of d with the absolute, so any straight line d through Y intersects a in two distinct points, q.e.d.

Corollary. *Any transformation T' in H' admits a restriction T to C, being itself* *a bijection of C. The restrictions of all transformations T' in H' make up a group $\tilde{H}$.*

We are now in the possession of a Kleinean Geometry (C, $\tilde{H}$) subordinated to the projective Geometry. The English mathematician Arthur Cayley discovered this Geometry and proved that *this Geometry constitutes a model of plane hyperbolic Geometry.* We shall prove this at the end of paragraph 5.

We close this paragraph with a commentary. The exposition in this paragraph and in the preceding one is correct but very little "geometrical". Such a situation is frequent since for most of Kleinean (m, G) Geometries a *standard mechanism* of precessing the figures and the invariants is valid. Looking for transformations T in G that should apply a figure $F \subset m$ on another figure $F' \subset m$, by gradually extending the figure F (and F', respectively) one can determine little by little the parameters of transformation T. At a given moment conditions of "compatibility" can appear, and they give conditions for figure F inducing *specific notions* of Geometry (m, G) and "numerical" characteristics for F which are transmitted to F' that is *invariants* of Geometry (m, G). This standard mechanism is rather algebraic than geometric and can lead to the false impression that Geometry is subordinated to Algebra.

But the above short presentation of the "standard mechanism" has neglected an essential aspect: "*the geometric interpretation*". In this respect, one can not suggest "receipts" anylonger, but the achievement of this aspect is what offers viability and elegance to a Klein space study. These geometric interpretations, "moments of truth" in edifying a Kleinean Geometry outline the organic connection of Geometry with other branches of Mathematics; in the case of "elementary" Geometries the link is done rather with Algebra, but other chapters of Geometry require above all topology or functional analysis; within "differential" Geometry a close contact with the study of differential equations and with the partial derivatives is achieved etc.

There appears then the necessity of re-iterating the exposition in this paragraph by inserting geometric interpretations into the adequate places. We shall satisfy this necessity in the next paragraph. Of course, one could have prepared the interpretations of projective Geometry by presenting first those of the affine Geometry; we shall obtain the plane affine Geometry by particularizing the projective one.

§4. Plane Projective Geometry

In the preceding paragraph we considered a set $P_2 = n / \gamma$, where n appeared as the numeric set: $n = \mathbb{R}^3 - \{(0, 0, 0)\}$. Now we can conceive $\mathbb{R}^3$ not only as a numeric set, but as set A_3 subjacent to a three-dimensional affine Geometry. The

edification of this Geometry can be achieved by following step by step the exposition in
§2 or by retaining only a part of the construction of model G of Euclidean Geometry in
Chapter V, §7 (the function of distance wich does not constitute an affine notion - must
be neglected). We shall distinguish in A_3 a point O playing the part of the triplet
$(0, 0, 0)$ in $\mathbb{R}^3$.

For an arbitrary point M different from O in A_3 we shall consider the straight
line $m = (OM)$. If M corresponds to point (m^1, m^2, m^3) in $\mathbb{R}^3$, the straight line m will
contain O and all other points M', to which the terns (km^1, km^2, km^3) correspond,
where k is in $\mathbb{R}^*$. The equivalence γ in $\mathcal{n}$ returns to a relation of co-linearity with O
for points in $A_3 - \{0\}$, so an equivalence class related to γ appears as a straight line
through O of A_3. We have thus achieved a first geometric interpretation of the points of
P_2: the set of *the straight lines through a point O of the affine space A_3* (which is also
called *sheaf of straight lines*).

But we should like more: to interpret the points of the projective plane as being
also "points" and not "straight lines". The idea consists in the choice of a surface in A_3
which should cut each straight line through O only in one point.

The choice is not a simple one. A sphere S of centre O cuts each straight line
through O, but in two points, not only in one! This finding out furnishes a second
geometric interpretation for P_2: *a sphere S within which the diametrically opposed points
identify themselves*. But we can still make another step towards a point correspondence
retaining only a closed hemisphere $\mathcal{E}$ of S, delimited by a large circle C. Thus, we obtain
the third interpretation for P_2: *a closed hemisphere $\mathcal{E}$ within which the points
diametrically opposed to "the equator of the hemisphere" identify themselves*.

We can try to intercept the straight line of A_3 through O with a plane π non-
incident to O. A straight line m cuts π in a unique point except the cases when $m \parallel \pi$.
There appears thus the idea of adding some more points "situated at the infinite" to the
plane π. To effectively achieve this completion we shall consider the set $\mathcal{L}_\pi$ of the
straight lines of the plane π and factorize it by the equivalence provided by the relation
of parallelism. We shall denote $\mathcal{L}_\pi / \parallel$ with Δ_π and the equivalence class of a straight line
d, that is the "direction" of the straight line d, with $[d]$. For the set $\pi \cup \Delta_\pi$ we shall use
the notation $\bar{\pi}$. We shall call *points* the elements $[d]$ in Δ_π, too, but when necessary we
shall distinguish them by the qualifier "improper". We finally obtain *the fourth
interpretation* of P_2 summarized by the formula $P_2 = \bar{\pi}$.

We have presented four geometric interpretations for P_2 in order to sustain the
impossibility of the existence of a "unique receipt" for this strictly necessary stage in

edifying the Geometry of a Klein space. On the other hand, certain notions of plane projective Geometry will be more easily to be intuited on some of these geometric hypostases of P_2. There also exists the eventuality that remarkable geometric ideas should be evidenced by passing from a "model" to another one!

The choice of P_2 as "geometric" set largely directs the researches towards the interpretations of group .

If we take P_2 as the sheaf of straight lines of A_3 through O it is natural to look for a group K of the transformations of A_3 that should generate group G on the projective plane P_2. We shall of course ask from a transformations S in K to preserve the distinguished point O. By taking again the "numeric" interpretations of A_3 and P_2 as base, we deduce that transformation S has to apply $X = (x^1, x^2, x^3)$ in $Y = (y^1, y^2, y^3)$, where

$$y^i = \sum_{j=1}^{3} s^i_j x^j , (i = 1, 2, 3)$$

the matrix (s^i_j) being not degenerated. We are suggested then to start from a sub-group K of the centro-affine transformations of A_3. A transformation S in K traces a straight line d into a straight line $d' = S(d)$. It is obvious that $O \in d \Rightarrow O \in d'$, so S induces a transformation $T = f(S)$ into the sheaf P_2. It is easy to find out on the expression that the map $f : K \to G$ is a homomorphism of groups. The matrices of the transformations S and $f(S)$ are identical and therefore f will be an isomorphism if the chose sub-group K has the property

$$\forall\, S \in K,\ \forall\, k \in \mathbf{R}^* : kS \in K \Rightarrow k = 1.$$

We can achieve this by imposing $\left|\det(s^i_j)\right| = 1$ which returns to say that the transformation S is also equi-affine. We can then interpret G *as the group of centro- and equi-affine transformations of* A_3 (on the condition on transferring by the isomorphism f the action of the transformations from the affine space A_3 onto the sheaf P_2).

We appreciate that the second and third interpretations of P_2 are not useful for the generic interpretation of G. But for the fourth interpretation of P_2 we shall find that the choice of the couple (O, π) is arbitrary outside the condition $O \notin \pi$. We shall analyze the significance of modifying the couple (O, π) through another couple (O', π') by supposing first that π and π' *are not parallel.*

For the moment, we keep point O and replace the plane π by π'. We thus determine a function f which associates to a point $M \in \bar{\pi}$ the point $M' = f(M)$ in $\bar{\pi}'$ determined so that O, M, M' should be co-linear. (The function f is characteristic for projective Geometry; but here we still consider it as insufficient because it does not have the same $\bar{\pi}$ as co-domain.) It has to be noticed that, without having added improper points to the plane π', we could not have defined f when $(OM) \parallel \pi'$; the improper points in $\bar{\pi}$ are applied by f in proper points of $\bar{\pi}'$. It is obvious that f is a bijection. Analogous considerations referreing to O' instead of O allow the introduction of a function $f' : \bar{\pi}' \to \bar{\pi}$, where $f'(M')$ is the point M_1 common to the straight line $O'M'$ and to the completed plane $\bar{\pi}$. We shall consider the function $F = f' \circ f$ which constitutes a bijection of the set $P_2 = \bar{\pi}$ in itself. Since F is a map resulted from the arbitrariness of choosing a "geometric" exemplar of the set P_2 the requirement that F should be element of the projective group G seems natural. (In the contrary case, the geometric interpretation of P_2 would not be compatible to the given group.) Given the importance of function F we shall try to characterize it in terms specific to the plane $\bar{\pi}$.

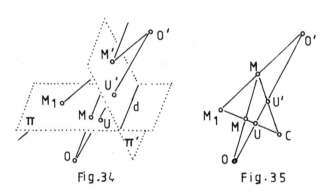

Fig.34   Fig.35

It is easy to prove that the straight line $d = \bar{\pi} \cap \bar{\pi}'$ contains the fixed points of the transformation F. Besides this straight line there exists another fixed point, namely the point U common to the straight line OO' and to the plane $\bar{\pi}$. (If $OO' \parallel \pi$, U is an improper point.) Let us consider an arbitrary plane μ through (OO'); figure 35 represents the intersection of the plane μ with the configuration illustrated by figure 34 and is meant to clarify the definition of point $M_1 = F(M)$. Let us also consider points $U' \in (OO') \cap \bar{\pi}'$, $C \in d \cap \mu$. By twice applying Menelaos' relation for the triangle

UCU', we deduce:

$$\overline{\frac{M'C}{M'U'}} \cdot \overline{\frac{MU}{MC}} \cdot \overline{\frac{OU'}{OU}} = 1 = \overline{\frac{M_1U}{M_1C}} \cdot \overline{\frac{M'C}{M'U'}} \cdot \overline{\frac{O'U'}{O'U}}.$$

We obtain from here

$$\overline{\frac{MU}{MC}} \cdot \overline{\frac{M_1C}{M_1U}} = \overline{\frac{OU}{OU'}} \cdot \overline{\frac{O'U'}{O'U}},$$

equality which is more concisely expressed by means of the cross ratio: $(U, C; M, M_1) = (U, U'; O, O')$. The second member is independent of point M in $\bar{\pi}$, so, for function F under consideration, it is a constant k. We would have reached the same conclusion (on ways that do not differ essentially from the above used one) even if U had been an improper point or incident to the straight line d.

When the straight line d and the point $U \notin d$ are given in the completed plane $\bar{\pi}$ and knowing the constant $k \in \mathbb{R}^*$, we have the possibility of re-defining the function $F: \bar{\pi} \to \bar{\pi}'$ on the basis of the formula

$$(4.1) \quad F(M) = M_1 \wedge C = d^\frown(UM) \Rightarrow (U, C; M, M_1) = k.$$

An algebraic calculus we are avoiding here given its too technical character ensures that the above transformations F under consideration (called *general homologies*) generate the group G. We thus have also *the geometric interpretation of group G*.

We are going on presenting the interpretations of the notions associated in the preceding paragraph to the Klein space (P_2, G).

The condition of co-linearity of points A, B, C returns within the first interpretation of P_2 to the condition of co-planarity of the straight lines $a = (OA)$, $b = (OB)$, $c = (OC)$. A straight line d of the projective space P_2 identifies then itself with a pencil of straight lines through O, situated in a plane δ. Within the second interpretation (on the sphere S within which the diametrically opposed points identify themselves) we conceive a straight line as a big circle of the sphere within which we identify the diametrically opposed points etc. Within the last interpretation, $P_2 = \bar{\pi}$, a straight line a is obtained by adding to the points of a straight line d its improper point $[d]$, but there also exists in "improper" straight line m appearing when $\delta \parallel \pi$; the points of this straight line m are the improper points of $\bar{\pi}$, that is $m = \Delta_\pi$.

The principle of duality also admits remarkable concretizations within the geometric interpretation $P_2 = \overline{\pi}$. For this purpose a conical Γ (non-degenerated, with centre) is considered in π. To each point M in π *the polar* straight line m of M related to Γ is associated.

A variable straight line u through M cuts Γ in points A, B and there is a point U so that $(A, B ; M, U) = -1$; when u varies, U is on the straight line m). If instead of M an improper point, $[d]$, is taken, the diameter conjugated to the direction $[d]$ (the geometrical locus of the midpoints of the chords of Γ parallel with d) is associated to it.

We thus attach to the conical Γ a map γ from P_2 to the set $\mathcal{L}$ of the straight lines of P_2 . We can then extend γ to $\mathcal{L}$, too, on the basis of the formula $\gamma(A \curlyvee B) = \gamma(A) \curlywedge \gamma(B)$. The thus extended map, γ, is a bijection of the set $P_2 \cup \mathcal{L}$ on itself called *polarity related to the conical* Γ. The principle of duality is now concretized this way:

By replacing in a proposition P of plane projective Geometry the points and straight lines with their polar ones and substituting the operations of union and intersection with each other we obtain a proposition P' ; P and P' are simultaneously true.

For the cross ratio $(A, B ; C, D)$ of four co-linear points the following interpretation was outlined in the course of this paragraph:

$$(A, B; C, D) = \frac{(A, B; C)}{(A, B; D)} \; .$$

We shall draw attention here only on a fact neglected in the preceding paragraph.

Theorem 4.1. *If A, B, C are distinct co-linear points and we associate to the arbitrary point X of the straight line $A \curlyvee B$ the cross ratio $f(X) = (A, B ; C, X)$, then f is a bijection from $A \curlyvee B$ to the set $\dot{R} = R \cup \{\infty\}$.*

The completion of $\ddot{R}$ with the "improper number" is necessary to seize the eventuality $X = A$; when X is the improper point of the (proper) straight line (AB), $(A, B ; C, X) = (A, B ; C)$ holds.

On the basis of the geometrical notion of cross ratio, Theorems 3.3 and 3.4 can benefit from geometrical proofs which, without being very simple, are preferable to the algebraic proofs suggested in the preceding paragraph.

As we had stated at the end of paragraph 3, we intend to deduce the affine Geometry by particularizing the projective one. In the real plane projective Geometry (P_2, G) we focus our attention on a projective straight line a. So that the projective transformation F given by (4.1) may preserve this straight line a (without making invariant each of its points) it is necessary that $U \in a$. By interpreting $P_2 = \overline{\pi}$, it is natural to suppose that a is the improper straight line and so to particularize figures 34 and 35 by considering $OO' \parallel \pi$. On figure 36 that overtakes the role of figure 35 we easily deduce the following characterization of the function $F : \pi \to \pi$:

$$(4.2) \quad F(M) = M_1 \Leftrightarrow C \in d \wedge MC \parallel OO' \wedge (M, M_1 ; C) = k,$$

where $k = (O, O'; U')$.

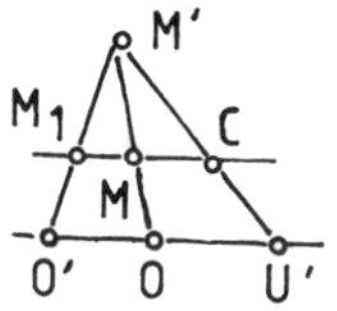

Fig.36 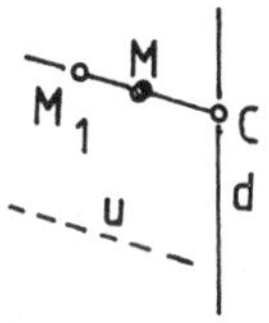   Fig.37

By renouncing to the elements outside the plane π, we shall re-define the transformation F by means of the straight line d, of a direction $[u]$ in π, where $d \notin [u]$, and of the non-null real number k. For the arbitrary point $M \in \pi$ we shall uniquely determine $C \in d$ so that $MC \parallel u$ and then determine only $M_1 = F(M)$ by condition $\overline{CM_1} = k \cdot \overline{CM}$. The transformation F obtained is described by the name: *affine dilatation to the direction* $[u]$, *of axis* d *and ratio* k. Of course, if $0 < k < 1$ holds, the term of *contraction* is more suggestive than that of dilatation, and for $k < 0$ we have to take into account that M_1 and M are in distinct semi-planes related to d.

It is immediately obtained:

Theorem 4.2. *The affine dilatation F of direction* $[u]$, *axis d and ratio k* :
1. *preserves the points of the straight line d;*
2. *preserves the straight lines of direction* $[u]$ *and axis d ;*
3. *turns straight lines into straight lines* ;

4. *preserves the simple ratio of three co-linear points.*

It can be proved without major difficulties the following result:

Any affine transformation of the plane π is obtained by composing such affine dilatations F .

§5. The Hyperbolic Plane

Let us consider an Euclidean plane E_2 in which we fix a straight line a we shall call *absolute* and an open semi-plane S delimited by a in E_2 . The elements of the set S are points ; to distinguish them from the elements in $E_2 - S$, we shall call them h-points. If d is a straight line perpendicular on a, the symmetry S_d of the plane E_2 related to the straight line d admits a restriction T_d which is a bijection of the set S in itself. Let C be a circle of centre C on the absolute a. The inversion I related to the circle C admits a restriction T_C which is a bijection of the set S an itself. For the transformations T_d and T_C we shall use the common name of h-symmetries, being able to distinguish their "type" according to the typographical character of the index. The presence of the prefix "h-" is meant to evoke the fact that the name refers to a notion in hyperbolic Geometry and not to one in Euclidean Geometry that provides the frame of the construction we are achieving.

Definition 5.1. *We call h-transformation a composition of a finite number of h-symmetries.*

Theorem 5.1. *The set H of the h-transformations and the composition constitutes a group. In H there also exist restrictions to S of the translations of direction [a] as well as of the homotheties whose centre is on the absolute a.*

Proof. H is a set of the family $B(S)$ constituted of the bijections of the set S, obviously stable related to the composition. The identical transformation $I = 1_s$ can be presented under the form $I = T_d \circ T_d$ or $I = T_C \circ T_C$ since both the symmetries related to a straight line and the inversions are involutive transformations. The inverse of the h-movement $T = T_\omega \circ ... \circ T_\beta \circ T_\alpha$ is obviously the h-transformations $T^{-1} = T_\alpha \circ T_\beta \circ ... \circ T_\omega$. (Here and from now on we shall be using Greek indexes not precise the "type" of an h-symmetry.) According to a result in Chapter V, §5, the compositions $T_d \circ T_c$ represent translations. Through direct calculus one can notice that

$I(0, k) \circ I(0, h) = H(0, \frac{k}{h})$, q.e.d.

Definition 5.2. *We call plane hyperbolic Geometry the Geometry of the Klein space* $[S, H]$.

Unlike the Kleinean Geometries studied in paragraphs 2 and 3, now there are no analytical expressions of h-transformations and a *synthetical* study of the h-transformations is necessary. We shall direct to study of the Geometry of Klein space (S, H) towards the defining certain interpretations for the primary notions and relations in Euclidean Geometry and the checking of the axioms in Hilbert's system. In paragraphs 3 and 8 of the preceding chapter we found out that the set made out of straight lines and circles is stable related to symmetries in relation with straight lines and inversions. This remark suggests to us the following definition:

Definition 5.3. *We call h-straight line the intersection with S of:*
1. *a straight line perpendicular on the absolute* a;
2. *a circle whose centre is on the absolute* a;

To also precise the "shape" of a straight line (related to the Euclidean space) we shall replace the prefix "h-" by "hs-" in the first case and by "hg-" in the second case; the letters s and g are abbreviations of the "special" and "generic", respectively; this names will sometimes accompany the term of h-straight line. The remark preceding the definition leads to the following strictly necessary specification.

Theorem 5.2. *The notion of h-straight line is made invariant by h-transformations.*

We shall denote by D the set of h-straight lines in (S, H). One can now understand effectively the basic idea of making the figures 1-12 which illustrated the first part of Chapter V.

Theorem 5.3. *For the triplet* $(S, D, \in)$ *the incidence axioms* I_1, I_2, I_3 referreing to the plane Geometry in Hilbert's axiomatic system are satisfied.

We shall argue only the validity of axiom I_1. Let A, B be h-points. If $A = B$, an h-straight line through A is, for example, provided by the perpendicular from A on the absolute a. If $A \neq B$, let m be the (Euclidean) mediatrix of the segment AB. If m cuts a in a point O, the circle $O(O, OA)$ intersects S along an h-straight line through

A, B. If $m \parallel a$, it results that the (Euclidean) straight line AB is perpendicular on a, thus intersecting S along an h-straight line.

We shall denote by $(AB)_h$ the only h-straight line passing through the distinct points A, B. If $(AB)_h$ is an hg-straight line, it can be obtained by interesting S with an circle $O_{A,B}$; we shall refer to the points U, V where the circle $O_{A,B}$ intersects the absolute a by naming them limit points of $(AB)_h$. It is obvious that U, V are not h-points. If $(AB)_h$ is an hs-straight line, we can evidence only a limit point $Z \in (AB) \cap s$.

Definition 5.4. For distinct h-points A, B, we consider the set $\text{Int}_h\, AB$ which we define as follows:

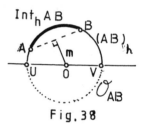

Fig. 38

- if $(AB)_h$ is an hs-straight line, then $\text{Int}_h\, AB = |AB| = \{X : A - X - B\}$;
- if $(AB)_h$ is an hg-straight line, then

$\text{Int}_h\, AB$ is that arc AB of Euclidean circle $O_{A,B}$ completely contained into S. For h points X satisfying $X \in \text{Int}_h\, AB$ we say that they are *interior to the h-segment AB* and note it which $A_h - X_h - B$ (it is read that X is h-between A and B). The set $\text{Int}_h\, AB$ will be called *the hyperbolic interior* of $\{A, B\}$.

Theorem 5.4. *The ternary relation of "to be -between" is made invariant by h-transformations, that is if A', B', X' are the images of A, B, X in an h-movement T, then $A_h - X_h - B$ and $A'_h - X'_h - B'$ hold simultaneously.*

The proof is simple since it is enough to find out the validity of the conclusion in the particular cases $T = T_d$ and $T = T_C$. For the second case we recommend Theorem 8.4 and Figure 32.

Theorem 5.5. *For the quartern $(S, D, \in, ?_h - ?_h - ?)$, group II of Hilbert's axioms is also verified.*

Proof. Propositions II_1, II_2, II_3 are immediate consequences of Definition 5.4 and of certain elementary remarks of Euclidean Geometry. To check axiom II_4 of Pasch, we recommend that one should previously find out that axiom B_7 of Birkhoff, equivalent to II_4, according to Theorem 2.6 in Chapter VI, were satisfied. The concept of "h-semi-plane delimited by $(AB)_h$" admits a natural interpretation:

- if $(AB)_h$ is an hs-straight line, then (AB) divides E_2 into two Euclidean semi-planes intersecting S along the two h-semi-planes;

- if $(AB)_h$ is an hg-straight line, the circle $O_{A,B}$ divides E_2 into two sets: the inside and the outside of the circle. The intersections with S of these are the h-semi-planes delimited by $(AB)_h$.

As a result of this theorem, into the hyperbolic Geometry can be introduced the notion of h-rays whose interpretations on the given model are:

- if $(AB)_h$ is an hs-straight line whose limit point is Z and $Z-A-B$, then the open h-rays of origin A situated on $(AB)_h$ are $|AZ|$ and $(A; B)$;

- if A is on an hg-straight line d whose limit points are U and V, then the open h-rays of origin A situated on d are the (Euclidean) arcs of circle $\widehat{AU}$ and $\widehat{AV}$ included into S.

We shall denote the open h-ray of origin O containing A, analogously as in Euclidean Geometry, by $(O; A)_h$. The set $(O; A)_h \cup \{O\}$ will be denoted by $[O; A)_h$ and will be called *(closed) h-ray*.

Lemma 5.1. *For any h-rays* $\beta = [B; M)$ *and* $\gamma = [C; N)$ *and any h-semi-planes* B, C *delimited by* $(BM)_h$ *and* $(CN)_h$, *respectively, there exists an h-movement* T *so that* $T(\beta) = \gamma$ *and* $T(B) = C$.

Proof. For the beginning we shall approach the problem in a particular case when instead of the h-ray γ there is an h ray $\delta = [D; P)$ figured by an inside of Euclidean segment: $\delta = |DZ|$. The role of the h-semi-plane C will be overtaken by $D = S \cap (DZ; B)$ (fig. 39).

Let us initially suppose that $(BM)_h$ is an hg-straight line with limit points U, V and $\beta = \widehat{BV}$. There exists a point E on $(U; B)$ so that (DE) should be parallel to the absolute a. Let X be the limit point of the hs-straight line containing E. Let us consider the h-movement T' which is restriction to S of the inversion of pole U and power $p = \overline{UB} \cdot \overline{UE}$. It is easy to find out that the

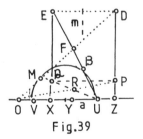

Fig.39

image through T' of the h-ray β is $|EX|$ and the h-semi-plane B', outside the circle $0_{B,V}$, is changed through T into the h semi-plane delimited by EX which contains U. If $B = B'$, the h-transformation T_1 which leads (β, B) into (δ, D) will be $T_1 = T'' \circ T'$, where T'' is (in our case, when (DZ) does not separate the points B, U) the restriction to S of the symmetry related to the mediatrix m of the (Euclidean) segment DE. If $B \neq B'$ (or B, U are in distinct semi-planes related to the straight line $\overrightarrow{DZ}$), instead of the transformation T'' we shall take the translation of vector $\overrightarrow{ED}$. For the latter case, if the power of the inversion T_1 had been $p' = \overrightarrow{UB} \cdot \overrightarrow{UF} \neq p$, we could have replaced the translation T'' by a homothety of centre $O \in a \cap (DF)$ and ratio $k = OF : OD$.

   If the hypothesis that $(BM)_h$ where a hg-straight line had not been achieved we could have reduced the problem to the analyzed case by first operating an inversion T_0 on (β, B) and taking in the end $T_1 = T'' \circ T' \circ T_0$.

   By fixing now the auxiliary element (δ, D) we shall determine, as in the preceding stage, an h-movement T_2 which should transform (γ, C) into (δ, D); the transformation T in the enunciation will be given by $T = (T_2)^{-1} \circ T_1$.

   Remark. The transformation T in the enunciation of the lemma is not unique: in diverse particular cases we can determine such an h-transformation composing less h symmetries, but we have here been interested only in *the existence* of such *h-transformations*.

   Definition 5.5. *Let A, B, C, D be four h-points not necessarily distinct; we shall say that the h-segments AB and CD are h congruent if there is an h-transformation T which applied $\{A, B\}$ into $\{C, D\}$, therefore: $(T(A) = C$ and $T(B) = D)$ or $(T(A) = D$ and $T(B) = C)$ should hold. We shall use in this case the notation $AB \underset{h}{=} CD$; a supplementary complication is not necessary because the presence of the index h evokes not only the "hyperbolic Geometry", but also distinguishes the binary relation $\underset{h}{=}$ from the equality $=$ of the sets.*

   Theorem 5.6. *For the introduced notions of hyperbolic Geometry axioms III_1 and III_2 are satisfied.*

Proof. For axiom III_1 we consider h-points A, B and h-ray $\alpha' = (A'; N)_h$. According to Theorem 5.1 there will exist an h-transformation T which should apply the h-ray $\alpha = (A; B)_h$ on α'. (We are not interested in the action of T on the semi-planes.) By noting $T(B) = B'$ we immediately deduce $B' \in \alpha'$ and $AB \underset{h}{=} A'B'$. Axiom III_2 is an immediate consequence of Definition 5.5. The theorem is thus completely proved.

Further on we shall consider as being known the real number $e = \lim(1 + \frac{1}{n})^n$ as well as the "natural logarithm" (or logarithm in base e) function $\ell n: \mathbb{R}_+ \to \mathbb{R}$, where $\ell n\, x = y \Leftrightarrow y = e^x$. After having done these preparations, we shall construct a function $d_h: S \times S \to \mathbb{R}_+ \bigcup \{0\}$ according to the following three conditions:

1. For any h-point A there holds:

(5.1) $\quad d_h(A, A) = 0$.

2. If $(AB)_h$ is an h-straight line whose limit point is Z, then:

(5.2) $\quad d_r(A, B) = |\,\ell n \dfrac{AZ}{BZ}\,|$.

3. If $(AB)_h$ is an hg-straight whose limit point are U, V, then

(5.3) $\quad d_h(A, B) = |\,\ell n(\dfrac{UA}{UB} \cdot \dfrac{VA}{VB})\,|$.

Theorem 5.7. *If T is an h-transformation and $T(A) = C$, $T(B) = D$, then* $d_h(C, D) = d_h(A, B)$.

The proof will successively approach the particular cases when T is an h-symmetry T_d and T_C, respectively. Taking into account the structure of group H, these two particular cases are sufficient.

For $T = T_d$, denoting by W the symmetric related to the straight line d of the limit point Z, respectively X, Y the symmetric related to the straight line d of U, V, we easily find out: $A = B \Leftrightarrow C = D$, $\frac{CW}{BW} = \frac{AJ}{BJ}$ (in case 2) and $\frac{XC}{XD} : \frac{YC}{YD} = \frac{UA}{UB} : \frac{VA}{VB}$ (for case 3), therefore $d_h(C, D) = d_h(A, B)$.

Let $T = T_C$ be the restriction to S of the inversion $I(O, p)$. If $(AB)_h$ is an hs-straight line, $(CD)_h$ is, according to Theorem 8.3, Chapter VIA, an hg-straight line with limit points O and $V = I(O, p)(Z)$. The equalities $AZ = \frac{p \cdot CV}{OC \cdot CV}$ and $BZ = \frac{p \cdot DV}{OD \cdot OV}$ ensures $\frac{AZ}{BZ} = \frac{CV}{DV} : \frac{OC}{OD}$, that is $d_h(A, B) = d_h(C, D)$ hold in this case.

If $(AB)_h$ is an hg-straight line of limit points U, V, let us suppose that $U \neq O \neq V$ (otherwise we would have to reverse the above calculus). In this case, the limit points of $(CD)_h$ are X, Y, the image through $I(O, p)$ of U and V, respectively. (See Theorems 8.4 and 8.2 in Chapter VIA.) We find out

$$AU = \frac{p \cdot CX}{OC \cdot OX}, \; BU = \frac{p \cdot DX}{OD \cdot OX}, \; AV = \frac{p \cdot CY}{OC \cdot OY}, \; BV = \frac{p \cdot DY}{OD \cdot OY}$$

and it immediately follows $d_h(A, B) = d_h(C, D)$, q.e.d.

Theorem 5.8. The necessaria and sufficient condition for $AB \underset{h}{=} CD$ is $d_h(A, B) = d_h(C, D)$.

The necessity results directly from Definition 5.5 and Theorem 5.7 after the simple study of the "symmetry" of the function d_h, that is:

(5.4) $\forall A, B \in S: d_h(A, B) = d_h(B, A)$.

For sufficiency, in the hypothesis $0 \neq d_h(A, B) = d_h(C, D)$ we shall consider an h-transformation T that applies $[A; B)_h$ in $[C; D)_h$ as in Lemma 5.1. We are not restraining the generality, supposing $[C; D)_h = |CZ|$ because by acting on the initially given h-points A, B, C, D with an h-transformation we do not alter the hypothesis. Of course, $T(A) = C$ holds and let be $E = T(B)$. It immediately follows $d_h(C, D) = d_h(A, B) = d_h(C, E)$. By means of the specifications we did the equality obtained is made explicit in $\frac{CZ}{DZ} = \frac{CZ}{EZ}$, from where $E = D$, so $D = T(C)$, that is $AB = CD$. q.e.d.

By adding some small elements to the above proof, we deduce:

Theorem 5.9. *For any non-negative real number x and the h-ray $[A; B)_h$, there exists a unique point X on this h-ray so that $d_h(A, X) = x$. If $x = 0$, then $X = A$.*

Theorem 5.10. *If $A \underset{h}{-} B \underset{h}{-} C$, then $d_h(A, C) = d_h(A, B) + d_h(B, C)$.*

Proof. According to the preceding theorem and to Lemma 5.2 we do not restrain the generality supposing $(AB)_h$ an hs-straight line of limit point Z so that $A-B-C$ and $B-C-Z$. It follows that $d_h(A, B)+d_h(B, C) = \ell n\frac{AZ}{BZ}+\ell n\frac{BZ}{CZ} = \ell n\frac{AZ}{CZ} = d_h(A, C)$, q.e.d.

As an immediate consequence we deduce:

Theorem 5.11. *Within the hyperbolic Geometry axiom* III_3 *in Hilbert's axiomatic system is valid.*

As in Euclidean Geometry, we call h-angle a pair of h-rays having the same origin. For such a pair $\{[O; A)_h, [O; B)_h\}$, we shall use the notation $\underset{h}{\sphericalangle}AOB$. The side $[O; A)_h$ of an h-angle is a curve in the Euclidean plane (ray, inside of segment of straight line or arc of circle); we can always associate to it the Euclidean ray of origin O, "tangent" to $[O; A)_h$. Associating to both sides of an h-angle $\underset{h}{\sphericalangle}AOB$ Euclidean rays $[O; X)$, $[O; Y)$, we attach thus to the h-angle $\underset{h}{\sphericalangle}AOB$ an

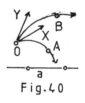

Euclidean angle $\widehat{XOY}$ to which we shall refer the notation

Fig. 40

$\widehat{XOY} = [\underset{h}{\sphericalangle}AOB]$. For the measure in radians of the Euclidean angle XOY we shall use the notation $(\underset{h}{\sphericalangle}AOB)$.

This convention regarding the notations is compatible with the Geometry of (concrete) Klein space (S, H) because the h-transformations preserve the angle of two curves. (See Theorems 3.1, 2.3 and 8.5 in Chapter VIA.) We define now in a natural way the congruence relation of the h-angles for which we shall use the same symbol $\underset{h}{=}$ as for the congruence of the h-segment:

$$\underset{h}{\sphericalangle}AOB \underset{h}{=} \underset{h}{\sphericalangle}A'O'B' \Leftrightarrow (\underset{h}{\sphericalangle}AOB) = (\underset{h}{\sphericalangle}A'O'B').$$

Theorem 5.12. *For the structure* $(S, D, \varepsilon, \underset{h}{?-?-?}, \underset{h}{=})$ *the axioms in groups I-IV of Hilbert and a negation of axiom V are checked.*

Proof. The axioms preceding axiom III_4 in Hilbert's system were checked by

Theorems 5.3, 5.5, 5.6 and 5.11.

To check axiom III_4 let us suppose now as given $\sphericalangle AOB$, $[O'; C)_h$ and an h-semi-plane C delimited by $(O'C)_h$. There are two distinct variants to assure ourselves about the existence of a unique h-ray $[O'; D)_h \subset C$ so that $\sphericalangle AOB = \sphericalangle CO'D$.

In the first variant we consider $\widehat{XOY} = [\sphericalangle AOB]$ and the Euclidean ray $[O'; U)$ tangent to $[O'; C)_h$. A semi-plane C' delimited by the (Euclidean) straight line $O'U$ containing the tangents to the h-rays through O' situated in C is evidenced and a unique ray $(O'; Y)$ included into C' so that $\widehat{XOY} \equiv \widehat{UO'V}$ is determined. If $O'V$ is perpendicular on the absolute a, then $[O'; Y) \cap S$ is the h-ray $[O'; D)_h$ we were looking for. Otherwise there exists a unique point Z on the absolute a so that $ZO' \perp O'V$; the circle of centre Z passing through O' intersects C along the h-ray we were looking for. (The proof is given in more details in [12], pp. 79-80).

In the present context we prefer the following variant of proving the existence of axiom III_4. Let B be the h-semi-plane delimited by $(OA)_h$ which contains B. According to Lemma 5.1 there exists an h-transformation which applies $[O; A)_h$ on $[O'; C)_h$ and B on C. Let be $[O'; D)_h = T([O; B)_h)$. It obviously holds $\sphericalangle AOB = \sphericalangle CO'D$. (For uniqueness we may appeal to the initial variant.) By means of this second proof variant we can immediately establish that the definition given for the congruence of h-angles is equivalent with *the existence of an h-transformation which applied the h-sides of the first h-angle on the h-sides of the second one.*

We establish the validity of axiom III_5. Let (A, B, C) and (A', B', C') be triplets of h-points which are not h-co-linear so that $AB = A'B'$, $AC = A'C'$ and $\sphericalangle BAC = \sphericalangle B'A'C'$. From $AB = A'B'$ we deduce the existence of an h-transformation T applying A, B in A', B'. Let be $T(C) = D$. We can suppose without restraining the generality that D is in that h-semi-plane delimited by $(A'B')_h$ in which C' is because, otherwise, we should replace T by $T' \circ T$, where T' is the h-symmetry related to $(A'B')_h$. We deduce that T applies $\sphericalangle BAC$ on $\sphericalangle B'A'D$ and, based on the above given

proof of axiom III$_4$, we conclude that $[A'$; $D)_h$ = $[A'$; $C')_h$. But there is $d_h(A', D) = d_h(A, C) = d_h(A', C')$; by Theorem 5.9 it results that $D = C'$. Let us notice now that the image through the h-transformation T of the h-ray $[B$; $C)_h$ is the h-ray $[B'$; $C')_h$. It immediately follows that $< ABC = < A'B'C'$, that is just the conclusion of $\quad\quad h\quad\quad h\ h$
axiom III$_5$.

We are giving here less importance to the axioms of continuity since Hilbert proved, [229], that they are consequences of the axioms in groups I, II, III, to which an axiom $\neg V$, (with a more restrictive formulation) is added.
Let us consider an hg-straight line b figured on the model by a semi-circle centred in $W \in a$ and limited in U, V by the absolute a. Let c be the (Euclidean) parallel (in Euclidean meaning) to a, and O the contact point of b and c. We shall define a function $f: b \rightarrow c$ by associating to the h-point $X \in b$ the point $Y = f(X) \in c$, where $Y \in c \cap (Z; Y)$. The function f is obviously bijective and compatible with the relations "to be between":

$$A_h - B_h - C \Leftrightarrow f(A) - f(B) - f(C).$$

By the existence of this equivalence it obviously appears that *Cantor's axiom*, IV$_2$, is simultaneously satisfied on the h-straight line and on the straight line c, its being satisfied on the Euclidean straight line c being directly ensured.

Using the notation $\overset{\frown}{VWX} = 2u$, it is easy to calculate:

$$d_h(O, X) = \left| \ln \frac{OU}{OV} : \frac{XU}{XV} \right| = \left| \ln \frac{XV}{XU} \right| = |\ln \cot u|.$$

We are constructing a new function $g : b \rightarrow [O ; T)$, where T and V in the same semi-plane delimited by the straight line OW, putting $g(X) = Z$ if $[W ; Z)$, is the inner bisector of the angle VWX. Taking the ray ZO as unit of measure of the (Euclidean) segments we obtain:

$$OZ = \cot u, \text{ hence } d_h(O, X) = [\ln OZ].$$

The established equality reduces the transfer of the validity of *Archimedes' axiom* from the Euclidean straight line c to the generic h-straight line b to a simple algebraic calculus.

We are at this moment assured that Hilbert's axioms in groups I-IV are satisfied in (S, H), as well as all their consequences that make up the absolute Geometry. It is particularly ensured, according to Theorem 6.1 in Chapter V, the existence of a canonical parallel b through A (non-situated on a to a. Taking into account that S is "plane" and not a three-dimensional space it is not necessary anylonger to specify that b is in the plane defined by A and a). Therefore, the negation of axiom V will be a proposition easy to precise, and we shall formulate it directly in terms specific to the model: $\neg.V$. There exist an h-point A and an h-straight line d non-incident with A since there exist through A at least two distinct h-straight lines parallel to d.

Indeed, for the generic h-straight line and an arbitrary h-point A there will exist

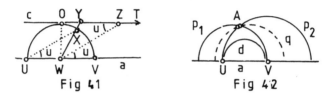

Fig 41   Fig 42

distinct h-straight lines, q, which passed through A and did not cut d. In Figure 42 there appear such "limit" h-straight lines p_1, p_2, too, each of them having a common limit point with d; any h-straight line situated "between" p_1 and p_2 will have the mentioned property. We thus find out that not axiom V, but its negation, $\neg.V$, is satisfied in hyperbolic Geometry.

As a consequence of Theorem 5.12, we remark

Theorem 5.13. *The axiom of parallels, V, is independent of the other axioms in Hilbert's system.*

Indeed, we have constructed in this paragraph a model of *plane* Geometry in which the only non-checked axiom is V. A spatial model would have been constructed in an analogous manner taking S as a semi-space delimited by a plane α, the h-planes being the intersections with S of the spheres or of the planes perpendicular on α; the h-straight lines are intersections of h-planes, and the remaining notions (and proofs) are kept within this ampler context.

Outlining the plane hyperbolic Geometry there appeared the possibility of recharacterizing and re-defining the h-transformations T in H in an "intrinsic" way (using only the Geometry constructed on the adjacent set S). Taking into account Definition 5.5 and Theorem 5.8 we find out that H is the group of the isometries for the metric space (S, d_h).

Let c be an hg-straight line obtained by intersecting S with a circle C of centre O (Fig. 43). Let be $M' = T_C(M)$, where M is an arbitrary h-point. Using Theorem 8.2 in Chapter VII, we easily deduce that any circle through M, M' is orthogonal to C, so the h-straight line (MM') is perpendicular on c. If we note the common point of these two h-straight lines with N, it is obvious that $T(N) = N$ holds, and from here $MN \underset{h}{=} M'N$. We can then characterize the h-transformation T_C as symmetry in the metric space (S, d_h) related to the "straight line" c, thus justifying the name (given at the beginning of this paragraph) of h-symmetry. An analogous result is also obtained for the h-symmetries T_d related to hs-straight lines. This re-interpretation of the h-symmetries enables us to re-actualize here Theorem 3.6 in the preceding chapter and present the specification offered by:

Theorem 5.14. *Any h-transformation T in H is a composition of at most three h-symmetries.*

This theorem is very important because it enables the re-establishing of a general analytical formula of h-movement, formula which does not have any interest for us here.

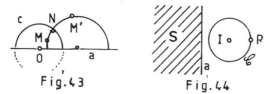

Fig. 43   Fig. 44

To conclude this paragraph, we shall also present two more concrete Klein spaces (I, H'), (I, H'') from the abstract Klein space for hyperbolic Geometry.

Let us consider the Euclidean plane $E_2 \supset S$ and a point $P \in E_2 - S$. Let J be an inversion of the pole P. The image of the absolute a through J is a circle C out of which point P would have to be excluded for the moment. The semi-plane S is applied by J on the interior I of the circle C. The inversion J admits then a restriction (which we shall denote identically) that establishes a bijection from S to I. We shall associate to an arbitrary h-transformation $T \in H$ a transformation $J(T) = J \circ T \circ J$ of I an itself and thus obtain an isomorphism J from the group H to a group we shall denote by H'. We obtained then a concrete Klein space (I, H') equivalent with (S, H). From the historic point of view the model (I, H') has preceded (S, H). For (I, H') we shall use the name of *Poincaré model of the plane hyperbolic Geometry*. Taking into account that (S, H) is obtained from (I, H') in a simple, elementary way, many authors use the name of Poincaré model for (S, H), too; but we shall prefer to say that (S, H) is the *Beltrami-Enneper model of the plane hyperbolic Geometry*.

The role of the h-straight lines d will be overtaken in (I, H') by sets $J(c)$, where c is h-straight line in (S, H). It is easy to find out that the straight lines d will now appear as arcs of circle orthogonal tot the absolute circle, C (and limited by C or by diameters of the circle C).

It is quite easy to find out that group H' is generate by the inversion T'_o related to circles O orthogonal to C and by symmetries T'_d related to straight lines d passing through the centre C of C (obviously, restrained to I).

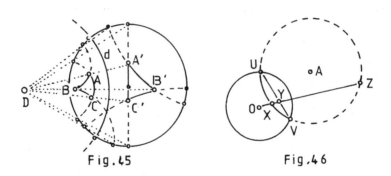

Fig. 45   Fig. 46

We shall introduce a new distance function $\delta: I^2 \to \mathbb{R}_+ \cup \{0\}$ analogous to function d_h, where $\delta(A, B) = d_h(J(A), J(B))$ If we denote the new limit points of the h-straight line AB with U, V, we easily obtain $\delta(A, B) = \left| \ell n \left(\frac{AU}{AV} : \frac{BU}{BV} \right) \right|$, too. Figure 45 illustrates a generic transformation of the model (I, H') which can be called h-symmetry related to the h-straight line d or restriction to I of the inversion T of pole D and power $p = (D)$. Taking into account that this model is "conformal" (that is the angle formed by the curve have equal measures with those of the Euclidean ones) the congruent relations are introduced as in (S, H). In figure 45 also appear "congruent triangles" ABC and $A'B'C'$.

Theorem 5.15. For any circle $C(O, r)$ having the interior denoted by I, there exists a bijection $F : I \rightarrow I$ which transform any arc of circle UV orthogonal to C and situated in I into the "chord" $|UV|$.

Proof. Let $A(A, R)$ be a circle orthogonal to C and let UV be their common chord. For an arbitrary point X on $UV = A \cap I$ we consider the points Y, Z in which $(O ; X)$ cuts (UV) and cuts again A, respectively. Points U and V are obviously the contact points of the tangents from O to A, so (UV) is the polar of O related to A. It follows that Y is harmonically conjugated to O related to XZ, so $\overline{OY} \cdot (\overline{OX} + \overline{OZ}) = 2\overline{OX} \cdot \overline{OZ}$. The power of the point O related to A is $A(O) = OV^2 = r^2 = \overline{OX} \cdot \overline{OZ}$. By eliminating OZ, we easily deduce:

$$(5.5) \qquad \overline{OY} \cdot (r^2 + OX^2) = 2r^2 \cdot \overline{OX}.$$

But this formula defines the point $Y = f(X)$ independently of the circle A under consideration. Obviously, $f(O) = O$. To find out that f is a bijection it will be sufficient to show that the restriction of f to an arbitrary "ray" $|OM|$ (with $M \in C$) is a bijection, that is that function $g : (O, r) \rightarrow (O, r)$ defined by $g(x) = \dfrac{2r^2 x}{r^2 + x^2}$ is bijective. An elementary modality of satisfying this desideratum consists in proving the fact that function $h : (O, r) \rightarrow (O, r)$ defined by $h(y) = \dfrac{r}{y}\left(r - \sqrt{r^2 - y^2}\right)$ is the inverse of g.

We shall transform Poincaré's model (I, H') by the bijection f, that is we shall replace group H' by H'', the elements of the latter group being transformations of the type $T' = f \circ T \circ f^{-1}$, where T is an arbitrary transformation in H'. An analytical calculus (or a synthetical one) long enough but lacking in principle difficulties enables us to establish that group H'' coincides (in the particular case $r = 1$) with the group H which we had considered in §3, retaining from among the transformations of the projective group G only the transformations that preserves C. We thus deduce that, leaving aside the notation, the Klein space (I, H'') coincides with the Cayley space (C, H'').

As in §3 we were not interested to introduce a distance in the Cayley space (C, H'') let us do it now.

For any distinct points A, B interior to the conical Γ the straight line (AB)

intersects Γ in exactly two "limit points" U, V. A distance $c: C \times C \to \mathbf{R}_+ \cup \{0\}$ is given by $c(A, B) = |\ell n (A, B; U, V)|$ and there are no difficulties to prove their independence related to the group H'' of motions.

Otherwise, the distance c above introduced is the same with that induced by δ through the Kleinean imbedding f.

By interpreting in (C, H'') the hyperbolic Geometry it results:

Theorem 5.16. *The plane hyperbolic geometry is subordinated to the projective one.*

§6. Complements of Absolute Geometry

All the considerations in this paragraph could have been exposed before paragraph 6 in Chapter V. But there are two principal reasons which determined the placing here of this paragraph.

The first reason comes from admitting the role of intuition in approaching elementary Geometry (even when this approach is based on a rigorously axiomatic system). We preferred thus to ensure the possibility of a double modality of intuiting the abstract results in this paragraph: the traditional modality and the one edified on the preceding paragraph.

The second reason is given by the fact that the results in this paragraph are paving the way for the considerations in the paragraph to immediately follow.

We have already stated that the term of *absolute Geometry* wad created by Janos Bolyai. His brilliant book, "Appendix", written and published in Transylvania constitutes one of the pillars of hyperbolic Geometry. Janos Bolyai shares the glory of having created the new Geometry with N.I. Lobatchevski (who had published his book concerning the "pan-Geometry" six years before) and with the great mathematician K.F. Gauss, who had ever since 1818 reached profound conclusions on the new Geometry, but had not considered as opportune to publish his results. We have evoked here the names of these fathers of hyperbolic Geometry, independent discoverers of a new "geometric world" because the expositions in this paragraph (and in the following ones) will present in turn ideas of Gauss, Lobatchevski and Bolyai.

Definition 6.1. *Let α, β be rays of origins A, B; we shall say that α is super-parallel with β and note $\alpha \uparrow \beta$ if:*
   1. they have the same support straight line and orientation, or
   2. are disjoint and any ray interior to the angle $\{[A; B), \alpha\}$ cuts β.

Remark. The coincidence of symbol ↑ of the super-parallelism relation with the symbol used in Chapter V, §3 for identically orientated rays is not completely accidental.

Theorem 6.1. *If A is not on the support straight line of the ray β, there is unique ray α through A so that α ↑ β.*

Proof. Let C be so that $B - A - C$ and $D \in \beta$. According to Pasch's axiom, in $\triangle BCD$, a straight line through A which does not cut (BD) will cut $|CD|$. We shall deal only with rays $[A; N)$ with $N \in |CD|$ without restraining thus the generality. Let f be a system of co-ordinates on (CD) so that $f(C) = 0$, $f(D) = 1$ (see Theorem 5.1, Chapter V). Let us denote by K the set of positive real numbers x having the property that the ray of origin A containing $f^{-1}(x)$ does not cut β. K is bordered in the upper part by 1. There exists the real number $m = \sup K$. Let be $M = f^{-1}(m)$ and $\alpha = [A; M)$.

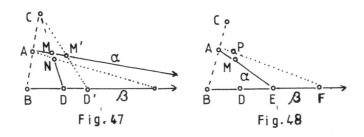

Fig. 47   Fig. 48

Let us suppose by absurd that α and β have a common point E (Fig. 48). Let F be so that $B - E - F$, $P \in (AF) \cap |CD|$ and $p = f(P)$. It is obvious that $p < m$. For any $x \in (p, m)$, the ray of origin A through $f^{-1}(x)$ will cut β (between E and F), so $x \notin K$, contrary to the supremum quality of m.

If γ is a ray inside the angle formed by α with $[A; B)$, it will cut $|BM|$ and, according to Pasch's axiom, it will also cut $|BD|$ or $|MD|$. The second variant keeps its interest. Let be $N \in \gamma \cap |MD|$ and $n = f(N)$. We deduce $n > m$ and, as in the preceding stage, we shall come to contradict the supremum quality of m.

For uniqueness, it is easy to find out that any super-parallel γ to β will contained the above considered point M.

Theorem 6.2. *If α ↑ β and γ ⊂ α or α ⊂ γ then γ ↑ β.*

Proof. Let be $\gamma \subset \alpha$ whose origin is C.

If, by absurd, $\gamma \uparrow \beta$ does not hold, it will exist a ray $\delta = [C; D)$ so that $\delta \uparrow \beta$. We shall

consider the inside of the angle formed by
$[A; B)$ and α . From $\alpha \cap \beta = \varnothing$ it easily
results $D \in (\alpha; B)$. As α is super-parallel with β
it follows that $[A; D)$ cuts β is a point E. As $|AB|$
can not be cut by (CD) it follows, according to
Pasch's axiom in $\triangle ABE$, that (CD) would cut
$|BE| \subset \beta$ in a point P, for which we establish
by means of the same axiom of Pasch (applied to
the triangle CPB) that C-D-P holds, so
$P \in \delta$, which is absurd.

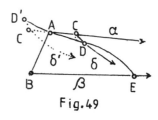

Fig. 49

The proof is practically the same when $\alpha \subset \gamma$. Now we should contradict $\gamma \uparrow \beta$
by the existence of a super-parallel δ' having the same origin C' as γ has and should
take D' on the "prolongation" of δ' etc.

By virtue of Definition 6.1 the following implication is obvious:

$$(6.1) \qquad \alpha \uparrow \beta \wedge (\delta \subset \beta \vee \beta \subset \delta) \Rightarrow \alpha \uparrow \delta.$$

Further on, we shall refer to oriented lines $(a, [\alpha])$, but we admit from the
notations the symbol $[\alpha]$ of the class of rays included into a, equivalent with α (the
equivalence $\sim$ was introduced in Chapter V, §3). We shall leave to the context the
specifications of the orientation whenever it is necessary; if the orientation is not specified
by the context it will be arbitrary but will be the same if the subjacent straight line appears
several times in the same context. We shall denote by $\vec{\mathscr{L}}$ the set of the oriented straight
lines.

Definition 6.2. *Let a and b be two oriented straight lines: we say that a is
super-parallel with b and note with $a \uparrow b$ if there exist rays α, β included into a and b,
respectively, positively oriented, so that $\alpha \uparrow \beta$. According to Theorem 6.1 and to
implication (6.1), if we replace α by α' and β by β', $\alpha' \uparrow \beta'$ also holds.*

Remarks. 1. From $a = b$, it follows $a \uparrow b$.

2. From $a \uparrow b$ and $a \neq b$, it follows, according to Theorem 6.2 and to
implication (6.1) that $a \cap b = \varnothing$ and therefore each of the straight lines a, b is included
into one of the semi-planes determined by the other one.

Usually, in the literature consecrated to hyperbolic Geometry a terminology distinct from the one suggested here is adopted. In [21], vol. III, p. 235, [335], p. 36, [521], p. 135, the term of parallels is used instead of that of super-parallels, and the non-secant (co-planar) straight lines are called "divergent". Bolyai chose the term of *asymptote* and used the notation ‖‖ instead of ↑. (see [47], p. 81); with Lobatchevski, [306] and in [29], p. 293, *the term limit straight line* is used, and in [347], p. 301, the quite analogous expression of *critical parallel* is introduced.

Theorem 6.3. *If $a \uparrow b$ and $a \neq b$, there is an equally slant secant related to (a, b), that is a straight line c secant to a, b in points A, B so that $\widehat{BAX} = \widehat{ABY}$ should hold for $A \prec X$ and $B \prec Y$.*

Proof. Let be $M \in a$, $N \in b$ and a point X on a so that $M \prec X$. The bisector of the angle NMX is inside this angle so it will cut b in a point P (of course, $N \prec P$). The interior bisector of the angle PNM cuts $|MP|$ in a point Q. Let A, B, C be the orthogonal projections of Q on a, b and (MN), respectively. From $Q \in [M; P]$ $QA \equiv QC$ results, and from $\widehat{MNQ} \equiv \widehat{QNP}$ it follows that $QC \equiv QB$. We deduce that $QA \equiv QB$, so $\widehat{QAB} \equiv \widehat{QBA}$ holds in the isosceles triangle QAB. We can suppose that we have chosen X so that $A \prec X$ and let be Y so that $B \prec Y$. We immediately find out that $\widehat{BAX} \equiv \widehat{ABY}$, so (AB) is an "equally slant" secant, q.e.d.

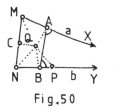
Fig. 50

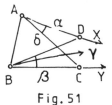
Fig. 51

Remarks. 1. The theorem does not state that the equally slant secant (AB) would be unique and this fact is not true, either; in the Corollary 1 of Theorem 7.2 the fact that each of the points A, B can be independently chosen will be proved.

2. If $a = b$, the conclusion of the above theorem, slightly reformulated, is a

common (but not false) one, the role of the equally slant secants being overtaken by the straight lines perpendicular on a.

Theorem 6.4. *The super-parallelism relation* ↑ *is symmetric.*

Proof. Let be a ↑ b. According to the preceding theorem we can evidence an equally slant secant AB (Fig. 51). Let $\alpha = [A; X)$ and $\beta = [B; Y)$ be rays which precise the orientations of the straight lines a, b that have been implied by a ↑ b. If, by absurd, β ↑ α does not hold, it will exist a ray γ of origin B so that γ ↑ α. We easily find out that $\gamma \subset \text{Int} \widehat{ABY}$. Let be a ray δ included into $(a; B)$ so that $(\widehat{\alpha, \delta}) \equiv (\widehat{\beta, \gamma})$.

One can immediately notice that $\delta \subset \text{Int} \widehat{BAX}$ and, according to the super-parallel quality of α, it will exist a point C common both to δ and β. Let us consider $D \in (A; X)$ so that $AD = BC$. It is easy to find out that $\triangle ABD \equiv \triangle BAC$ (the LUL case) and it follows that $\gamma = [B; D)$, contrary to the hypothesis γ ↑ α, etc.

Theorem 6.5. *The super-parallelism relation* ↑ *is an equivalence in the set* $\vec{\mathcal{L}}$ *of the oriented straight lines.*

Proof. The reflexivity has been noticed immediately after Definition 6.2, and the preceding theorem has ensured the symmetry. So the only one to be proved is transitivity.

Let us suppose then that a ↑ b and b ↑ c hold (Fig. 52). For the beginning, we shall also adopt the supplementary hypothesis that a, c are in distinct semi-planes delimited by b. Let $\alpha = [A; X)$ and $\gamma = [C, Z)$ be positively oriented rays included into a and c respectively, If, by absurd, α ↑ γ does not hold, it will exist, according to Theorem 6.1, a ray α' of origin A so that α' ↑ γ. Since α ↑ β held, it easily follows that there exist a point B common to b and to α'. We distinguish the ray $\beta = \{Y: B \preceq Y\}$ of the straight line b and the ray $\alpha'' \subset \alpha'$ of origin B, which is inside the angle CBY. It follows β ↑ γ and it results that α'' ↑ γ can not hold, in contradiction with the hypothesis α' ↑ γ or with Theorem 6.2. The contradiction we obtained proves that α ↑ γ holds, that is a ↑ c, what we had intended to prove.

Further on, we shall give up the supplementary hypothesis that a, c are "separated" from b and analyze the remaining case: a and c are included into the same semi-plane delimited by b (Fig. 53). We choose "positive"rays α, β, γ included into a, b, c; let A, B, C be their origins. One of the rays $[B; A)$, $[B; C)$ will form a smaller angle with β; let us suppose it is the first one. In this case it follows that $[B; A)$ is secant to the straight line c in a point C'. By eventually replacing C by C', we can then suppose points A, B, C as being co-linear and there will exist one of them to be among the other ones. Consequently, from hypotheses we have adopted, one of given straight lines a, b, c separates them from the other two ones; let this be c.

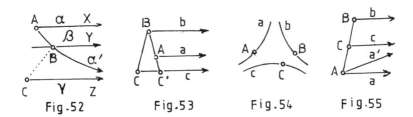

Fig.52   Fig.53   Fig.54   Fig.55

We draw attention that this stage was strictly necessary because we could not *a priori* know that the straight lines a, b, c were not disposed as in Figure 54.

Supposing now (by absurd) that $\alpha \uparrow c$ does not hold, we should find another oriented straight line a' through A so that $a' \uparrow c$. For $a' \uparrow c$ and $c \uparrow b$ where c separates a' from b, we shall deduce, as in the above stage, that $a' \uparrow b$. But by means of Theorem 6.1, it follows that $a = a'$ etc.

Definition 6.3. *Let a be an oriented straight line and a point A non-incident to it. It is considered $B \in a$ so that $AB \perp a$ and the ray $\beta \subset a$ of origin B which represents the orientation given for a. We call parallelism angle for the couple (a, A) the real number* $\sup \widehat{BAX}$. *For the beginning, we shall note this angle with $U_{a,A}$; we imply the measuring of angles in radians.*

Theorem 6.6. *If the distance of points A, A' up to the straight lines a and a', respectively, are congruent, then $U_{a,A} = U_{a,A'}$. $U_{a,A} < \frac{\pi}{2}$ holds.*

Proof. Let be, by absurd, $U_{a',A'} < U_{a,A}$. By comparing Definitions 6.1 and 6.3, we notice that a ray $[A; Y)$ situated in the semi-plane limited by (AB) which contains the ray β and satisfying $BAY = U_{a',A'}$ is super-parallel to . For a ray $[A; Z)$ included into the same semi-plane so that $\widehat{BAZ} = U_{a',A'}$, it thus follows that it admits a point Z common with a (Fig. 56). We are considering Z' on a' so that $B' \preceq Z'$ and $B'Z' \equiv BZ$. We immediately find out that $\triangle BAZ = \triangle B'A'Z'$ (the SAS case) and it follows that $\widehat{B'A'Z'} = U_{a',A'}$, which is absurd etc.

We shall particularize this result 1. for the case when $A' = A$, and the straight lines a, a' coincide as sets, but have distinct orientations. We obtain the conclusions that the parallelism angle $U_{a,A}$ is independent of the orientation of the straight line a.

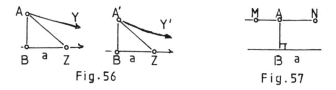

$$\text{Fig.56} \qquad\qquad \text{Fig.57}$$

According to Theorem 6.1 in Chapter V, there exists a straight line c through A parallel with a (Fig. 57). Let be points M, N on c so that $M - A - N$. In order not to contradict the supremum quality of $U_{a,A}$ it is necessary that $U_{a,A} \leq \widehat{BAN}$ should hold; when adding member by member the two inequalities, we also deduce conclusion 2., q.e.d.

As a consequence of this theorem, the parallelism angle $U_{a,A} = L(AB)$. Let us consider the set $\Sigma^* = \{(A, B) ; A \neq B\}$ of the non-null segments. We have thus introduced a function $L : \Sigma^* \to \left(0, \dfrac{\pi}{2} \right]$. It is easy to see that, if the axiom V of the parallels holds, the function L will be constant and its considerations lacking in importance. The brilliant Russian mathematician N.I. Lobatchevski was the first to grasp the importance of this function in hyperbolic Geometry; in his honour we shall use here the notation L (instead of that more frequently used in mathematical literature, $\sqcap$) and the name of Lobatchevski's function. Important properties of this function will be evidenced in this paragraph and in the following two ones.

Definition 6.4. *We call defect of the triangle ABC the real number* $\delta = \pi - (A + B + C)$, *where* A, B, C *are the values of the angles of the triangle ABC (measured in radians).*

Theorem 6.7. *The defect of a triangle is a non-negative and strictly smaller than the real number* π.

Proof. The statement $\delta(ABC) < \pi$ directly results from the definition. We suppose by absurd that there exists a triangle ABC having the negative defect $\delta(ABC) \leq -u$, where u is an angle. We may suppose that we have noted the vertexes so that $A \leq B \leq C$. Let D be the midpoint of BC and E the symmetric of A related to D (Fig. 58). The triangles DAC and DEB are congruent (the SAS case) and we easily find out that

$$\delta(ABE) = \pi - (\widehat{EAB} + \widehat{ABE} + \widehat{BEA}) = \pi - (\widehat{DAB} + B + \widehat{CBE} + \widehat{DAC})$$
$$= \pi - (A + B + C) = \delta(ABC)$$

We find out then that we can that we can replace the triangle ABC with a triangle $A_1 B_1 C_1$ having the same defect as ABC and its minimum angle A_1 satisfying the inequality $2A_1 \leq A$. (It is obvious thaz A_1, B_1, C_1 can be taken as A, B, E.) Iterating this stage we shall find a triangle $A_n B_n C_n$, where $2^n \cdot A_n \leq A$ and $\delta(A_n B_n C_n) \leq -u$. We can suppose that we have chosen the natural number n being big enough so that $A_n \leq u$, and the inequality $\delta(A_n B_n C_n) \leq -u$ is made explicit by $B_n + C_n > \pi$ which would state that the interior angle B_n is bigger that the exterior angle C_n, which is absurd etc.

Theorem 6.8. *The function* δ *satisfies the following "adivity" properties:*
1. $B - M - C \Rightarrow \delta(ABM) + \delta(AMC) = \delta(ABC)$;
2. $M \in \text{Int}_\Delta ABC \Rightarrow \delta(ABM) + \delta(BMC) + \delta(CMA) = \delta(ABC)$.

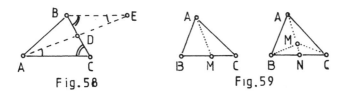

Fig.58   Fig.59

The proof of the first implication consists in making explicit in a simple way the defect of the triangles under consideration. The premise $M \in \text{Int}_\Delta ABC$ must be understood as follows: there exists $N \in |BC|$ so that $M \in |AN|$ and then, according to 1., the equalities

$$\delta(ABM) + \delta(BMC) + \delta(CMA) = \delta(ABM) + [\delta(BMN) + \delta(NMC)] + \delta(CMA) =$$
$$= [\delta(ABM) + \delta(BMN)] + [\delta(NMC) + \delta(CMA)] = \delta(ABN) + \delta(NCA) = \delta(ABC)$$

hold, q.e.d.

Theorem 6.9. *Let be* $(AB) \perp a$ *with* $B \in a$. *The following three propositions are equivalent:*

1. *there exist through* A *at least two distinct parallels to* a;
2. *there exists* $M \notin a$ *so that* $\delta(ABM) > 0$;
3. $L(AB) < \frac{\pi}{2}$.

Proof. The equivalence of propositions 1. and 3. has already been evidenced; for their equivalence with 2. we shall adapt an idea of the mathematician Legendre.

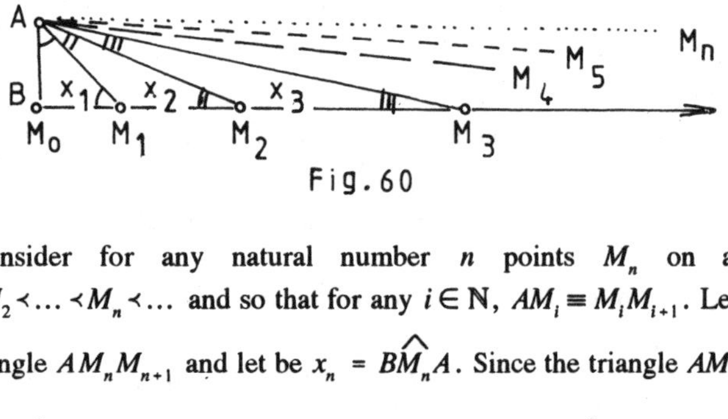

Fig. 60

We consider for any natural number n points M_n on a so that $B = M_0 \prec M_1 \prec M_2 \prec \ldots \prec M_n \prec \ldots$ and so that for any $i \in \mathbb{N}$, $AM_i \equiv M_i M_{i+1}$. Let d_n be the defect of the triangle $AM_n M_{n+1}$ and let be $x_n = \widehat{BM_n A}$. Since the triangle $AM_i M_{i+1}$ is an isosceles one, $M_i \widehat{A} M_{i+1} = x_i$ holds and, for $i \in \mathbb{N}^*$, $M_{i+1} \widehat{M_i} A = \pi = x_i$. We thus deduce:

$$(*) \quad x_i - 2x_{i+1} = d_i.$$

Totalizing these equalities for $i = 1, \ldots, n-1$, we shall obtain $(x_1 - x_n) - (x_2 + x_3 + \ldots + x_n) = d_1 + d_2 + \ldots + d_n$. Using the preceding theorem, as well, and totalizing the angles in A we offer this equality the following form:

$$(2x_1 - x_n) - B\widehat{AM}_n = \delta(AM_1 M_n).$$

Since we had noted the defect of the triangle ABM_1 with d_0, we also deduce $\frac{\pi}{2} - 2x_1 = d_0$

and the above equality becomes $\frac{\pi}{2} - x_n - B\widehat{AM} = \delta(ABM_n)$, and therefore

$(**)$ $\quad B\widehat{AM}_n = \frac{\pi}{2} - \delta(ABM_n) - x_n.$

From the relation $(*)$ we also deduce $2x_{i+1} \leq x_i$, so $0 < x_n \leq 2^{-(n+1)}\pi$ and $\lim_{n \to \infty} x_n = 0$. The sequence of real numbers $u_n = \delta(ABM_n)$ *is monotonously non-decreasing* and bordered in the upper part, so $\lim_{n \to \infty} u_n$ exists. It is easy to find out that, for n tending to the infinite, the angle BAM_n tends to the parallelism angle $L(AB)$ and we then obtain from the relation (**):

(6.2) $\quad L(AB) = \frac{\pi}{2} - \lim_{n \to \infty} \delta(ABM_n).$

On the basis of this equality, the equivalence of propositions 2. and 3. in the enunciation is immediate.

Theorem 6.10. *The axiom* $\neg V$ *is equivalent with the following propositions:*

1. *the co-domain of Lobatchevski's function L contains at least an element* $u < \frac{\pi}{2}$;

2. *there exists an right-angled triangle ABC of positive defect;*

3. *any right-angled triangle DEF has the positive defect;*

4. *any triangle MNP has the positive defect;*

5. *for any x in* $\mathbb{R}_+$, *$L(x) < \frac{\pi}{2}$;*

6. *for any point Q and straight line d so that $Q \notin d$ there exist through Q at least two parallels to the straight line d.*

Proof. The axiom $\neg V$ (see the preceding paragraph) states that there exists a point A and a straight line a (with $A \notin a$) so that *there exist through A at least two distinct parallels to the straight line a*. Let B be the foot of the perpendicular drawn down from A on a. According to the preceding theorem the final statement in $\neg V$ is equivalent with $L(AB) < \frac{\pi}{2}$ and with the existence of point $C \in a$, respectively, so that $\delta(ABC) > 0$. So $\neg V \Leftrightarrow 1.$ and $\neg V \Leftrightarrow 2..$

We prove 2. ⇒ 3. by initially considering right-angled triangles DEF so that $DE = AB$ and $EF > BC$. We consider $G \in |EF|$ so that $EG = BC$; it follows $\triangle ABC = \triangle DBG$ and $\delta(DEF) \geq \delta(DEG) \geq \delta(ABC) \geq 0$. The proof is immediately extended when instead of $DE \equiv AB$ there is $D'E > AB$ by repeating the same reasoning.

We go on proving the implication 2. ⇒ 3. keeping the notations but supposing now that $DE < AB$ (and, eventually $EF < BC$). Let us suppose by absurd that $\delta(DEF) = 0$ and construct point H so that $\widehat{FDH} \equiv \widehat{DFE}$ and $DG \equiv EF$. It follows that

$\triangle DEF \equiv \triangle FHD$ (the ASA case), so $DH \perp HF$, $HF \equiv DE$ and $\widehat{DFH} \equiv \widehat{EDF}$. The

hypothesis we adopted, $\delta(DEF) = 0$, leads to $EF \perp FH$. We shall refer to the quadrangle $DEFH$ with the provisory name of *rectangle* (such a name is correct in Euclidean Geometry, but contradictory in hyperbolic one). By laying one over the other one congruent copies of the rectangle $DEFH$ as in Figure 62 we shall obtain a rectangle D_nEFH_n, number n being chosen so that $AB \leq ED_n$ (see axiom IV_1 of Archimedes).

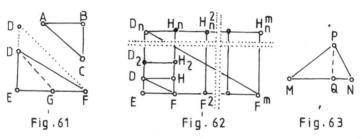

Fig.61   Fig. 62   Fig.63

In the eventuality that $EF < BC$ we shall lay laterally congruent copies of the rectangle D_nEFH_n to obtain the "wider" rectangle $D_nEF^mH_n^m$ with $BC \leq EF^m$. (This "paving" of the plane with rectangles is possible only on the basis of the hypothesis that the defect of the triangle DEF is null.) We trace D_nF^m and obtain the congruent triangles (according to the cases SAS or SSS) ED_nF^m and $H^mF^mD_n$. The sum of the angles of these two triangles is obviously 2π, so the right-angled triangle D_nEF^m has the null defect. But its catheti are respectively bigger than those of *ABC* and we proved in the preceding stage that the defect of D_nEF^m was strictly positive. The absurdity comes from the hypothesis that $\delta(DEF) = 0$, then 2. $\Rightarrow$ 3..

We are proving that 3. $\Rightarrow$ 4.. We suppose that the vertexes M, N, P of the triangle *MNP* are noted so that $M \le N \le P$. In this case M and N are acute angles and the foot Q of the height in P is on the side $|MN|$. The right-angled triangles PQM and PQN have (according to hypothesis 3.) positive defects and so we immediately get $\delta(MNP) = \delta(MQP) + \delta(PQN) > 0$. We thus proved that 3. $\Rightarrow$ 4., but the inverse implication is obvious, so 3. $\Leftrightarrow$ 4..

Let *DE* be a segment of length x and an arbitrary point F on the perpendicular in E on (DE). According to Theorem 6.9, the condition $\delta(DEF) > 0$ is equivalent with $L(DE) < \frac{\pi}{2}$, so 3. $\Leftrightarrow$ 5..

We do not restrain generality by supposing that, in Proposition 6., point Q coincides with the above considered D and $d = (EF)$. According to the same Theorem 6.9, 3. $\Leftrightarrow$ 5. $\Leftrightarrow$ 6. holds.

One can immediately notice that 6. $\Rightarrow$ $\neg V$.

Out of the above proofs we retain the implications summarized in the adjacent diagram. Obviously, the propositions "in the corners" are equivalent. We have also proved that 1. $\Leftrightarrow$ $\neg V$ and 3. $\Leftrightarrow$ 4. $\Leftrightarrow$ 5. and the theorem is completely proved.

§7. Plane Hyperbolic Geometry

Within this paragraph we shall admit Hilbert's axioms in groups I-IV and the axiom $\neg V$. The considerations in §5 ensure the consistence of such a Geometry. To deduce and prove the hyperbolic Geometry propositions we have now two distinct modalities of reasoning:

- abstractly;
- on a specific model (Beltrami-Enneper, Poincaré or Cayley-Klein).

We do not know yet whether the axiomatic system we adopted (to which we shall refer with the symbol $\mathcal{H}$) is categorical or not and can not state that the two modalities described lead to the same results. There exists the theoretical possibility that one can prove on a certain model certain propositions P_n which should not be logical consequences of the system $\mathcal{H}$. In this case, the study on the model would not be equivalent with the logical development of the axiomatic system $\mathcal{H}$, but with that of an ampler axiomatic system $\mathcal{H} = \mathcal{H} \cup \{P_n\}$. This is the main reason that privileges the method of the abstract reasoning related to that of the study on a model. But here the axiomatic system is *categorical* and the above presented theoretical eventuality is not achieved.

Unfortunately, to prove the categoricity of the axiomatic system $\mathcal{H}$ it is necessary to have an ampler space than we have here at our disposal, and the deduction of some important propositions in hyperbolic Geometry by abstract reasoning is difficult and sophisticated enough.

This state of affairs determine us to group in this paragraph the deduction by abstract reasoning of only some consequences, and, in the following one, *use also the Beltrami-Enneper model* to obtain as quickly as possible some central propositions of hyperbolic Geometry.

Definition 7.1. *A convex quadrangle ABCD is called Saccheri quadrangle if $AB \perp BC$, $BC \perp CD$ and $AB \equiv CD$. The order in which we shall present the vertexes of a Saccheri quadrangle is not completely arbitrary (Fig. 64).*

Theorem 7.1. *If ABCD is a Saccheri quadrangle, then:*

1. *DCBA is a Saccheri quadrangle;*

2. *the diagonals BD and AC are congruent;*

3. *the angles BAD and CDA are congruent and acute;*

4. $\widehat{BDC} \equiv \widehat{CAB} < \widehat{ACD} \equiv \widehat{ABD}$.

Proof. For 1. we shall find out that the three conditions in Definition 7.1 are simultaneously satisfied for the convex quadrangles $ABCD$ and $DCBA$. 2. The triangles BAC and CDB are congruent (the SAS case), so $AC \equiv BD$. We also retain from here that $\widehat{BAC} \equiv \widehat{CDB}$. For 3. we shall notice (according to the SSS case) that $\triangle CAD \equiv \triangle BDA$, so $\widehat{CAD} \equiv \widehat{BDA}$. It follows that $\widehat{BAD} \equiv \widehat{CDA}$. The sum of the angles of the quadrangle $ABCD$ is equal to the sum of the angles of the triangles ABD and BCD and the latter sum is strictly inferior to 2π. It follows that $\pi + 2 \cdot \widehat{BAD} < 2\pi$, so $\widehat{BAD} < \frac{\pi}{2}$. 4. Because the defect of the triangle ABC is positive, we immediately deduce $\widehat{BAC} < \pi - \frac{\pi}{2} - \widehat{ACB} = \widehat{DCB} - \widehat{ACB} = \widehat{DCA}$. It immediately follows the sequence of equalities and inequalities in the enunciation.

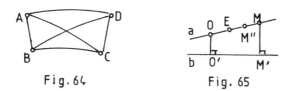

Fig. 64   Fig. 65

Remarks. We have not proved that if two triangles have inequal angles placed between respectively congruent sides, then the sides opposed to the inequal angles will satisfy the same inequality. If we had proved this propositions, we should have applied it to the triangles BAC and DCA in order to deduce from the inequality 4. that $BC \le AD$ holds. We shall prove this inequality in Theorem 8.6. The names of *small base* and *large base*, respectively, for the sides BC and AD of the Saccheri quadrangle $ABCD$ are thus justified.

Let us consider two distinct points O, E and a straight line b (Fig. 65). Further on, we shall associate to the triplet (O, E, b) a function $g: \mathbb{R} \to \mathbb{R}_+$ which we shall call *Gauss' function*. Let $f: (OE) \to \mathbb{R}$ be the Cartesian system of co-ordinates satisfying $f(O) = 0, f(E) = 1$ (see Theorem 5.1, Chapter V). For any real number x there exists a unique point M on (OE) so that $f(x) = M$. Let M' be the foot of the perpendicular drawn down from M on b and the point $M'' \in [O; E)$ so that $MM' \equiv OM''$. We define the function g of Gauss by $g(x) = f(M'')$. There are two particular cases when function g of Gauss is expressed simply but lacks in interest; when (OE) coincides with b, g is the identically null function, and when $OE \perp b$ there will be $k \in \mathbb{R}$ so that $g(x) = |x+k|$. But in other cases Gauss' function will prove to be very important in edifying the hyperbolic Geometry.

Definition 7.2. *We shall say that the straight lines a, b are divergent if they do not have common points and do not admit such orientations that they should super-parallel.*

Theorem 7.2. *Let be $O \ne E$ and Gauss' function g associated to the triplet (O, E, b). Let O' be the orthogonal projection of O on b. We shall note with g_+ the restriction of g to $\mathbb{R}_+$.*

1. *If $O = O'$ or $O \ne O'$ and $\widehat{O'OE}$ is obtuse or right, then g_+ will be monotonously increasing, continuous and unbounded.*

2. *If $O \neq O'$ and $O'\widehat{}OE = L(OO')$, then g_+ will be monotonously decreasing to zero and continuous.*

3. *If (OE) and b are divergent there will exist $x_0 \in \mathbb{R}$ so that, for $x > x_0, g$ should be monotonously increasing and unbounded and $g(x_0 - u) = g(x_0 + u)$.*

Proof. We shall approach the statement 1. in the generic case $O \neq O'$, supposing therefore that $O'\widehat{}OE \geq \frac{\pi}{2}$. Let be the arbitrary positive real number x, $M = f^{-1}(x)$ and the orthogonal projection M' of M on b (Fig. 66). There is a point A on $(M'; M)$ so that $OO' \equiv M'A$. We can notice that the quadrangle $OO'M'A$ is a Saccheri one, so $O'\widehat{}OA$ is acute. It immediately follows that $M - A - M'$.

Let us then consider the arbitrary c in $\mathbb{R}_+$ and let be $N = f^{-1}(x+c)$. We are also noting with N' the orthogonal projection of N on B. There is $B \in (N'; N)$ so that $N'B \equiv MM'$. In the Saccheri quadrangle $MM'N'B$ there results $M'\widehat{}MN \leq \frac{\pi}{2}$. We also find out that $M'\widehat{}MN$, as an angle exterior to the triangle OAM, is bigger than the obtuse angle $O\widehat{A}M = \pi - OAM'$. Then $M'\widehat{}MN$ is obtuse and $(M; B)$ is a ray inside it. This ray cuts then $|M'N|$. According to Pasch's axiom, it follows that $N' - B - N$, so $N'B < N'N$, that is $g(x) = MM' < NN' = g(x+c)$. The arbitrary of the positive real numbers x and c ensure that the restriction of g to $\mathbb{R}_+$ should be monotonously increasing.

In the triangle MBN the side MN opposed to an obtuse angle is the longest one, so $NB < MN$. We deduce from here $g(x+c) - g(x) < c$, inequality that ensures the continuity of function g on $\mathbb{R}_+$.

Let be $P = f^{-1}(x+2c)$ and its orthogonal projection $\mathcal{P}'$ on b. We are considering points C, D defined as follows:

$$C \in (N'; N) \wedge N'C \equiv P'P, \ B - N - D \text{ and } BN \equiv ND.$$

The triangles MBN and PDC are congruent (the SAS case) so $N\widehat{D}P \equiv N\widehat{B}M = \pi - M\widehat{B}N' > \frac{\pi}{2}$. So $N\widehat{D}P$ is obtuse. From the Saccheri quadrangle

$CN'P'P$ it follows that $N'CP$ is an acute angle. So that the theorem of the exterior angle should not be contradicted it is necessary that $N - D - C$ held, that is $NB \equiv ND < NC$. We have thus found out: $g(x+c) - g(x) < g(x+2c) - g(x+c)$. By induction, we find out that after the natural number $n \neq 0$, $g(x+nc) - g(x) > (n-1) \cdot [g(x+c) - g(x)]$ holds. This inequality ensures unboundness of function g on $\mathbb{R}_+$ and the statement 1. is completely proved in the case $O \neq O'$.

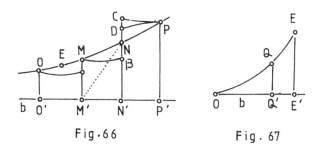

Fig.66   Fig. 67

Let be $O = O'$ (Fig. 67). If $OE \perp b$ the statement 1. will result from the remark that has preceded the theorem; we shall suppose that OE is not perpendicular on b. Let E' be the orthogonal projection of E on b. Obviously, $\widehat{EOE'}$ is acute. For arbitrary q in $(0, 1)$ let be $Q = f^{-1}(q)$ and its orthogonal projection Q' on b. $\widehat{Q'QE}$ is obviously obtuse, as exterior angle to the right-angled triangle $Q'OQ$. According to the above given proof in the generic case, for $x > q$, Gauss' function g satisfies the enunciated conclusions. Given the arbitrary of q the statement 1. results.

2. The hypothesis $\widehat{O'OE} = L(OO')$ expresses that for naturally chosen directions on (OE) and b, respectively, $OE \uparrow b$ holds. By choosing arbitrary $r \in \mathbb{R}_+$, evidencing $R = f^{-1}(r)$ and its orthogonal projection R' on b, we shall find out that $\widehat{ORR'}$ (as a supplement of parallelism angle) is obtuse (see Figure 68). Let g_r be the Gauss function associated to the triplet (R, O, b); by interpreting for g the statement 1.

referring to g_r we assure ourselves that g_+ is monotonously decreasing and continuous (for values $x < r$, but the liberty of choosing r annulus the necessity of this supplementary caution). It has thus remained to prove only that g_+ decreases to zero.

For arbitrary y in $\mathbb{R}_+$ [but supposed to be smaller than $g(O)$] we shall find $D \in |OO'|$ so that $O'D = y$. Let d be the super-parallel through D to OE (so oriented that $O \prec E$). Obviously, d will be super-parallel with b (with ω orientation in which O' precedes the points R'). Choosing the opposed orientation, $-\omega$, on b we shall determine another

Fig. 68

super-parallel to b which will cut (OE) in a point F. Let us consider M so that $O\text{-}F\text{-}M$, $FM \equiv FD$ and let F' be the foot of the perpendicular drawn down from F on b. Let H be the symmetric of O' related to F'. The angles $F'FD$ and $F'FM$ are congruent as parallelism angles. By *SAS* case we deduce $F'D \equiv F'M$ and $\widehat{DF'F} \equiv \widehat{MF'F}$.

It then follows (according to the same *SAS* case of congruence of the triangles) that $\widehat{MHF'} \equiv \widehat{DOF'} = \frac{\pi}{2}$. Therefore, the point H under considerations coincides with the orthogonal projection M' of M on b and $MM' \equiv DO' = y$. Taking $x = f(M) > 0$, we shall obtain $g(x) = y$. But y was chosen as arbitrary in $\mathbb{R}_+$ so g_+ takes values as near to zero as possible (and obviously can not take negative values).

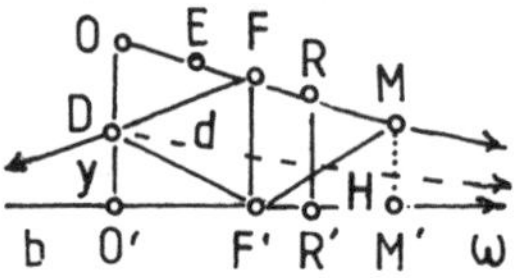

3. Let us suppose the generic case when $\widehat{O'OE} \neq \frac{\pi}{2}$. We are fixing the ideas supposing that $\widehat{O'OE} < \frac{\pi}{2}$. Let us choose the orientation from O' towards E' $(EE' \perp b)$ on b. Let c be the straight line through O super-parallel with b. The points M of the ray $(O; E)$ and the points $M' \in b$ are obviously separated from the straight line c; we shall suppose that $f(M) = m$ and $MM' \perp b$. There exists then $M_1 \in |MM'| \cap c$. We consider the orthogonal projection M_2 of M on c. We easily find out that $g(m) = MM' > MM_1 > MM_2 = g_c(m)$, where we noted the Gauss function associated to the triplet (O, E, c) with g_c. But, according to 1., $g_c(x)$ is unbounded (for $x \in \mathbb{R}_+$), so $g_+(x)$ is also unbounded. We retain from this stage that exists $m \in \mathbb{R}_+$ so that $g(m) > OO' = g(O)$.

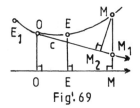

Fig. 69

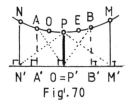

Fig. 70

Let now be E_1 so that $E - O - E_1$ and the Gauss function g_1 associated to the triplet (O, E_1, b). According to 1., there exists the real number $n > 0$ so that $g_1(n) = g(m)$, that is a point N so that $NN' \equiv MM'$. The continuation of this reasoning is illustrated in Figure 70.

Obviously, $NN'M'M$ is a Saccheri quadrangle. Let P, Q be the midpoints of the bases NM and $N'M'$. The triangle $NN'P$ and $MM'P$ are congruent (SAS), so $N'P \equiv PM'$ and moreover $\widehat{NPN'} \equiv \widehat{MPM'}$. In the isosceles triangle $PN'M'$ the median $|PQ|$ is also height (so Q coincides with the orthogonal projection $\mathcal{P}$ of P on b) and bisector, so $\widehat{N'PP'} \equiv \widehat{P'PM'}$. Adding up this congruence of angles to that previously evidenced we deduce $\widehat{NPP'} \equiv \widehat{P'PM} = \frac{\pi}{2}$, so $P'P \perp OE$. (Given its intrinsic importance we shall re-formulate this intermediate result in the Corollary 2.)

Let be $x_0 = f^{-1}(P)$. For the (positive) arbitrary real number u we shall determine the points A, B so that $f(A) = x_0 - u$, $f(B) = x_0 + u$. Obviously, P will be the midpoint of AB. Let A', B' be the orthogonal projections of A, B on b. Using the triangles PAP' and PBP' we deduce $AP' \equiv BP'$ and $\widehat{AP'P} \equiv \widehat{BP'P}$. It follows the congruence of the complementary angles $AP'A'$ and $BP'B'$ and then, according to the SAS case, $AA' \equiv BB'$, so $g(x_0 - u) = g(x_0 + u)$.

We had initially considered $\widehat{O'OE} \neq \frac{\pi}{2}$. If $O'O \perp DE$ we should have directly taken $P = O$ resuming only the last line of the proof. The concluding part referring to the monotony of function g (for $x < x_0$ and for $x > x_0$) is obtained as with 1. substituting the point O by P. The proof is thus completed.

Corollary 1. *If $a \uparrow b$, an equally slant secant (AB) related to (a, b) will pass through any point $A \in a$.*

Indeed, let be $AA' \perp b$ with $A' \in b$ and $A'A'' \perp a$ with $A'' \in a$. Obviously, $AA' > A'A''$, according to the statement 2. in the theorem that there exists a unique point $B \in b$ whose distance BB' to a should be equal to AA'. One can find out that the right-angled triangles $A'BA$ and $B'AB$ are congruent, so $A'\widehat{AB} \equiv B'\widehat{BA}$. We also notice that

$$A'\widehat{AB}' = L(AA') = L(BB') = B'\widehat{BA}' \text{ etc.}$$

Corollary 2. *Any two divergent straight lines a, b admits a common perpendicular AB with $A \in a$, $B \in b$. If d is a distance in the hyperbolic plane and $M \in a$, $N \in b$ are considered as variable points, then $d(M, N) \geq d(A, B)$, the equality being satisfied only when $M = A$, $N = B$.*

Indeed, a common perpendicular $|AB|$ has been evidenced (with the notation $|PP'|$) in the proof of statement 3.. Fixing M arbitrary on a, the distance $d(M, N)$ will be minimum when N coincides with the orthogonal projection M' of M on b. We conceive $d(M, N')$ as a Gauss function admitting a unique minimum when $M = A$ (and then $N = M' = B$). If, by absurd, besides the common perpendicular $|AB|$ under consideration, existed another one, $|MM'|$, the quadrangle $ABM'M$ would have the sum of the angle 2π and a diagonal AM' would divide it into two triangles of null defect, which is absurd.

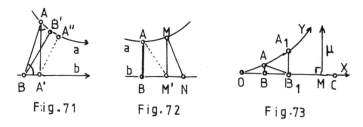

Fig.71   Fig.72   Fig.73

Remark. This Corollary 2 also admits a converse which constitutes in fact the re-formulation of Theorem 6.1 in Chapter V: *if two straight lines a, b admit a common perpendicular, they will be divergent.*

Theorem 7.3. *For any acute angle XOY a unique ray μ perpendicular on OX and super-parallel with $[O; Y)$ exists.*

Proof. Let be $A \in (O; Y)$ and its orthogonal projection B on OX. Let us consider the symmetric B_1 of O related to B. The defect of the triangle OAB is a strictly positive number d (Theorem 6.10, statement 3.). The triangle ABB_1, congruent with ABO, will of course have the same defect d. Let us suppose that the perpendicular drawn up in B_1 on OX cuts OY in A_1 (Fig. 73). For the defect d_1 of the triangle ABC, equal with the sum of the defects of the triangles ABO, ABB_1 and AA_1B there will then be $d_1 > 2d$. We can consider points B_n for any n in $\mathbb{N}^*$, continuing their being defined recursively with: B_n is the symmetric of O related to B_{n-1}. If, by absurd, the perpendicular drawn up in any point B_n on OX cuts OY in a point noted with A_n, we shall find out that the defect d_n of the triangle OA_nB_n exceeds $2^n \cdot d$. But this supposition is absurd because for a big enough n it would follow that $d_n > \pi$, that is the sum of the angles of the triangle OA_nB_n would be negative.

We retain from this proving stage that there exist points C on OX so that the perpendicular in C on OX should not cut OY. By means of Pasch's axiom and by reductio ad absurdum, one can notice that if C has the mentioned property and $O-C-C'$, then C' will also have this property.

Let f be the Cartesian system of co-ordinates, $f:(OX) \to \mathbb{R}$ with $f(O) = 0$, $f(B) = 1$. The set P of points $Z \in (O; X)$ with the property that the perpendicular in Z on OX cuts OY is applied by f into a subset P' of $\mathbb{R}_+$, upperly bounded by values $f(C)$ for points C whose existence has already been established. So $m = \sup P'$ exists; let be $M = f^{-1}(m)$ and μ the ray of origin M perpendicular on OX on that side of OX where Y is situated.

Let us state that μ is super-parallel with OY. Indeed, if μ cut OY in N we should consider T so that $O-N-T$ (Fig. 74). The orthogonal projection T' of T on OX would be in P, so $t = f(T') \in P'$. But $m < t$ would hold, in contradiction with the quality of m as a majorant of the set P'.

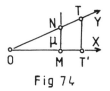

Fig 74

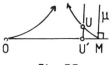

Fig 75

If the super-parallel through M to $[O; Y)$ were distinct from μ it would contain

a point $U \in \text{Int}\,\widehat{OMN}$ (Fig. 75). Let be $UU' \perp OX$ with $U' \in OX$. But (OY) can not cut

the segment $|UU'|$ because, according to Pasch's axiom, it would be obliged to cut $|MU|$, too. On OY no points V can exist so that $U' - U - V$ because these points V are on the other side of (MU) related to that where the points of the straight line OY are. Consequently, the real number $u = f(U')$ is a majorant for P' and the obvious inequality $u < m$. The existence of the super-parallel μ has been proved. Its uniqueness returns to the uniqueness of M which is easy to prove be reductio ad absurdum.

No main difficulties appear to adapt the above proof in order to prove:

Theorem 7.4. *For any non-perpendicular straight lines a, b just two perpendiculars on a, c_1, c_2 exist so that, by conveniently orienting them, orientations on b should exist on each of them so that $c_i \uparrow b$ $(i = 1, 2)$.*

Theorem 7.5. *If a, b are secant straight lines in O, for any of the four modalities of choosing orientations on a and b, there will exist one unique straight line c_i $(i = 1, 2, 3, 4)$ for each on which convenient orientations could be taken so that $c_i \uparrow a$ and $c_i \uparrow b$, respectively.*

Proof. Let $\alpha \subset a$ and $\beta \subset b$ be rays of origins O which precise the choices of

orientations on a and b, respectively. Let γ be the bisector of $(\widehat{\alpha, \beta})$. According to

Theorem 7.3, a point $M \in \gamma$ exists so that the perpendicular drawn up in M on γ (taken as ray σ situated on that side of (OM) on which α is) should be super-parallel with α. Out of "symmetry" reasons one can find out that the "prolongation" σ' of this ray is super-parallel with β. The union of the rays σ and σ' is one of the common super-parallels c_i in the enunciation.

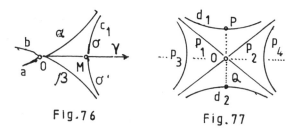

Fig. 76   Fig. 77

To find out that $c_1 = \sigma \bigcup \sigma'$ is the only super-parallel common with α and β we use the same above Figure 76, but supposed to be achieved in another order which we are presenting here. Let d (illustrated by c_1) be supposed super-parallel with α and β. Let M be the orthogonal projection of O on d; we note $\gamma = [O; M)$. The situations of super-parallelism impose $(\widehat{\alpha, \gamma}) = L(OM) = (\widehat{\gamma, \beta})$ so γ is the (interior) bisector of $(\widehat{\alpha, \beta})$.

The point M in which a perpendicular super-parallel with α can be drawn up is unique on γ. It thus immediately results that $d = c_1$, q.e.d.

Theorem 7.6. *If d_1, d_2 are divergent straight lines, for any of the four modalities of choosing orientations on d_1 and d_2, there exists a straight line p_i ($i = 1, 2, 3, 4$) for each, on which convenient orientations could be taken on so that $p_i \uparrow d_2$ and $p_i \uparrow d_2$, respectively.*

The proving reasoning is illustrated in Figure 77. Let $|PQ|$ be the perpendicular common to the straight lines d_j and its midpoint, O. There are two super-parallels through O to d_1 which we shall note with p_1, p_2. We draw attention that relations $p_1 \uparrow d_1$ and $p_2 \uparrow d_1$ imply opposed choices of orientations on d_1. From $L(OP) = L(OQ)$, using angles opposed at the vertex, we can establish that (after having changed the orientations on the straight lines p_i) $p_i \uparrow d_2$ holds (with $i = 1, 2$ but implying in the two cases distinct choices of orientations on d_2). Applying the preceding theorem for (p_1, p_2) instead of (a, b) there will exist four straight lines c_i super-parallel with p_1, p_2 ($i = 1, 2, 3, 4$). Two of them, let us say c_1, c_2, coincide with d_1 and d_2, respectively, and the other two will be re-noted with p_3, p_4 to obtain the very notation in the enunciation; the transitivity of the relation of super-parallelism is also used.

Theorem 7.7. *Lobatchevski's function $L: \mathbb{R}_+ \to (0, \frac{\pi}{2})$ is monotonously decreasing and surjective.*

Proof. Let XOY be a right angle and a Cartesian system of co-ordinates $f: (OY) \to \mathbb{R}$ so that $f(O) = 0$ and $f(Y) > 0$. Let be positive real numbers a, b so that $a < b$. We consider points $A = f^{-1}(a)$ and $B = f^{-1}(b)$ on $(O; Y)$. Let be $u = L(a)$ and the ray $[A; Z)$ so that $\widehat{OAZ} = u$. Of course, $[A; Z)$ is super-parallel with OX. Let also be $U \in \text{Int} \widehat{XOY}$ so that

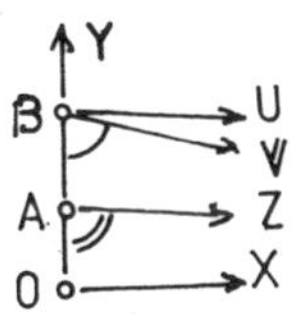

Fig. 78

$\widehat{OBU} = u$. The points of the straight line BU are on that side of (AZ) on which O is not (or any other point of the straight line OX). We deduce that $BU \cap OX = \varnothing$. Consequently, the super-parallel $[B; V)$ to $[O; X)$ is inside the angle OBU. We deduce that the angle $v = L(b) = \widehat{OBV}$ is strictly smaller than $u = L(a)$, that is function L is monotonously decreasing.

The surjectivity of function L constitutes a re-formulation of Theorem 7.3.

Remarks. 1. We had initially conceived (in §6) Lobatchevski's function L as being defined by the set of the segments and associating the same value to congruent segments. According to that first acception, function L has a geometric character. In the enunciation of Theorem 7.7 we interpreted L as a function defined on $\mathbb{R}_+$; we are re-noting it with $\tilde{L}$ to be able to analyze it in comparison with the initial function L, without any dangers of confusions. As we have found out, the presence of a Cartesian system of co-ordinates $f: (OY) \rightarrow \mathbb{R}$ enables us to consider $\tilde{L}$ as function defined on $\mathbb{R}_+$ according to the following scheme: we associate a point $A = f^{-1}(a) \in [O; Y)$ to the real number a, the segment OA to the point A, and conclude in the end that $\tilde{L}(a) = L(OA)$. This way, but starting from another system of co-ordinates, $g: (OY) \rightarrow \mathbb{R}$, we shall obtain another function, $\check{L}$, where, obviously, $\tilde{L} \circ f = \check{L} \circ g$. Since the Cartesian systems f and g under consideration satisfy $f(O) = g(O)$, their specification can be achieved only by fixing certain points of unity. We easily find out that this returns to the choice of a constant k in $\mathbb{R}_+$ and it follows that $\check{L}(x) = \tilde{L}(kx)$. We thus retain that the functions $\tilde{L}$ do not have a geometric character (or that they do not have it *yet*).

2. In absolute Geometry we can arbitrarily choose a non-null segment which served as unit of measure of the segments (Theorem 5.2, Chapter V). This statement keeps being valid when specializing the absolute Geometry in the Euclidean one. But in hyperbolic Geometry, using for example the function L to determine (up to a congruence) a segment OE so that $L(OE) = \frac{\pi}{4}$. We thus get a canonical procedure of choosing a standard OE to measure the lengths within hyperbolic Geometry. Gauss knew this fact in 1824 as it results from a letter of his to Taurinus.

3. The properties mentioned in the enunciation of Theorem 7.7 for function L also ensures the continuity of this function. The analytical expression of function L will appear in the following paragraph.

4. Let us retain the fact that for lower values of the distance a, function L has values quite near $\frac{\pi}{2}$. Let us suppose that the psychical space we live in would be homogeneous and that we should like to experimentally establish whether axiom V of the parallels were satisfied or not, by measuring a parallelism angle. If the standard for measuring the length were the "light-year", for example (that is about 10^{16} meters), the angle of parallelism for a distance of a few centimetre in a copy-book would differ in hyperbolic Geometry so little from the right angle in Euclidean Geometry that no apparatus to achieve a measurement of a small enough tolerance existed. The German geometrician Schweikart noticed in 1818 (in a short statement) that, besides Euclidean Geometry, there *existed* an "*astral*" Geometry in which the defect of the triangles was positive and variable and Theorem 7.3 (using other formulations) held. This is perhaps *the first document* stating the existence of a non-Euclidean Geometry.

As for the physical space we live in, the things are but a little more complicated because by virtue of Einstein's generalized theory of relativity this space is not homogeneous, admitting in different points different "curvatures" (depend on the way the matter is disposed). The geometric concept adequate to the description of such a Universe is that of *differentiable variety* (Riemannian or pseudo-Riemannian). This fact does not undermine the role of Euclidean Geometry that proves to be a simple approximation very good for the objective reality and, moreover, that permits other Geometries to be edified within it: hyperbolic (as in the models we suggested here) or elliptic ones.

§8. Hyperbolic Trigonometry

As we have shown at the beginning of the preceding paragraph, we shall further on use the Beltrami-Enneper model to deduce some powerful results of hyperbolic Geometry. The trigonometric functions of sine, cosine, tangent, exponential function and its reverse, the "natural logarithm", will prove to be useful to formulate these results, but the following two hyperbolic functions, as well:

- *hyperbolic cosine*, cosh: $\mathbb{R} \to [\,1,\ \infty\,)$, where $\cosh x = \frac{1}{2}(e^x + e^{-x})$;

- *hyperbolic sine*, sinh: $\mathbb{R} \to \mathbb{R}$, where $\sinh x = \frac{1}{2}(e^x - e^{-x})$.

The following formulae analogous to those in elementary Trigonometry can be proved without any difficulties:

$$(8.1) \quad \begin{aligned} \cosh(x \pm y) &= \cosh x \cdot \cosh y \pm \sinh x \cdot \sinh y, \\ \sinh(x \pm y) &= \sinh x \cdot \cosh y \pm \cosh x \cdot \sinh y, \end{aligned}$$

(the signs in the two members coinciding each time);

(8.2) $\cosh(-x) = \cosh x, \ \sinh(-x) = -\sinh x$

(8.3) $0 < x < y \Rightarrow \cosh x < \cosh y,$
$x < y \Rightarrow \sinh x < \sinh y.$

In §5 we have also considered the function of "hyperbolic distance" $d_h: S^2 \to \mathbb{R}_+ \bigcup \{0\}$. It is useful here to consider a little more general distance: for an arbitrary positive real number k we shall consider the function $d: S^2 \to \mathbb{R}_+ \bigcup \{0\}$, where $d(A, B) = \frac{1}{k} \cdot d_h(A, B)$; in fact, the specifying of this constant k returns to fixing a standard OE for the measurement of the lengths (with the condition $d(O, E) = 1$).

Theorem 8.1. *Lobatchevski's function L admits the following analytical expression:* $L(x) = 2 \arctan e^{-kx}$.

Proof. Let d be an oriented straight line figured on the model by open Euclidean ray $(O; X)$, where O is, of course, on the absolute a. Let A be an h-point so that $[O; A)$ should form an acute angle u with a. The super-parallel through A to d will be an Euclidean ray $[A; Y)$. The perpendicular from A on d is figured by a semi-circle of centre O which contains A, limited by the points U, V. For the hyperbolic distance we shall easily find out:

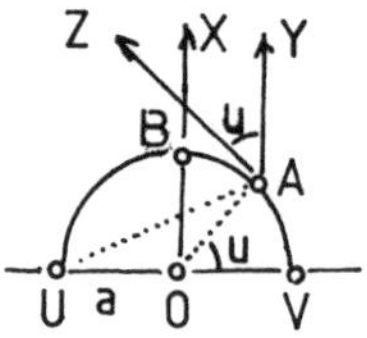

Fig.79

$$d(A, B) = \frac{1}{k} \cdot d_h(A, B) = \frac{1}{k} \ln \frac{AU}{AV} = \frac{1}{k} \ln \frac{2 \cdot OA \sin \frac{1}{2}(\pi - u)}{2 \cdot OA \sin \frac{1}{2} u} = \frac{1}{k} \ln \cot \frac{u}{2} = -\frac{1}{k} \ln \tan \frac{u}{2}$$

The angle of parallelism $L(AB)$ is then $\widehat{ZAY}$, where $[A; Z)$ is the tangent in A to $(AB)_h$. We immediately obtain $\widehat{ZAY} = u$. According to the above formulae the equality $x = d(A, B)$ returns to $\tan \frac{u}{2} = e^{-kx}$, so $u = 2 \arctan e^{-kx}$, q.e.d.

Remark. The established formula enables us to concretize the conclusion in Remark 4. at the end of the preceding paragraph. The choice suggested there, $k = 10^{-16}$,

$x = 10^{-1}$ lead to a value $\frac{\pi}{2} - u$ of the angle of corresponding parallelism, where u is an angle of about 10^{-27} radians, angle inferior to that under which we should see the orbit of an electron on the Moon from the Earth!

Further on, we shall establish some formulae of hyperbolic Trigonometry. For this purpose, for an arbitrary h-triangle ABC, we shall agree to note with $\dot a = d(B, C)$, $\dot b = d(A, C)$, $\dot c = d(A, B)$ the distances between the vertexes, the presence of the superior point being meant to withdraw the possible confusions with an eventual Euclidean triangle of vertexes ABC. Analogously, we shall note $\dot A = (\underset{h}{\sphericalangle} BAC)$ etc.

Theorem 8.2. *In the h-triangle ABC the following "formula of the side cosine" holds:*

$$\cosh k\dot a \cdot \sin \dot B \cdot \dot C = \cos \dot A + \cos \dot B \cdot \cos \dot C.$$

Proof. By means of an h-symmetry we can suppose that $(BC)_h$ is an hs-straight line whose limit point U. Let V, Z be the centres of the circles $O_{A,B}$ and $O_{A,C}$, respectively. We easily find out that $\dot A = \widehat{VAZ}$, $\dot B = \widehat{BVU}$, $\dot C = \widehat{CZU}$. We can then express:

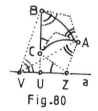

Fig.80

$$VA = VB = \frac{BU}{\sin \dot B}, \quad ZA \doteq ZC = \frac{CU}{\sin \dot C}, \quad VU = BU \cot \dot B,$$
$$UZ = CU \cdot \cot \dot C, \quad VZ = BU \cdot \cot \dot B + CU \cdot \cot \dot C.$$

The cosine theorem (in Euclidean Geometry) for the triangle AVZ, $VZ^2 = VA^2 + ZA^2 - 2VA \cdot ZA \cdot \cos A$, becomes thus:

$(*)\quad 2 \cdot BU \cdot CU \cdot (\cos \dot A + \cos \dot B \cdot \cos \dot C) = (BU^2 + CU^2) \cdot \sin \dot B \cdot \sin \dot C.$

But $k\dot a = d_h(B, C) = \ln \frac{BU}{CU}$; therefore

$$\frac{BU^2 + CU^2}{2 \cdot BU \cdot CU} = \frac{1}{2}\left(\frac{BU}{CU} + \frac{CU}{BU}\right) = \frac{1}{2}(e^{k\dot a} + e^{-k\dot a}) = \cosh k\dot a.$$

Substituting this equality in (*) one obtained the enounced formula.

Corollary. *The "case AAA of congruence of the triangles" is valid in hyperbolic Geometry: if the triangles ABC and $A'B'C'$ have their angles respectively congruent, then $\triangle ABC \underset{h}{\equiv} \triangle A'B'C'$.*

Indeed, applying the theorem we deduce $\cosh k\dot{a} = \cosh k\dot{a}'$. According to the definition of the hyperbolic cosine function, taking into account the a the real numbers $k\dot{a}$ and $k\dot{a}'$ are positive, we deduce from the above equality that $\dot{a} = \dot{a}'$, that is $BC \underset{h}{=} B'C'$. In the end we make appeal to the *ASA* case of the congruence of the triangles.

Remarks. 1. The name given in the enunciation of the theorem of "side cosine" formula is arbitrary in principle because it permits that $\cosh k\dot{a}$ and $\cos\dot{A}$ be made explicit; the name evokes that the formula express a *side* by the *angles* of the triangle.

2. Cyclicly permuting the enounced formula we also get

$$\cosh k\dot{b} \cdot \sin\dot{A} \cdot \sin\dot{C} = \cos\dot{B} + \cos\dot{A} \cdot \cos\dot{C};$$
$$\cosh k\dot{c} \cdot \sin\dot{A} \cdot \sin B = \cos\dot{C} + \cos\dot{A} \cdot \cos B.$$

3. The inequality $\cosh k\dot{a} > 1$ leads to $\cos\dot{A} + \cos(\dot{B} + \dot{C}) > 0$, that is $\dot{A} + \dot{B} + \dot{C} < \pi$, or $\delta(ABC) > 0$.

4. We draw attention that before precising the value of k, the formula of side cosine does not offer the effective possibility to determine a fourth element of the triangle ABC in accordance with other three ones.

5. Let $u < \frac{\pi}{3}$ be an arbitrary angle and $m = \cos u$; the h-points B, C will be vertexes of an "equilateral h-triangle of h-angles u" if and only if $d_h(B, C) = \frac{m}{1-m}$ holds. Particularizing for example $u = \frac{\pi}{4}$ a canonical procedure of choosing a unity h-segment BC will result. Gauss had probably in view such a possibility to fix the length unit in hyperbolic Geometry.

6. Owing to the presence of the *AAA* case of the congruence of triangles it is obvious that it is not possible that congruent similar triangles should exist in hyperbolic Geometry.

Theorem 8.3. *In the hyperbolic triangle ABC the following "sines formula" holds:*

$$\frac{\sinh k\dot{a}}{\sin \dot{A}} = \frac{\sinh k\dot{b}}{\sin \dot{B}} = \frac{\sinh k\dot{c}}{\sin \dot{C}}.$$

Proof. It is easy to prove that $\sinh^2 x = \cosh^2 x - 1$. Using the formula of side cosine we obtain:

$$\sinh^2 k\dot{a} \cdot \sin^2 \dot{B} \cdot \sin^2 \dot{C} = (\cosh^2 k\dot{a} - 1) \cdot (\sin \dot{B} \cdot \sin \dot{C})^2 = (\cos \dot{A} + \cos \dot{B} \cdot \cos \dot{C})^2 -$$
$$-(1 - \cos^2 \dot{B}) \cdot (1 - \cos^2 C) = \cos^2 \dot{A} + \cos^2 \dot{B} + \cos^2 \dot{C} +$$
$$+2 \cdot \cos \dot{A} \cdot \cos \dot{B} \cdot \cos \dot{C} - 1.$$

The final expression we reached is independent of the order in which the vertices A, B, C are taken. By also taking into account the expression from which we have started we deduce that the final expression could be re-noted with Δ^2, the real number Δ being associated to the triangle ABC. We transcribe the obtained equality under the formula $\sinh k\dot{a} \cdot \sin \dot{B} \cdot \sin \dot{C} = \Delta$. After cyclic permutations one can obtain:

$$\sinh k\dot{a} \cdot \sin \dot{B} \cdot \sin \dot{C} = \sinh k\dot{b} \cdot \sin \dot{A} \cdot \sin \dot{C} = \sinh k\dot{c} \cdot \sin \dot{A} \cdot \sin \dot{B} = \Delta.$$

After having divided by the non-null factor $\sin \dot{A} \cdot \sin \dot{B} \cdot \sin \dot{C}$ we obtain

$$\frac{\sinh k\dot{a}}{\sin \dot{A}} = \frac{\sinh k\dot{b}}{\sin \dot{B}} = \frac{\sinh k\dot{c}}{\sin \dot{C}} = \frac{\Delta}{\sin \dot{A} \cdot \sin \dot{B} \cdot \sin \dot{C}}$$

formula which also contains the equalities in the enunciation.

Remarks. 1. The invariant under considerations Δ also admits other symmetric expression related to the angles of a triangle; we shall deduce such a formula in which the "semi-defect" $s = \frac{1}{2}\delta(ABC)$ also figures.

$$\Delta^2 = (\cos\dot{A} + \cos\dot{B} \cdot \cos\dot{C})^2 - (\sin\dot{B} \cdot \sin\dot{C})^2 = [\cos\dot{A} + \cos(\dot{B} + \dot{C})] \cdot [\cos\dot{A} + \cos(\dot{B} - \dot{C})] =$$

$$= 4\cos\frac{\dot{A} + \dot{B} + \dot{C}}{2}\cos\frac{-\dot{A} + \dot{B} + \dot{C}}{2}\cos\frac{\dot{A} - \dot{B} + \dot{C}}{2}\cos\frac{\dot{A} + \dot{B} - \dot{C}}{2} =$$

$$= 4 \cdot \sin s \cdot \sin(s + \dot{A}) \cdot \sin(s + \dot{B}) \cdot \sin(s + \dot{C}).$$

2. One can find out that $\lim_{x \to 0}\frac{\sinh ax}{x} = a$. In the sines formula we divide by k and then pass to the limit for k tending to zero. The theorem of the sines in Euclidean Geometry in thus obtained. There appears then a supplementary argument to interpret the Euclidean Geometry as "limit of hyperbolic Geometry" when the value of the constant k is diminished, operation that returns to the increasing of the standard for the measure of the segments.

3. We have deduced here the sines formula from the cosine formula of the side; one can do it anti-clockwise, too.

4. It is possible that the deduction of the sines formula by synthetical Geometry and Euclidean Trigonometry on the model (using on Figure 80) should present interest. Our preference expression the intention of restraining the role of the reasoning on the model as much possible.

Theorem 8.4. *In an h-triangle ABC the following "formula of the angle cosine" holds:*

$$\cosh k\dot{a} = \cosh k\dot{b} \cdot \cosh k\dot{c} - \sinh k\dot{b} \cdot \sinh k\dot{c} \cdot \cos\dot{A}.$$

Proof. Using the three formulae of the side cosine we obtain:

$$[\cosh k\dot{b} \cdot \cosh k\dot{c} - \cosh k\dot{a}]\sin^2\dot{A} \cdot \sin\dot{B} \cdot \sin\dot{C} = (\cos\dot{B} + \cos\dot{A} \cdot \cos\dot{C}) \cdot (\cos\dot{C} +$$
$$+ \cos\dot{A} \cdot \cos\dot{B}) - (\cos\dot{A} + \cos\dot{B} \cdot \cos\dot{C})(1 - \cos^2\dot{A}) = \cos\dot{A} \cdot (\cos^2\dot{B} +$$
$$+ \cos^2\dot{C} + \cos^2\dot{A} - 1) + 2\cos^2\dot{A} \cdot \cos\dot{B} \cdot \cos\dot{C} = \Delta^2 \cdot \cos\dot{A} =$$
$$= [\sinh k\dot{b} \cdot \sin\dot{A} \cdot \sin\dot{C}] \cdot [\sinh k\dot{c} \cdot \sin\dot{A} \cdot \sin\dot{B}]\cos A =$$
$$= [\sinh k\dot{b} \cdot \sinh k\dot{c} \cdot \cos A]\sin^2 A \cdot \sin\dot{B} \cdot \sin\dot{C}.$$

Remarks. 1. Permuting the vertexes A, B, C we obtain two other formulae of the "angle cosine".

2. Starting from the three formulae of the angle cosine we can, by simple algebraic calculuses, deduce the formulae of the side cosine.

3. One can find out $\lim_{x \to 0}(\cosh ax - 1) = 0$ and $\lim_{x \to 0}\frac{\cosh ax - 1}{x^2} = \frac{1}{2}a^2$. By provisorily

using the notations: $\alpha = k^{-2}(\cosh k\dot{a} - 1)$, $\beta = k^{-2}(\cosh k\dot{b} - 1)$, $\gamma = k^{-2}(\cosh k\dot{c} - 1)$ and $\varepsilon = k^{-2} \cdot \sinh k\dot{b} \cdot \sinh k\dot{c}$ we may transcribe the enounced formula as

$$\alpha = \beta + \gamma + k^2\beta\gamma - \varepsilon\cos A.$$

Passing to the limit for $k \to 0$ we get $a^2 = b^2 + c^2 + 0 - 2bc\cos A$, that is *the cosine theorem in Euclidean Geometry*.

Theorem 8.5. *For any positive real numbers* a, b, c *so that* $|b - c| < a < b + c$ *an* h-*triangle exists so that* $\dot{a} = a$, $\dot{b} = b$, $\dot{c} = c$.

Proof. The enunciation refers to the notation convention $\dot{a} = d(B, C) = \frac{1}{k}d_h(B, C)$, implying then the specified constant k. Let us consider (for $k > 0$) the real number

$$\alpha = \frac{\cosh kb \cdot \cosh kc - \cosh ka}{\sinh kb \cdot \sinh kc}.$$

Using (8.1) and (8.3) one deduces $\alpha < 1 \Leftrightarrow \cosh kb \cdot \cosh kc - \cosh ka < \sinh kb \cdot \sinh kc \Leftrightarrow \cosh ka > \cosh k|b - c| \Leftrightarrow ka > k|b - c| \Leftrightarrow a > |b - c|$. Hence, the premise $|b - c| < a$ ensures $\alpha < 1$. One can analogously find out that $-1 < a$ derives from $a < b + c$. Therefore, an angle $u \in (0, \pi)$ exists so that $\cos u = \alpha$.

We shall choose an arbitrary h-point A and the h-rays $[A; X)$, $[A; Y)$ which from the angle u between them. We shall take on these h-rays the point B and C, respectively, so that $d_h(A, B) = kc$, $d_h(A, C) = kb$. For the h-triangle ABC we obtained, the equalities $\dot{b} = b$ and $\dot{c} = c$ obviously hold. Calculating by the cosine formula we obtain of course $\cosh k\dot{a} = \cosh ka$, so $\dot{a} = a$ also holds, q.e.d.

Remarks. 1. We appreciate that it is necessary to underline the role of the constant k (whose choice was equal to fixing an h-segment of "hyperbolic length" 1). The angle u we constructed *effectively depends on* k; for example, it is possible that for a certain k we should obtain $A = \frac{\pi}{2}$, that is a right-angled triangle ABC, and for other values k' a triangle $A'B'C'$ which should not be right-angled in A anymore. In Euclidean Geometry the lengths of the sides of a triangle do not effectively determine it, either, except after the standard has been fixed, but they determine *the values of angles*.

2. Theorem 8.5 belongs to *the absolute Geometry* since it is valid independently of the accepting of the parallels postulate or of its negation. The validity of the theorem

was proved here in the supplementary hypothesis that axiom $\neg V$ held, hypothesis that permitted the development of some techniques specific to hyperbolic Geometry and concurred in obtaining the proof. In Euclidean Geometry Theorem 8.5 benefits from another proof necessitating techniques specific to Euclidean Geometry such as the introduction of the concept of circle and the study of the relative positions of two circles. Of course, the non-Euclidean Geometry can adopt (taking the due precautions) a good part of the techniques of Euclidean Geometry (and conversely).

Theorem 8.6. *Let ABCD be a Saccheri quadrangle; AD > BC holds.*

Proof. Let be $d(A, B) = a$, $d(B, C) = b$. We are going to calculate $d(A, C) = m$, $\widehat{ACB} = u$ and then $d(A, D) = x$. We apply the formula of the cosine of the angle B into the triangle ABC and deduce: $\cosh km = \cosh ka \cdot \cosh kb$. In the same triangle, the sines formula ensures $\sin u \cdot \sinh km = \sinh ka$, which is transcribed under the form $\cos ACD \cdot \sinh km = \sinh ks$. The formula of the cosine of the angle ACD leads us to:

$$\cosh kx = \cosh km \cdot \cosh ka - \sinh km \cdot \sinh ka \cdot \cos ACD = \cosh^2 ka \cdot \cosh kb - \sinh^2 ka =$$
$$= \cosh kb + \sinh^2 ka \cdot (\cosh kb - 1).$$

Given its intrinsic importance, we shall retain the obtained formula:

$$(*) \quad \cosh kx - \cosh kb = \sinh^2 ka \cdot (\cosh kb - 1).$$

The second member is obviously positive, so $\cosh kx > \cosh kb$, from where, according to (8.3), $x > b$ results, q.e.d.

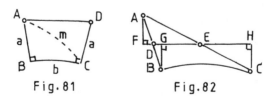

Fig. 81   Fig. 82

Corollary. *If we note with D, E the midpoints of sides AB, AC in triangle ABC, the length of the middle line $d(D, E)$ is strictly smaller than the half of the base $a = d(B, C)$.*

Proof. Let F be the foot of the perpendicular drawn down from A on (DE) and G, H the symmetrics of point F related to points D, E. According to the *SAS* case, we establish $\triangle DAF \equiv \triangle DBG$ and $\triangle EAF \equiv \triangle ECH$. It follows that $BG \perp GH$, $GH \perp HC$ and the equality $BG \equiv HC$ holds. Therefore, $BGHC$ is a Saccheri quadrangle and $d(B, C) > d(G, H) = 2 \cdot d(D, E)$, q.e.d.

Problem

Let $S(O, R)$ be a sphere and A, B, C points on it, any two of them being not "antipodal". We can join each two of them by the small arc of big circle to get a spherical triangle $\measuredangle ABC$.

Let a, b, c the lengths of its sides $\overset{\frown}{BC}$, $\overset{\frown}{CA}$, $\overset{\frown}{AB}$. They rays tangent in A to $\overset{\frown}{AB}$, $\overset{\frown}{AC}$ form an angle whose radians measure we denote by $\dot{A}$; analogous definitions introduce the others angles of $\measuredangle ABC$.

Prove the following three formulas:

1. $\cos\dot{A}\sin\dfrac{b}{R}\sin\dfrac{c}{R} = \cos\dfrac{a}{R} - \cos\dfrac{b}{R}\cos\dfrac{c}{R}$.

2. $\cos\dfrac{a}{R}\sin\dot{B}\sin\dot{A} = \cos\dot{A} + \cos\dot{B}\cos\dot{C}$.

3. $\dfrac{\sin\dot{A}}{\sin\dfrac{a}{R}} = \dfrac{\sin\dot{B}}{\sin\dfrac{b}{R}} = \dfrac{\sin\dot{C}}{\sin\dfrac{c}{R}}$.

CHAPTER IX

BACHMANN'S AXIOMATIC SYSTEM
(Chapter written by Prof. Dr. Francisc Radó)

§1. The Isometries of the Absolute Plane

The isometries of the Euclidean plane (see Chapter VII) and the h-transformations of the hyperbolic plane (Chapter VIII) present a series of common properties. The explanation resides in the fact that both types of transformations are isometries of the absolute plane.

It can also be noticed that many proofs regarding the isometries of the Euclidean plane are based only on the first three groups of Hilbert's axioms. It is therefore necessary a study of the isometries within this more general frame.

According to the classical acception, introduced by J. Bolyai, the term of absolute plane concerns the first four groups of Hilbert's axiom. We shall use here the notion of *absolute plane* in a more general acception, concerning only the axioms of incidence (in the plane), those of ordering and those of congruence.

Let $\mathcal{P}$ be the set of the points of such an absolute plane. A map $T: \mathcal{P} \to \mathcal{P}$ is called *isometry* if

$$(1.1) \quad \forall M, N \in \mathcal{P}: MN \equiv T(M)T(N).$$

It immediately results from this definition that T is an injective map.

Remark. If we admit the axioms of continuity, too, condition (1.1) will be equivalent with $\forall M, N \in \mathcal{P}: d(M, N) = d(T(M), T(N))$, and our definition returns to that in Chapter VII, §2.

Theorem 1.1. *The image, through an isometry T, of a straight line is a straight line, and two perpendicular straight lines change into perpendicular straight lines.*

Proof. Let d be any straight line. Let us take two points A and B symmetric related to $d(A \neq B)$ and not $A': = T(A)$, $B': = T(B)$, $d': =$ the mediatrix of $A'B'$. If $M \in d$, from $MA \equiv MB$ we shall deduce $T(M)A' \equiv T(M)B'$, so $T(M) \in d'$ and $T(d) \subset d'$. Let be now $N \in d'$; it result from the axioms of congruence that the triangles AM_1B, AM_2B, $A'N_1B'$ there exist so that $M_1 \neq M_2$, $N \neq N_1$ and $\triangle AM_1B \equiv \triangle AM_2B \equiv \triangle A'N_1B' \equiv \triangle A'NB'$. Taking account of (1.1) we obtain $T(M_1) = N$ or $T(M_1) = N'$. In the second case it necessarily follows that $T(M_2) = N$, so $N \in T(d)$

in both cases, that is $d' \subset T(d)$. It result $T(d) \subset d'$. Applying this result to the straight line (AB) we obtain $T((ab)) = (A'B')$ and the theorem is proved.

Theorem 1.2. *Any isometry is a bijective map. The set of the isometries of the absolute plane $\mathcal{P}$ form a group $\underline{G}$ with respect to the composition.*

Proof. As T is any of the injective isometries, we have only to show that $\operatorname{Im} T = \mathcal{P}$. Let d be a fixed straight line and d' its image. Let us consider an any point $N \in \mathcal{P}$. The perpendicular d_1 on d, traced through N, cuts d' in a point P. According to Theorem 1.1, there exist $Q \in d$ so that $T(Q) = P$ and the image of the perpendicular through Q on d is d_1. Therefore, $N \in \operatorname{Im} T$ and $\mathcal{P} = \operatorname{Im} T$.

Obviously, together with T_1, T_2 and $T_1 \circ T_2$ it is an isometry; $1_{\mathcal{P}}$ is an isometry. Let P, Q be any points; replacing $M = T^{-1}(P)$, $N = T^{-1}(Q)$, in (1.1), one can obtain: $T^{-1}(P)T^{-1}(Q) \equiv PQ$ so T^{-1} is an isometry.

$T_1 \circ T_2$ will be also noted with $T_1 \cdot T_2$ or $T_1 T_2$ and will be called the product of the isometries T_1 and T_2.

The (axial) symmetry related to an straight line d, noted with $\widetilde{d}$, and the (central) symmetry related to a point A, noted with $\widetilde{A}$, are defined as in Chapter VII, §3. The proofs of the following theorems are analogous to those in Chapter VII, §3.

Theorem 1.3. *Let a be a straight line and $\widetilde{a}$ the symmetry related to a. Then:*
1) $\widetilde{a}$ *is an isometry.*
2) $\widetilde{a}(M) = M \Leftrightarrow M \in a$.

Theorem 1.4. 1) *Let be $\triangle ABC \equiv \triangle A'B'C'$. Then $T \in G$ exists so that $T(A) = A'$, $T(B) = B'$, $T(C) = C'$.*

2) *Any isometry is a product of at most three axial symmetries.*

3) *If $a = (AB)$ and A and B are fixed points of the isometry T, then $T = 1_{\mathcal{P}}$ or $T = \widetilde{a}$ (depending on the existence or non-existence of a fixed point non-situated on a).*

4) *If $T \in G$ has a unique fixed point, then T is a product of two axial symmetries.*

As a corollary we obtain: if a and b are distinct straight lines, then

$$(1.2) \qquad \widetilde{a}\widetilde{b}(M) = M \Leftrightarrow \{M\} = a \cap b.$$

Indeed, let be $\widetilde{a}\widetilde{b}(M) = M$, that is $\widetilde{b}(M) = \widetilde{a}(M) =: N$; if $M \neq N$, then a

and b will be the mediatrixes of the same segment MN, that is $a = b$, contrary to the hypothesis. So $M = N$ and, by virtue of Theorem 1.3, we deduce from $\tilde{a}(M) = M$, $\tilde{b}(M) = M$ that $M \in a$, $M \in b$. The contrary implication results immediately.

Let us note with I the set of the involutive (non-identical) isometries: $I: = \{T \in G : T^2 = 1_p \wedge T \neq 1_p\}$. Obviously, the axial and central symmetries are involutive.

Theorem 1.5. *Let a and b be two straight lines; then:*

1) $a \perp b \Rightarrow \tilde{a} \circ \tilde{b} \in I$.

2) $a \perp b$, $\{C\} = a \cap b \Rightarrow \tilde{C} = \tilde{a} \circ \tilde{b}$.

Proof. 1) Let us suppose that $a \perp b$ and $\{C\} = a \cap b$. Let be $T = \tilde{a} \circ \tilde{b}$, $A \in a \setminus \{C\}$, $B \in b \setminus \{C\}$. As $\tilde{b}(A) \in a$, there is $T(A) = \tilde{a}(b(A)) = b(A)$, so $T^2(A) = t[T(A)] = \tilde{a}\tilde{b}\tilde{b}(A) = \tilde{a}(A) = a$ and, analogously, $T^2(B) = B$. Taking also that $T^2(C) = C$, in Theorem 4.3, into account, we obtain $T^2 = 1_p$. But $T = 1_p$ is not possible since there would be $\tilde{a} = \tilde{b}$, in contradiction with $a \perp b$, so $T \in I$.

Let be $\tilde{a} \circ \tilde{b} \in I$. Then $a \neq b$. Let us suppose that a is not perpendicular on b and $B \in b \setminus a$; it results $\tilde{a}(B) \notin b$ (because, in the contrary case, $a \perp b$), so $\tilde{b}\tilde{a}(B) \neq \tilde{a}(B) = \tilde{a}\tilde{b}(B)$, which implies $\tilde{b} \circ \tilde{a} \neq \tilde{a} \circ \tilde{b}$ and $\tilde{a} \circ \tilde{b} \circ \tilde{b} \circ \tilde{a} \neq 1_p$, in contradiction with $\tilde{a} \circ \tilde{b} \in I$.

2) Let be $a \perp b$, $\{C\} = a \cap b$ and $T = \tilde{a} \circ \tilde{b}$. By virtue of (1.2) C is the only fixed point of T. As T is involutive, for any $M \in \mathcal{P} \setminus \{C\}$, the segment $MT(M)$ changes into itself (its ends change into each other), from where it results that the midpoint of $MT(M)$ is a fixed point, that is it coincides with C. One can see then that $T(M) = \tilde{C}(M)$ and thus $T = \tilde{C}$.

Corollary 1.1. *Any involutive isometry T is an axial or central symmetry.*

Indeed, according to the above reasoning, T has at least a fixed point. If it has two fixed points, then, from Theorem 1.4, 3), it results that T will be an axial symmetry. If T has only one fixed point, we deduce from Theorem 1.4, 4) that T has the form $\tilde{a} \circ \tilde{b}$, and from Theorem 1.4, 1) it results that $a \ldots b$, so T is a symmetry of centre belonging to $a \cap b$.

Corollary 1.2. *For any straight lines a, b, c there is $\tilde{a} \circ \tilde{b} \circ \tilde{c} \neq 1_p$.*

Indeed, let us admit that $\tilde{a} \circ \tilde{b} \circ \tilde{c} = 1_p$, that is $\tilde{a} \circ \tilde{b} = \tilde{c}$. It results that $\tilde{a} \circ \tilde{b} \in I$, so, by virtue of Theorem 1.5, $\tilde{a} \circ \tilde{b} = \tilde{C}$, where $C \in a \cap b$. Then $\tilde{C} = \tilde{c}$ and this equality is absurd because $\tilde{C}$ has only one fixed point, while $\tilde{c}$ has an infinity of fixed points.

Theorem 1.6. *If a is a straight line and A a point, then $A \in a \Leftrightarrow \tilde{A}\tilde{a} \in I$* $\Leftrightarrow \tilde{a}\tilde{A} \in I$.

Proof. Let b be the perpendicular in A on a. If $A \in a$, then, according to Theorem 1.5, $\tilde{A} = \tilde{a} \circ \tilde{b} = \tilde{b} \circ \tilde{a}$, then $\tilde{a} \circ \tilde{A} = \tilde{A} \circ \tilde{a} = \tilde{b} \in I$. Let be now $\tilde{A} \circ \tilde{a} \in I$ and let us suppose that $A \notin a$. Then $A' := \tilde{a}(A) \neq A$, so $\tilde{A}(A') \neq A$, that is $\tilde{A}\tilde{a}(A) \neq \tilde{a}(A) = \tilde{a}\tilde{A}(A)$. It result that $\tilde{A} \circ \tilde{a} \neq \tilde{a} \circ \tilde{A}$ and $(\tilde{A} \circ \tilde{a}) \circ (\tilde{A} \circ \tilde{a}) \neq 1_p$, in contradiction with $\tilde{A} \circ \tilde{a} \in I$. Therefore, $\tilde{A} \circ \tilde{a} \in I$ (and so does $\tilde{a} \circ \tilde{A} \in I$) implies $a \in A$.

Theorem 1.7 (The theorem of the three symmetries). *Let be the straight lines a,* b, c.

1) *If a, b, c have a common point P, then $\tilde{a} \circ \tilde{b} \circ \tilde{c} = \tilde{d}$, where d is a straight line containing the point P.*

2) *If a, b, c have a common perpendicular ℓ, then $\tilde{a} \circ \tilde{b} \circ \tilde{c} = \tilde{d}$, where d is a straight line perpendicular on ℓ.*

Proof. If $a = b$, both statements are obvious. Let further on be $a \neq b$ and $T = \tilde{a} \circ \tilde{b} \circ \tilde{c}$.

1) As P is the unique fixed point of $\tilde{a} \circ \tilde{b}$, taking a point $C \in c \setminus \{P\}$ it results that $T(C) = \tilde{a}\tilde{b}(C) \neq C$. The mediatrix d of the segment $CT(C)$ passes through P (because $PC = T(P)T(C) = PT(C)$). The isometry $\tilde{d}T$ has P and C as fixed points, so, according Theorem 1.4, 3), $\tilde{d}T = 1_p$ or $\tilde{d}T = \tilde{c}$, In the second case, $\tilde{d} = \tilde{d}\tilde{c}\tilde{c} = \tilde{d}\tilde{d}T\tilde{c} = T\tilde{c} = \tilde{a}\tilde{b}\tilde{c}\tilde{c} = \tilde{a}\tilde{b}$, therefore $\tilde{a}\tilde{b}\tilde{d} = 1_p$ which contradicts Corollary 1.2. It results that $T = \tilde{d}$.

2) We are noting $\{Q\}$; $c \cap \ell$. As $a \cap b = \varnothing$, the relation (1.2) implies that $Q' : T(Q) = \tilde{a}\tilde{b}\tilde{c}(Q) = \tilde{a}\tilde{b}(Q) \neq Q$. Let d be the mediatrix of the segment QQ'. Then $\tilde{d}T(\ell) = \tilde{d}\tilde{a}\tilde{b}\tilde{c}(\ell) = \ell$ (because a, b, c, $d \perp \ell$) and $\tilde{d}T(Q) = \tilde{d}(Q') = Q$; it results that the perpendicular through Q on ℓ, that is the straight line c, is also fixed to the isometry $\tilde{d}T$. Since $\tilde{d}T = \tilde{d}\tilde{a}\tilde{b}\tilde{c}$, preserves the two semi-planes limited by ℓ, it results $\tilde{d}T = \tilde{c}$

or $\tilde{d}T = 1_{\mathcal{g}}$. The first case leads to a contradiction, as above; therefore $T = \tilde{d}$.

§2. The Embedding of the Absolute Plane Into the Group of Its Isometries

Since the maps $a \to \tilde{a}$ and $A \to \tilde{A}$ from the set of straight lines (points) in the group G are injective, we can identify the any straight line a with the symmetry $\tilde{a}$ and the arbitrary point A with the symmetry $\tilde{A}$. The set $\mathcal{P}$ of the points and the set D of the straight lines of the absolute plane become thus sub-sets of G, for which $\mathcal{P} \cup D = /$ and $\mathcal{P} \cap D = \varnothing$. According to Theorem 1.4, 2), the set D is a system of generators for G: $G = \langle D \rangle = \{d_1 d_2 d_3 \mid d_1, d_2, d_3 \in D\}$. The neuter element $1_{\mathcal{g}}$ of D will be noted with 1.

The Theorems 1.5 and 1.6 enable us to express the fundamental geometric relations of incidence and perpendicularity by a group property: $A \in a \Leftrightarrow Aa \in /$, and $a \perp b \Leftrightarrow ab \in /$. Thus, the possibility of solving geometric problems by means of the group calculus, that is a new algebraic method in Geometry (besides the analytical one), appears.

After $\tilde{A}$ had been identified with A, the notation AB will mean a product in G; to avoid any confusions, the straight line passing through A and B will be noted with $\overline{A, B}$ (the segment AB does not appear anylonger).

Theorem 2.1. *For any* $a \in D, A \in \mathcal{P}$ *and* $T \in G$ *there are*

(2.1) $T(a) = TaT^{-1}$ and $T(A) = TAT^{-1}$.

Proof. Since $TaT^{-1} \neq 1 = TaT^{-1}TaT^{-1}$, it results $TaT^{-1} \in /$ and, analogously, $TAT^{-1} \in /$. If $M \in T(a)$, that is $M = T(N)$, $N \in a$ then the equality $TaT^{-1}(M) = Ta(N) = = T(N) = M$ holds. Therefore, any point of the straight line $T(a)$ is fixed to the involutive isometry TaT^{-1}, so TaT^{-1} coincides with the symmetry related to $T(a)$, which is now noted with $T(a)$. The second equality (2.1) is analogously proved.

One can obtain from (2.1):

(2.2) $\begin{cases} b(a) = bab, & B(A) = bAb, \\ B(a) = BaB, & B(a) = BAB. \end{cases}$

The relation $b(a) = c$ means that b is an axis of symmetry for a and c; $b(A) = C$ means that b is the mediatrix of the pair of points A, C; $B(A) = C$ means that B is the midpoint of A, C. Therefore, there is:

Theorem 2.2. *If a, b, $c \in D$, and A, B, $C \in \mathcal{P}$, then*

1) *b will be axis of symmetry for a, $c \Leftrightarrow ab = bc$;*
2) *B will be the midpoint of A, $C \Leftrightarrow AB = BC$;;*
3) *b will be the mediatrix of A, $C \Leftrightarrow Ab = bC$.*

One can then see that the properties of the absolute plane are translated into group terms and that this plane can be studied with the help of group G. We are led to a new generalization, if we retain only the essential properties of G, what we shall do in the next paragraph.

§3. The Group Plane

Let Γ be a group and Δ a system of generatrices of Γ made up of involutive elements, that is any $\gamma \in \Gamma$ should be written as $a_1 a_2 \ldots a_n$, with $a_i \in \Delta$ and

$$(3.1) \quad \Delta \subseteq \mathrm{Inv}\Gamma : = \{\gamma \in \Gamma \backslash \gamma^2 = 1 \neq \gamma\}.$$

Let us suppose that the system Δ is made invariant:

$$(3.2) \quad a \in \Delta \Rightarrow a\Delta a \subseteq \Delta.$$

Let be $\Delta\Delta : = \{ab \backslash a, b \in \Delta\}$ and

$$(3.3) \quad \Pi : = \Delta\Delta \cap \mathrm{Inv}\Gamma.$$

The elements of Δ and Π will called *straight lines* and *points*, respectively. The elements of Δ, Π, Γ will be, respectively, noted with a, b, c, ...; A, B, C, ...; α, β, γ, ...

The pair (Γ, Δ) is called *a group generated by symmetries* or an *AGS group* (for the initials of the title of Bachmann's book: "Aufbau der Geometrie au dem Spiegelungsbegriff" (The construction of Geometry Out of the Notion of Symmetry), first edition, Berlin, 1959) if the following *axioms of Bachmann* are satisfied:

B_1. *If A, $B \in \Pi$ are distinct points, then a unique $c \in \Delta$ will exist so that $Ac \in \mathrm{Inv}\Gamma$. The straight line c will be noted with $\overline{A, B}$.*

B_2. If aA, bA, $cA \in \text{Inv}\Gamma$ or ad, bd, $cd \in \text{Inv}\Gamma$, then $abc \in \Delta$.

B_3. There exist g, h, $j \in \Delta$ so that $gh \in \text{Inv}\Gamma$ and jg, jh, $j(gh) \notin \text{Inv}\Gamma$.

A geometric structure $\mathcal{P}(\Gamma, \Delta)$: $= (\sqcap, \Delta, I, \perp)$, where
I: $= \{(A, a) \backslash A \in \sqcap, a \in \Delta, aA \in \text{Inv}\Gamma\}$ and $\perp$: $= \{(a, b) \backslash a, b \in \Delta, ab \in \text{Inv}\Gamma\}$
is associated to the algebraic structure (Γ, Δ). The structure $\mathcal{P}(\Gamma, \Delta)$ is called *group plane* or *metric plane*.

The relation $(A, a) \in I$ is more briefly noted this way: AIa or aIA and is read "A is incident to a". The relation $(a, b) \in \perp$ is also noted $a \perp b$ and read "a is perpendicular on b".

From $ab \in \text{Inv}\Gamma$ results that $ba = (ab)^{-1} \in \text{Inv}\Gamma$, so $a \perp b$ implies $b \perp a$. If $a \neq b$ and there is C so that CIa, CIb, we say that the straight lines a and b are "concurrent".

The map $\tilde{c}$: $\Gamma \to \Gamma$, defined by $\tilde{c}(\gamma) = c\gamma c$ for any $\gamma \in \Gamma$, called *the symmetry related to the straight line* c, is associated to any $c \in \Delta$. One can find out without any difficulties that $\tilde{c}$ is bijective and, by virtue of (3.2), if $a \in \Delta$, then $\tilde{c}(a) \in \Delta$. Also,

(3.4) $\alpha \in \text{Inv}\Gamma \Rightarrow \tilde{c}(\alpha) \in \text{Inv}\Gamma$,

since $[\tilde{c}(\alpha)]^2 = c\alpha cc\alpha c = 1$ and $c\alpha c \neq 1$. It results that $\tilde{c}$ transforms point into point. Therefore $\tilde{c}$ preserves the straight lines and points and, taking (3.4) into account, it results that it preserves the incidences and perpendicularities.

Let be $\tilde{\Delta}$: $= \{\tilde{c} \backslash c \in \Delta\}$ and let $\tilde{\Gamma}$ be the group of transformations of Γ, generated by $\tilde{\Delta}$. We make so that $\tilde{\gamma}$: $\Gamma \to \Gamma$, $\tilde{\gamma}(\alpha) = \gamma\alpha\gamma^{-1}$ should correspond to an arbitrary element $\gamma \in \Gamma$. Since γ may be expressed under the form $\gamma = a_1 a_2 ... a_n$, $a_i \in \Delta$, there will be $\tilde{\gamma}(\alpha) = a_1 ... a_n \alpha a_n ... a_1 = \tilde{a}_1 ... \tilde{a}_n(\alpha)$, so $\tilde{\gamma} = \tilde{a}_1 ... \tilde{a}_n$. It results that the bijection $\tilde{\gamma}$ also preserves the straight lines, the points, the incidences and the perpendicularities. The restriction of $\tilde{\gamma}$ on $\Delta \bigcup \sqcap$ is called an *isometry*, and $\tilde{\gamma}$ itself a *movement* of the plane $\mathcal{P}(\Gamma, \Delta)$. Since Δ is a system of generatrices of Γ, a movement will be determined by its restriction on Δ; there is no essential distinction between an isometry and the movement determined by it. The group of transformations $\tilde{\Gamma}$ is called *the group of movement* of $\mathcal{P}(\Gamma, \Delta)$.

The absolute plane is a group plane. Indeed, taking $\Gamma = G$ and $\Delta = D$, condition (3.1) is obviously satisfied, and condition (3.2) results from the first formula (2.1). In this case $\sqcap = \mathcal{P}$ because the symmetries are the only involutive isometries, and, according to Theorem 1.5, any central symmetry is a product of two axial symmetries. B_1 results from Hilbert's axioms of incidence, B_2 is the theorems of the three symmetries and B_3 is verified with the following elements: g is a straight line in the absolute plane, A a point non-incident to g, h the perpendicular from A on g, $C := h \cap g$, $B \in g \setminus \{C\}$ and, finally, $j := $ the straight line $\overline{A, B}$.

One can then see that the group plane is a generalization of the notion of absolute plane. In the modern meaning, one understand by "absolute Geometry" the study of the group plane and of certain structures obtained by weakening the axioms of the group plane.

But let us remain at the notion of group plane, which, besides the (hyperbolic and elliptic) classical non-Euclidean Geometries, also contains many other Geometries. No suppositions of ordering or concurrence of the straight lines or of homogeneity exist in the group plane, that is points A and B can exist so that no movements should apply A in B. Then the pair A, B has no mediatrix.

We stated that the classical non-Euclidean planes are group planes. Of course, the statement is true for the hyperbolic plane, because this is an absolute plane. We want to sketch the notion of classical elliptical plane by means of one of its models to show that it is also a group plane. (We mention that, in fact, the axiomatic definition of the elliptic plane differs from Hilbert's axiomatic system by the axioms of ordering and parallelism that impose that any two straight lines should have a common point. This system of axioms is also a categorical one, so the elliptical plane can be studied on a model.)

Let S be a sphere of centre O in the classical Euclidean space. By identifying the diametrally opposed points of S, one can obtain *the points of the model* or the *e*-points. Therefore, an *e*-point is a pair of points of S. *The straight lines of the model* or *the e-straight lines* are the big circles of S, whose diametrally opposed points have been identified. It immediately results that two distinct *e*-straight lines always have only a common *e*-point. Two *e*-straight lines are called *perpendicular* if their planes are perpendicular. Every *e*-straight line perpendicular on an *e*-straight line p passes through a fixed *e*-point P (the intersection of S with the perpendicular through O on the plane of p); the *e*-point P is called *the pole of p* and the *e*-straight line p *the polar of P*. Two distinct *e*-straight lines have always a common perpendicular, namely the polar of their *e*-point of intersection. If the *e*-point A is not the ole of b, then a unique *e*-perpendicular through A on B exist; if A is the pole of B, any *e*-straight line through A will be perpendicular on B.

The movements of the model are induced by the isometries of the Euclidian space which preserve the sphere S. The symmetry of centre O changes the diametrally opposed points of S with each other, so it induces the identical map on the model. Therefore, any movement of the model can be obtained from the Euclidean rotations round the axes passing through O. The axis of a rotation ρ cuts the sphere S in an e-point P, so ρ induces *a rotation* $\tilde{\rho}$ of the model round P. The rotation $\tilde{\rho}$ will be involutive if the angle of ρ is of $180°$; then ... will be an e-*symmetry of centre* P. But the same rotation $\tilde{\rho}$ can be also induced by the symmetry related to the plane of the polar p of P and then $\tilde{\rho}$ will appear as an e-*symmetry of axis* p. Therefore, any central e-symmetry is also an axial e-symmetry, and conversely.

Let now Γ be the group of the movements of the model and Δ the set of the involutive movements. One can immediately see that (Γ, Δ) is an *AGS* group, and the corresponding group plane can be identified with the spheric model. In this example, $\Delta = \Pi$ (the straight lines coincide with the points), in opposition with the absolute plane in which $\Delta \cap \Pi = \varnothing$.

§4. The Consequences of Bachmann's Axioms

Proposition 4.1. 1) *A straight line a will be a fixed straight line of the symmetry $\tilde{c}$ if and any if $a = c$ or $a \perp c$.*

2) *Point A will be fixed to $\tilde{c}$ if and only if $A|c$ or $A = c$.*

Proof. 1) There is $\tilde{c}(a) = a \leftrightarrow cac = a \leftrightarrow (ac)^2 = 1$. If $ac \neq 1$, then $ac \in \operatorname{Inv}\Gamma$, so $a \perp c$; and if $ac = 1$, then $a = c$.

2) $\tilde{c}(A) = A \leftrightarrow (cA)^2 = 1 \leftrightarrow cA \in \operatorname{Inv}\Gamma$ or $A = c$.

Remark. In the case of the absolute plane the equality $A = c$ will not be possible; in the case of the elliptic plane any point A will also be a straight line.

Proposition 4.2. *Let P be a point and a, b straight lines. There will be $P = ab$ if and only if $a \perp b$ and $P|a$, b.*

Proof. 1) Let be $a \perp b$ and $P|a$, b. Then $ab \in \operatorname{Inv}\Gamma$ and $ab \in \Pi$; on the other hand, $a(ab) = b \in \operatorname{Inv}\Gamma$ and $b(ab) \in \operatorname{Inv}\Gamma$, so $ab|a$, b. As $P|A$, b, too, from B_1 it results that $P = ab$.

2) Let be $P = ab$; then $ab \in \operatorname{Inv}\Gamma$, so $a \perp b$; as $aP = b \in \operatorname{Inv}\Gamma$, there will be $P|a$ and, analogously, $P|b$.

Corollary 4.1. *Two perpendicular straight lines, a and b, are incident to a single point, to ab.*

Proposition 4.3. 1) *For any point A and straight line a, there will exist a straight line b so that Alb and $a \perp b$.*

2) *The straight line b is unique if and only if $A \neq a$.*

3) *If Ala, then $A \neq a$ and $b = Aa = aA$.*

Proof. The following three cases are possible:

i) *Ala*. Then $A \neq a$, since $aa = 1 \notin \operatorname{Inv}\Gamma \cdot A = ef$ and e, f, aIA, so, according to the axiom B_2, $b: = efa$ is a straight line. From $ba = ef = A \in \operatorname{Inv}\Gamma$ it results that $a \perp b$, and from $bA = bba = a \in \operatorname{Inv}\Gamma$ we deduce that *Alb*. If b' is a straight line with $a \perp b'$ and Alb', then, according to Proposition 4.2: $A = b'a$ so $b'a = ba$ and $b' = b$.

ii) $A = a$. We write A under the form $A = ef$, then $ea = f = \operatorname{Inv}\Gamma$, $af = e \in \operatorname{Inv}\Gamma$, so $a \perp e$, f and *Alae*, f. Since $e \neq f$, there will exist several straight lines incident to A and perpendicular on a.

iii) $(Aa)^2 \neq 1$. Then $A': \tilde{a}(A) = aAa \neq A$. Let be $b: \overline{A, A'}$. Since $A, A'Ib$, there will be $\tilde{a}(A), \tilde{a}(A')I\tilde{a}(b)$, that is $A', Alaba$, so $\overline{A', A} = aba$. Therefore, $(ba)^2 = 1$. But $b \neq a$ (since *Alb* and non *Ala*), so $ba \in \operatorname{Inv}\Gamma$ and $b \perp a$. If for the straight line b', there is: Alb', $a \perp b'$ then $\tilde{a}(b') = b'$; one can notice that $\tilde{a}(A)I\tilde{a}(b)$, $\tilde{a}(b')$, so $A, A'Ib$, b', which, by virtue of B_1, implies $b = b'$.

Corollary 4.2. *For any straight line a, there will exist at least one point incident to a.*

Indeed, according to B_3, there exists at least a point P (gh, for example). So, by virtue of Proposition 4.3, there will be a straight line b perpendicular on a; but then ab will be incident to a.

Proposition 4.4. *The straight lines a, b, c will be perpendicular two if and only if $abc = 1$.*

Proof. Let be $a \perp b \perp c \perp a$. Then ab will be a point and $abla$, b. Taking also $a, b \perp c$, $a \neq b$ into account, Proposition 4.3.2) shows us that $ab = c$, so $abc = 1$. Conversely, if $abc = 1$, then $ab = c \in \operatorname{Inv}\Gamma$, so $a \perp b$ and, analogously, $b \perp c$, $c \perp a$.

If $abc = 1$, the set $\{a, b, c\}$, will be called *a polar triangle*. If (Γ, Δ) has a polar triangle, then the group plane $\mathcal{P}(\Gamma, \Delta)$ will be called *elliptic*.

Proposition 4.5. *Let be $abc = d$. If a, b, cIA, then dIA; if $a, b, c \perp e$, then $d \perp e$.*

Proof. Let us suppose that a, b, cIA. We have to show that $abcA = Aabc$ and

$abc \neq A$. The first relation results from the fact that A is permutable with a, b and c. If we had $abc = A$, then, from $ab = Ac \in \mathrm{Inv}\Gamma$, we should deduce $a \perp b$ and $A = ab$; it would follow from here that $Ac = abc = A$, so $c = 1$, which is false. We are leaving to the readers the proof of the second statement.

Proposition 4.5 completes the axiom B_2. The following proposition is a converse of B_2.

Proposition 4.6. 1) $a \neq b \wedge Pla$, $b \wedge abc = d \Rightarrow Plc$.

2) $a \neq b \wedge e \perp a$, $b \wedge abc = d \Rightarrow e \perp c$.

Proof. 1) According to Proposition 4.3, there exists $\ell \in \Delta$ with $Pl\ell$ and $\ell \perp c$. Then $\ell c = : Q \in \Pi$ and $f: \ell ba \in \Delta$ (by virtue of B_2). Using Proposition 4.5, there will be Plf. On the other hand, $Q = \ell c = \ell baabc = fd$, so, according to Proposition 4.2, Qlf and thus P, Q. If, ℓ. It results that $P = Q$ or $f = \ell$. In the second case we deduce from $f = \ell ba$ that $a = b$, which is not possible. So $P = Q$ and Plc.

2) By virtue of the above corollary, P exists so that Plc. If $P = e$, then, obviously, $e \perp c$. Let further on be $P \neq e$ and c' the perpendicular from P on e. Let us suppose that $c \neq c'$. According to axiom B_2, $d': abc \in \Delta$ and according to Proposition 4.5, $d' \perp e$. Since $cc'd' = cc'd'^{-1} = cc'c'ba = cba = d^{-1} = d$, c' and $c \neq c'$, from part 1) of this Proposition it results that Pld'. As $P \neq e$, Pld', c' and d', $c' \perp e$, using proposition 4.3.2), one obtains: $d' = c'$, from where one can deduce that $a = b$, in contradiction with the hypothesis. Therefore, $c = c'$ and thus $e \perp c$.

Let be α, β, $\gamma \in \Delta \cap \Pi$. The following relations will play an important part further on:

(4.1) $\alpha \gamma \beta \in \Delta$,

(4.2) $\alpha \beta \gamma \in \Pi$.

Proposition 4.7. *Relations (4.1) and (4.2) are symmetric (they keep being valid for any permutations of α, β, γ) and reflexive (if two of the elements are equal and the third one is a straight line in case (4.1) and a point in case (4.2), respectively, the relation will hold).*

Proof. Let us suppose that relation (4.1) is true. Then $\gamma \beta \alpha = (\alpha \beta \gamma)^{-1} = \alpha \beta \gamma \in \Delta$ and $\gamma \alpha \beta = \gamma (\alpha \beta \gamma) \gamma = \tilde{\gamma}(\alpha \beta \gamma) \in \Delta$. Since any permutation of α, β, γ is obtained by successively applying these two permutations, relation (4.1) will be symmetric. If $\beta \in \Delta$, there will obviously be $\alpha \alpha \beta \in \Delta$ and $\beta \alpha \alpha \in \Delta$, and $\alpha \beta \alpha = \widetilde{\alpha}(\beta) \in \Delta$, so relations (4.1) will be reflexive. The same proof is also valid in the case of relation (4.2).

Proposition 4.8. *The relation $AbC \in \Delta$ will hold if any only if there exists a straight line d so that A, Cld and $b \perp d$.*

Proof. The case $A = C$. Then, for any b, the relation $AbA \in \Delta$ will be true (according to Proposition 4.7) and there will also exist d with Ald, $b \perp d$ (according to Proposition 4.3).

The case $A \neq C$. Let be $d : \overline{A, C}$, $a : \neq Ad$, $c : dC$ (Figures 83). Then $a \neq c$ (since in the contrary case there would be $Ad = Cd$ and $A = C$) and

$$AbC \in \Delta \Leftrightarrow ACb \in \Delta \Leftrightarrow (ad)(dc)b \in \Delta \Leftrightarrow acb \in \Delta.$$

Taking axiom B_2, Proposition 4.6.2) and $d \perp a$, c into account, the last relation will be equivalent with $b \perp d$.

Proposition 4.9. *The relation $aBc \in \bigcap$ will be true if and only if there exist $d \in \Delta$ so that a, $c \perp d$ and Bld.*

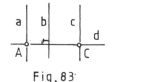

Fig. 83 Fig. 84

Proof. i) Let us suppose that $aBc = P \in \bigcap$. If Bla, then, noting $d : aB$, there will be $d \in \Delta$, a, $c \perp d$ and Bld. Let us suppose then that B is not incident to a (figure 84). Since $a = PcB$, applying Proposition 4.8, we shall find a straight line d for which P, Bld, $c \perp d$. Therefore, d is permutable with P, B, c and thus $da = d (PcB) = (PcB)d = ad$ and $(ad)^2 = 1$. But $ad \neq 1$ (otherwise it would result Bla), so $a \perp d$.

ii) Let now be a, $c \perp d$, Bld. Noting $b : Bd$, there will be a, b, $c \perp d$, so $\ell : abc \in \Delta$ and $\ell \perp d$. Therefore, we can write $aBc = abdc = abcd = \ell d \in \bigcap$.

§5. The Theorem of the Perpendiculars
and Its Applications

If a and b are two distinct straight lines, the set of straight lines $[ab]$: $\{x \in \Delta \mid abx \in \Delta\}$ will be called a *pencil (of straight lines)* with a, b as base. The relation $abc \in \Delta$ is read: "the straight lines a, b, c belong to a pencil".

When a, b are concurrent straight lines in a point P, the axiom B_2 shows that any straight line incident to P will belong to $[ab]$, and Theorem 1.6 shows that $x \in [ab] \Rightarrow P/x$. Therefore, in this case, $[ab]$ is the set of the straight line incident to P; there is a *central pencil*. Analogously, if a, b have a common perpendicular d, $[ab]$ will coincide with the set of the straight lines perpendicular on d and will be called *orthogonal pencil*. In the Euclidean plane, two distinct straight lines are either concurrent ones or have a common perpendicular, so any pencil is either central or orthogonal. In the elliptic plane, each pencil is a central one. But, in general, a pencil has not to be central or orthogonal.

The importance of the relation of belonging to a pencil resides in the fact that many Euclidean theorems in which the concurrence of three straight lines occurs, they are generalized if we replace the concurrence by the belonging to the pencil. To obtain such theorems we need a criterion of belonging to the pencil. This is given by the following theorem, established by J. Hjemslev:

Theorem 5.1 (The theorem of the perpendiculars). *If $aa' = A$, $cc' = C$ and $a'bc' = d$, then $abc \in \Delta \Leftrightarrow \exists \ell \in \Delta$: A, $C\ell\ell$, $d \perp \ell$.*

Proof. Since $abc = aa'a'bc'c'c = AdC$, Theorem 5.1 will result from Proposition 4.8 (Figure 85).

Remark. In the particular case of the Euclidean plane, when a and c, and also a' and c', are concurrent, Theorem 5.1 easily results from the property of the angles inscribed into circle, because in this case $a'bc' = d$, which is equivalent with $a'b = dc'$, it means that by rotating a' on b, the straight line d will take the position of c'.

Theorem 5.2. *For any pencil $[ac]$ and point P, not simultaneously incident to a and c, there will exist a unique straight line b incident to P in $[ac]$.*

Proof. Let a' and c' be perpendicular on a and c, respectively, incident to P (which exists by virtue of Proposition 4.3) and A: $= aa'$, C: $= cc'$, and ℓ a straight line incident to A and C (Figure 86). Let d be a straight line incident to P, perpendicular on ℓ. According to axiom B_2, $b:\, = a'dc' \in \Delta$, and according to Proposition 4.5, *Plb*. There is $d = a'bc'$ and from Theorem 5.1 it results that $abc \in \Delta$, so $b \in [ac]$.

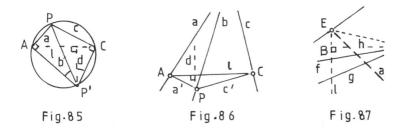

Fig·85   Fig·86   Fig·87

To prove the uniqueness, let us first establish that $P \neq \ell$. When supposing that $P = \ell$, there will be a', $c' \perp \ell$ and since only one perpendicular on a' exists through A (incident to a'), it results that $a = \ell$ and, analogously, $c = \ell$, which is false, because $a \neq c$. Let us suppose that $b_1 \in [ac]$ and Plb_1. According to Theorem 5.1, the straight lines $a'bc'$ and $a'b_1c'$ are perpendicular on ℓ and incident to P. As $P \neq \ell$, we deduce from Proposition 4.3 that $a'bc' = a'b_1c'$, so $b = b_1$.

Theorem 5.3 (The reduction theorem). *A product of an even and an odd number of straight lines, respectively, can always be written under the form ab and aB (also Ab), respectively.*

Proof. i) Given the straight lines e, f and the point P, there will exist a, $b \in \Delta$ so that $efP = ab$. Indeed, if $e = f$, the statement is obvious. If $e \neq f$, we can determine, by virtue of Theorem 5.2, the straight line $g \in [ef]$ so that Plg. Then $b:\, = gP \in \Delta$, $a:\, = efg \in \Delta$ and $efP = (efg)(gP) = ab$.

ii) For any straight lines e, f, g there will exist $a \in \Delta$ and $B \in \Pi$ for which $efg = aB$. Indeed, we can suppose that $f \neq g$ (Figure 87). We know that there exist Ele and $e' \in [f, g]$, Ele'; then $h:\, = e'fg \in \Delta$. Let ℓ be a perpendicular on h, incident to E, $B:\, \ell h \in \Pi$ and $a:\, ee'\ell \in \Delta$. It results $efg = (ee'\ell)\ell(e'fg) = a\ell h = aB$.

iii) A product $efgh$ can be written, according to ii), under the form ejB and then, according to i), under the form ab. Applying this procedure several times, any product of an even number of straight lines can be written as a product of two straight lines. But then any add product of straight lines takes the form efg, which is then transformed, by procedure ii), into the form aB. By slightly modifying the construction ii), aB can take the form Ab.

Corollary 5.1. *Any element of Γ is equal with a product of two involutive elements.*

The elements of Γ of the form aB can also be written as abc so that a, $b \perp c$. Indeed, if $c \perp a$, B/c and $b \perp c$, then $B = bc$. The movement $a\tilde{B}$ is called *symmetry with sliding*. If $aB \neq 1$, there will exist a unique perpendicular on a, incident to B, called *the axis* of aB.

Theorem 5.4. Inv $\Gamma = \Delta \cup \sqcap$.

Proof. Let be $\alpha \in$ Inv Γ. If $\alpha = ab$, then, according to the definition, α will be a point. If $\alpha = aB$, then a/B and, according to Proposition 4.3, aB will coincide with the perpendicular on a, incident to B.

Corollary 5.2. *The centre of group Γ is equal with $\{1\}$.*

Indeed, let γ be an element of the centre. We can write $\gamma = \alpha\beta$, where $\alpha, \beta \in$ Inv Γ, so $(\alpha\beta)^2 = (\alpha\beta) \cdot \alpha \cdot \beta = \alpha \cdot (\alpha\beta) \cdot \beta = (\alpha\alpha)(\beta\beta) = 1$. So $\gamma = 1$ or $\gamma \in$ Inv Γ. In the second case, γ is a straight line a or a point A; but, according to B_3, there exists B non-incident to a and b non-incident to A, respectively; it results that $aB \neq Ba$ and $Ab \neq bA$, respectively, so neither a nor B can belong to the centre. Therefore, only the case $\gamma = 1$ is possible.

Theorem 5.5. *There exists an isomorphism of group Γ on the group $\tilde{\Gamma}$ to which the image of Δ is $\tilde{\Delta}$ (the structures (Γ, Δ) and $(\tilde{\Gamma}, \tilde{\Delta})$ are isomorphic).*

Proof. One can notice that the map $\varphi : \Gamma \to \tilde{\Gamma}$, defined by $\varphi(\alpha) = \tilde{\alpha}$, is a surjective morphism of groups. The nucleus of φ is

$$\text{Ker } \varphi = \{\gamma \backslash \tilde{\gamma} = 1_\Gamma\} = \{\gamma \backslash \forall \xi \in \Gamma : \gamma \xi \gamma = \xi\} = \{\gamma \backslash \forall \xi \in \Gamma \backslash \gamma \xi = \xi \gamma\},$$

hence Ker φ coincides with the centre of Γ. From the above corollary it results Ker $\varphi = \{1\}$, so φ is an isomorphism. By definition, $\tilde{\Delta} = \varphi(\Delta)$.

Theorem 5.6. *If (Γ, Δ) is elliptic, then :*

i) *any $\gamma \in \Gamma$ can be written as ab* ;

ii) $\Delta = \sqcap$;

iii) *any two straight lines will be incident to a point.*

Proof. i) Since (Γ, Δ) is elliptic there will exist a_0 , b_0 , $c_0 \in \Delta$ so that $a_0 b_0 c_0 = 1$. If $\gamma \in \Gamma$ is of the form $\gamma = efg$, then $\gamma = efga_0 b_0 c_0$ and, according to the reduction theorem, γ is a product of two straight lines. So all the elements of Γ have the form *ab*.

ii) In particular, any straight line d can be written as $d = \ell \cdot m$ and since $\ell \cdot m \in$ Inv Γ, $d \in \sqcap$. Conversely, let $P = a_1 b_1$ be any of the points; we have seen that a_1 is a point and as $a_1 b_1 \in$ Inv Γ, $a_1 I b_1$, so c_1 exists so that $a_1 = c_1 b_1$. We finally obtain $P = c_1 b_1 b_1 = c_1 \in \Delta$.

iii) It is sufficiently to consider two distinct straight lines a and b. As a, b are also points, there will exist the straight line $c := \overline{a, b}$ so that a, bIc. As c is also a point, the Theorem is proved.

Theorem 5.7. *If (Γ, Δ) is not elliptic, then the elements of the form ab make up a normal sub-group in Γ of the index 2 and $\sqcap \cap \Delta = \varnothing$.*

Proof. One can notice by means of the reduction theorem that $\Delta \Delta = \{ab \setminus a, b \in \Delta\}$ is a subgroup of Γ. No element of the form abc belongs to $\Delta \Delta$, since from $abc = ef$ it result $f = eabc = gh$ and $fgh = 1$, contrary to the hypothesis. Therefore, $\Delta \Delta \cap \Delta \Delta \Delta = \varnothing$. One can easily see that for any $a \in \Delta$ there will be $a \Delta \Delta = \Delta \Delta \Delta$, so $\Delta \Delta$ has two cosets on the left: $\Delta \Delta$ and $\Delta \Delta \Delta$, from where it results that $\Delta \Delta$ is a normal sub-group. As $\sqcap c \Delta \Delta$ and $\Delta \subset \Delta \Delta \Delta$, there will be $\sqcap \cap \Delta = \varnothing$.

We are giving now two theorems without proof:

Theorem 5.8. *If the distinct straight lines c and d belong to the pencil [ab], then $[ab] \equiv [cd]$.*

Theorem 5.9. *If $(abc)^2 \neq 1$, $e \in [bc]$, $e \perp a$, $f \in [ca]$, $f \perp b$, $g \in [ab]$, $g \perp c$ then the straight lines e, f, g will belong to a pencil.*

We put an end to the introduction into the Geometry of the group plane with the following remarks:

1) There exists the theorem which generalizes the concurrence of the mediatrixes, bisectors and medians, respectively, of a triangle, as Theorem 5.9 generalizes the concurrent property of the heights of a triangle.

2) From Theorem 5.6 it results that the elliptic group plane is a projective plane.

3) If the group plane is not elliptic and has the property that for A and a being nonincident there are at most two straight lines incident to A, which are not secant to a and do not have a perpendicular common with a, either, then it is called "hyperbolic group plane" and it is a generalization of the classical hyperbolic plane. One can prove that a hyperbolic group plane $\mathcal{P}(\Gamma, \Delta)$ can be embedded in a projective plane, the points of the latter one being the pencils of straight lines of $\mathcal{P}(\Gamma, \Delta)$, (the central pencils identify themselves with their centres), and that only one improper straight line must be added to the straight lines of $\mathcal{P}(\Gamma, \Delta)$.

4) A group plane is called "Euclidean" if there exists a "rectangle", that is four distinct straight lines a, b, c, d so that a, $b \perp c$, d ; it generalizes the classical Euclidean plane.

CHAPTER X

HINTS

CHAPTER I

1. Let us note for a binary relation ρ by F_ρ, S_ρ its first and second base sets. There exists $\sigma \circ \rho$ iff $S_\rho = F_\sigma$. It follows $F_{\sigma \circ \rho} = F_\rho$, $S_{\sigma \circ \rho} = S_\sigma$. Then

$$x[(\alpha \circ \beta) \circ \gamma]y \Leftrightarrow \exists u \exists v \ x\gamma u \wedge u\beta v \wedge v\alpha y.$$

2. The map $f: Z \rightarrow N$ defined by

$$f(x) = \begin{cases} 2x & \text{if } x \geq 0, \\ -1-2x & \text{if } x < 0 \end{cases}$$

is a bijection.

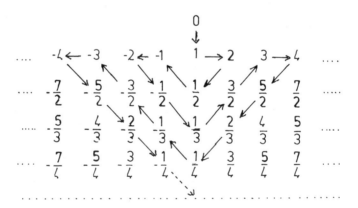

We can present Q in a plane configuration in the following way. On the first line we put only 0. On the second line we put $Z^* = Z - \{0\}$ (1 under 0). On the third line we put all the simplified fractions with the denominator 2 ($\frac{1}{2}$ under 1)... On the $(n+1)$-line we put all the simplified fractions with the denominator n. A way through all these rational numbers is suggested by arrows: only an arrow is vertical (from 0 to 1), some are horizontal on the second line (from $2x-1$ to $2x$; to the right if $x \geq 0$, to the left if $x < 0$) and several have the slope ± 1. A bijection $g: N \rightarrow Q$ appears if we take $g(n)$ the n-th rational way touched by this trip.

3. Let $\varepsilon(n)$ be the number we are looking for. We agree (*) $\varepsilon(0) = 1$, $\varepsilon(1) = 1$. To count the number $\varepsilon(n+1)$ let α be an equivalence of $\{1, 2, ..., n, n+1\}$ and X_α be the equivalence class of $(n+1)$, $X_\alpha = \{n+1\}\bigcup Y_\alpha$, where $Y_\alpha \subset \{1, ..., n\}$. Let be $k = Y_\alpha$. If we fix k there are C_n^k sets Y_α; for one of these, Z, we can reconstitute α by knowing an equivalence of the set $\{1, n\}\setminus Z$ and there are $\varepsilon(n-k)$ such equivalences. We deduce the recursive identity.

$$(**)\quad \varepsilon(n+1) = \sum_{k=0}^{n} C_n^k \varepsilon(n-k)..$$

Now (*) and (**) define ε. For example $\varepsilon(2) = 2$, $\varepsilon(3) = 5$, $\varepsilon(5) = 15$ and so on.

4. The condition is:

$$x\varepsilon x' \wedge y\varepsilon y' \Rightarrow (x \perp y)\varepsilon(x' \perp y').$$

5. To point out $1+\omega = \omega$ we interpret the first member as the ordinal of $\{-1\}\bigcup N$ and a similitude f from this set to N is given by $f(x) = x+1$.

We interpret $\omega+1$ as the ordinal of $N\bigcup\{+\infty\}$ and we remark that this ordered set, having a maximal element can not be similar to N.

We interpret $\omega \cdot 2$ as the ordinal of $\{0, 0^*, 1, 1^*, 2, 2^*, ..., n, n^*, ...\}$ obviously equal to $\omega = \mathrm{ord}\,N$. We interpret $2 \cdot \omega$ as the ordinal of $\{0, 1, 2, ..., 0^*, 1^*, 2^*, ...\}$ and in this ordered set there exist no predecessor of 0^*.

6. Let U be the universe of the sets A, $C = \mathrm{card}\,U$ and $M = \{m: X \to C\}$. The function $f: U_X \to M$ defined by $f(A, \alpha) = m$ and $m(x) = \alpha^{-1}(x)$ establishes a bijection between $\mathrm{card}\,U_X$ and M.

CHAPTER II

1. No, because now $a+x = b+x$ does not imply $a = b$.

2. Let $X = (x_n)_{n \in N}$ be a Cauchy sequence. For $i = 1, 2$ subsets $T_i(X)$ of Q are defined by:

$t \in T_1(X) \Leftrightarrow \exists A \in \mathbb{N}: n > A \Rightarrow t \leq x_n$

$t \in T_2(X) \Leftrightarrow \exists B \in \mathbb{N}: n > B \Rightarrow x_n < t.$

There are no difficulties to prove that the couple $t(X) = (T_1(X), T_2(X))$ is a Dedekind cut, hence $t: \mathbb{C} \to T(Q)$. If we take another Cauchy sequence Y, it follows $t(X) = t(Y) \Leftrightarrow X - Y \in \underline{N}$ bijection $\tilde{t}$ from $\mathbb{R} = \mathbb{C} / \underline{N}$ to $T(Q)$.

3. Obviously $r = \sqrt{a^2 + b^2}$; there exists only one t, $0 \leq t < \pi$ so that $\cos t = \dfrac{a}{r}$ and $\sin t = \dfrac{b}{r}$.

CHAPTER III

1. A binary relation ε on A satisfies the system E iff ε is an equivalence, hence E is consistent. An order relation ρ on A satisfies $E1$ but not $E2$. For $A = \{1, 2, 3\}$ let us take ω so that $x \rho y \Leftrightarrow (x, y) \in \{(1, 2), (2, 1)\}$; now $E2$ is verified but $E1$ is not (for $x = 3$). Hence E is minimal.

2. The given system specifies the distances on A and it is consistent.
For $A = \mathbb{N}$, $d(x, y) = 0$ the independence of $D1$ is ensured.
For $A = \mathbb{Z}$, $d(x, y) = x - y$ the independence of $D2$ is ensured.
For $A = \mathbb{Z}$, $d(x, y) = (x - y)^2$ the independence of $D3$ is ensured, too.

3. There results $f(a, b) = a^b$.

CHAPTER IV

There are no difficulties to verify the conditions of the Definition 1.1 (p. 63) for $(V \times G, \oplus, \otimes)$.

CHAPTER V

1. a) Proposition 1.9. Let B, C be distinct points of A. The plane $\beta = (A, B, C)$ satisfies the conditions for α and it is the only one which does it.
b) Proposition 1.10. Let α be the given plane. I_4 ensures the existence of a point A in α. By I_8 we deduce the existence of a point M not belonging to α, and after that of a point N not belonging to AM. By I_5 there exists $\beta = (A, M, N)$, distinct of α. We

apply I_7 for α and β to deduce the existence of $B \neq A$ in α. Using again I_8, we find $P \notin \beta$. We take $\gamma = (A, M, P)$ and we find $C \in \alpha \cap \gamma$. There are no difficulties to point out the absurdity of the supposition that A, B, C would be co-linear.

2. Let A, A' be distinct points. Using II_2 we find D, D' so that $A' - A - D$ and $A - A' - D'$. By twice applying the Theorem 2.1 we find M, M' so that $A - M - A'$ and $M - M' - A'$. Using I_8 we deduce the existence of a point N non incident to AA'. Using again the axiom II_2 and the Theorem 2.1 we find: E, E', F, F' so that $N - M - E$, $N - M' - E'$, $N - F - M$ and $N - F' - M'$. (Draw yourself a figure.) It is easy to state that DEF and $D'E'F'$ are triangles, $a \in \operatorname{Int}\triangle DEF$, $A' \in \operatorname{Int}\triangle D'E'F'$, and the interiors of these two triangles are disjoint.

3. Enounce: The interiors of tetrahedrons constitute two neighbourhoods of a Hausdorf topology of the space.

Proof: The above considerations are continued taking $P \notin (A, A', N)$ and Q so that $P - N - Q$. Pash's axioms ensures the existence of $G \in PF \cap [EQ]$ and $G' \in PF' \cap E'Q$. The tetrahedrons will be $DEGP$ and $D'E'G'P$.

4. The Archimedes' axiom is valid in U' but not in $U \cup U'$ (if $u \in U'$ and $v \in U$ there is no n so that $v < n \cdot u$).

5. Let E_3 be a model of Hilbert's axiomatics and $S(O, R)$ a sphere in this model: $M \in S \Leftrightarrow OM = R$. Let $f: [0, \infty) \to [0, R)$ be a bijection; $f(x) = \frac{Rx}{R+x}$ for example. We get a bijection $T: E_3 \to \operatorname{Int} S$, by defining: $T(0) = 0$ and, for $X \neq O$, $T(X) = X'$ so that $X' \in (A, X)$ and $AX' = f(AX)$. This bijection T transfer the geometric structure of E_3 to $\operatorname{Int} S$ and this set is bounded (in the initial acception given by E_3 structure).

CHAPTER VI

Using the techniques expressed in Chapter V and VI there are no difficulties to prove the enounced theorems.

CHAPTER VII

1. *First solution.* It is easy to prove that the symmetry S related to AB transform H in N on the circle C. An injective correspondence $f : C \to C$ ensures $f(M) = N$, $f(N) = M$. Since $H = S(N)$, the locus of H is included into the circle $C' = S(C)$. But M can not touch A, B; consequently N can not touch $A' = f(A)$ and $B' = f(B)$ and H can not touch $A'' = S(A')$ and $B'' = S(B')$. (Draw yourself the figure: $ABB'A'$ and $ABB''A''$ are equal rectangles.) The locus we are looking for is $C' \setminus \{A'', B''\}$.

Second solution. Let us consider the midpoint D of AB and the symmetry U related to D. It is easy to prove that $P = U(H)$ is on C and $P = g(M)$, where $g : C \to C$ associates to a point the diametrally opposed related to C, $g(M)$. We find again the locus $C' \setminus \{A'', B''\}$, defining now: $C' = U(C)$, $A'' = U \circ g(B)$, $B'' = U \circ g(A)$.

Third solution. Let us consider B' the diametrally opposed of A in C. (For M in C the orthocentre H is in B.) It is easy to prove that $B'BHM$ is a parallelogram. Let T be the translation specified by $T(B') = B$; the locus is $C' \setminus \{A'', B''\}$, where $C' = T(C)$, $A'' = T(A)$, $B'' = T(B)$.

2. Draw yourself a figure: ABC equilateral and parallel straight lines a, b, c through A, B, C so that b cuts (AC). Let R be the rotation of centre A which leads B in C, and let us consider $b' = R(b)$. Obviously $C \in b' \cap c$.

To prepare the calculus, let us consider equilateral triangles AMN, ANP so that $M \in b$, $N \in b$, $P \in a$. Let D be the midpoint of MN, $E \in AD \cap c$, $F \in AN \cap c$. Obviously, $b' = NP$, $C \in NP \cap c$, $CF = 2n/\sqrt{3}$, $EF = (m+n)/\sqrt{3}$, $EC = |m-n|/\sqrt{3}$. Therefore,

$$\ell^2 = AC^2 = AE^2 + EC^2 = (m+n)^2 + \frac{1}{3}(m-n)^2 = \frac{4}{3}(m^2 + mn + n^2).$$

3. The major difficulties of the problem arises from the orthocentres. But the homothety H of centre O and ratio $\dfrac{1}{3}$ leads the orthocentre of a triangle inscribed in C in the barycentre of the same triangle. Let G, G_a, G_b, G_c, G'_a, G'_b, G'_c the barycentres of the sever considered triangles, images by H of H, H_a, H_b, H_c, H'_a, H'_b, H'_c. Now the problem is to prove the co-linearity of G, G_a, G_b, G_c, G'_a, G'_b, G'_c.

To pass from ABC to $A'BC$ (and from G to G_a) we consider the midpoint D of BC. A homothety of centre D and ratio $\frac{1}{3}$ transform A, A' into G, G'. Therefore, GG' has the direction δ and so on.

4. Let M, N, P be the contact points of (A, C), (C, A), (A, B). Let I be the inversion of pole P and power $-4ab$ (the sign is chosen to have less points on the figure). Let D, E be the diametrally opposed points to P in A, B. Obviously, $I(D) = E$ and $I(E) = D$. Circles A, B are transformed by I into straight lines d, e perpendicular on DE: $D \in e$ and $E \in d$; the distance between d, e is $2(a+b)$. Let F be $I(C)$; it will be a circle tangent to d and e; its radius is $a+b$.

There are not difficulties to construct circles $G(G, a+b)$, $H(H, a+b)$ tangent to d, e and F. Obviously, X, Y are images of G, H in the inversion I. (The degeneration stipulated in the enunciation comes to $P \in G \bigcup H$.) We deduce

$$4ab(a+b)(\frac{1}{x}+\frac{1}{y}) = |G(P)| + |H(P)|.$$

By evaluating the powers in the second member we quickly find

$$|G(P)| + |H(P)| = 4(a+b) \cdot 2 \cdot FF',$$

where FF', the distance from F' to DE is given by

$$FF' = \frac{a+b}{c} \cdot CC' = \frac{2}{c} \cdot \sigma(ABC) = 2 \cdot \sqrt{\frac{ab}{c}(a+b+c)}$$

and we find the enounced equality.

5. Let I be an inversion of pole A and power p. Obviously, I transforms the circumcircle (without the point A) into a straight line to which belongs the points $B' = I(B)$, $C' = I(C)$, $D' = I(D)$. The following equality holds

$$B'D' = B'C' + C'D'$$
$$AB'^2 \cdot C'D' + AD'^2 \cdot B'C' = (AC'^2 + B'C' \cdot C'D') \cdot B'D'.$$

In these equalities we are only to replace

$$B'D' = \frac{p \cdot BD}{AB \cdot AD}, \; AB' = \frac{p}{AB}$$

and the analogous ones to get the enounced equalities.

CHAPTER VIII

1. Let us consider the trihedron of the vertex O and its edges passing through A, B, C. Obviously, $\sphericalangle BOC = \frac{a}{R}$ and so on. The plane perpendicular in A to OA cuts (!) OB, OC in D, E. (If it is not convenient to accept the eventuality that one or both of these be improper the special cases $\frac{b}{R} = \frac{\pi}{2}$, $\frac{c}{R} = \frac{\pi}{2}$ may be separately analyzed.) By expressing DE^2 by generalized Pithagoras' theorem in triangles ODE and ADE, we get

$$OD^2 + OC^2 - 2 \cdot OD \cdot OC \cdot \cos\frac{a}{R} = AD^2 + AC^2 - 2 \cdot AD \cdot AC \cdot \cos\dot{A}.$$

Using Pythagoras' theorem in triangles OAD and OAE this equality becomes

$$OD \cdot OC \cdot \cos\frac{a}{R} = R^2 - AD \cdot AC \cdot \cos\dot{A}.$$

We divide this equality by $OD \cdot OC$ and we replace

$$\frac{R}{OD} = \cos\frac{c}{R}, \; \frac{R}{OE} = \cos\frac{b}{R}, \; \frac{AD}{OD} = \sin\frac{c}{R}, \; \frac{AE}{OE} = \sin\frac{b}{R}.$$

We get

$$\cos\frac{a}{R} = \cos\frac{b}{R}\cos\frac{c}{R} - \sin\frac{b}{R}\sin\frac{c}{R}\cos\dot{A}$$

the same equality as the enounced one.

2. Let us consider the "polar" spherical triangle $\dot{\triangle}DEF$ of $\dot{\triangle}ABC$; that means that D is in the same semisphere delimited by (OBC) as A is, $OD \perp (OBC)$ and analogously to define E, F. We quickly point out the equalities

$$\dot{D} = \pi - \frac{a}{R}, \ \dot{E} = \pi - \frac{b}{R}, \ \dot{F} = \pi - \frac{c}{R};$$

$$\frac{d}{R} = \pi - \dot{A}, \ \frac{e}{R} = \pi - \dot{B}, \ \frac{f}{R} = \pi - \dot{C}.$$

By these equalities, 1 becomes 2.

3. Let us project A on (OBC), OB, OC in M, N, P. By expressing AM in the rectangle triangles AMN and AMP we point out:

$$AN \cdot \sin \dot{B} = AP \cdot \sin \dot{C}.$$

But $AN = R \sin \dfrac{c}{R}$ and $AP = R \sin \dfrac{b}{R}$ lead this equality to

$$\frac{\sin \dot{B}}{\sin \dfrac{b}{R}} = \frac{\sin \dot{C}}{\sin \dfrac{c}{R}}.$$

There are no difficulties to finish the proof.

Bibliography

Section A: Books

1. Abian, A., *La teoria degli insieme e l'aritmetica tensfinita*, Feltrineli, Milano, 1972.
2. Abramescu, N., *Analytical Study of the Non-Euclidean Geometries* (in Romanian), University of Cluj, 1926.
3. Addison, J. W. et. al., *The Theory of Models*, Proceedings of the 1963 Int. Symp., Berkely, North Holland, Amsterdam, 1965.
4. Adler, C.F., *Modern Geometry; an Integrated First Course*, Mc. Graw-Hill, New York, 1958.
5. Adler, I., *A New Look at Geometry*, John Day, New York, 1966.
6. Albu, A.C., *Lectures on the Foundations of Mathematics* (in Romanian), University of Timişoara, 1973 and 1983.
7. Albu, A.C., *Introduction in the Foundation of Mathematics* (in Romanian), University of Timişoara, 1989; Eurobit, Timişoara, 1995.
8. Albu, A.C., et. al. *Geometry for Teacher's Improvement* (in Romanian), EDP, Bucureşti, 1983.
9. Alexandroff, P.S. et. al. (editors), *Enziklopädie der Elementarmathematik*, 4 vols., DVW, Berlin, 1966.
10. Allendoerfer, C.B.; Oakley, C.O., *Principles of Mathematics*, Mc Graw-Hill, New York, 1956.
11. Artin, E., *Geometric Algebra*, Interscience, New York, 1957.
12. Artzy, R., *Linear Geometry*, Addison-Wesley, Reding, Massachusetts, 1974.
13. Bachmann, K., *Aufbau der Geometrie aus dem Spiegelungsbegriff*, Springer, Berlin, 1973.
14. Baker, H. F., *An Introduction to Plane Geometry*, Univ. Press, Cambridge, 1943.
15. Bal, L., *Synthetic Geometry* (in Romanian), EDP, Bucureşti, 1960.
16. Baldus, R., *Nichteuklidische Geometrie*, 2 vols., Göschen, Berlin, 1933-1934.
17. Baldus, R.; Lobell, F., *Nichteuklidische Geometrie*, Gruyter, Berlin, 1984 (or Göschen, Berlin, 1964).
18. Barbarin, P., *La géométrie non euclidienne*, Gauthier-Villars, Paris, 1928.
19. Barbilian, D., *Algebra* (in Romanian), EDP, Bucureşti, 1985.
20. Barbilian, D., *Axiomatics of the Geometry* (in Romanian), Faculty of Mathematics, Bucureşti, 1949.
21. Barbilian, D., *Didactic works* (in Romanian and German), 3 vols., Ed. Tehn., Bucureşti, 1968-1973.
22. Bârliba, D.M. et. al., *Logic, Formal truth and Interpretative Revelence* (in Romanian), Ed. Şt. Enc., Bucureşti, 1988.
23. Barry, E.H., *Introduction to Geometrical Transformation*, Prandl, Weber & Schmidt, Boston, 1966.

24. Barwise, J., *Admissible Sets and Structures*, Springer, Berlin, 1975.
25. Barwise, J. (editor), *The Syntax and Semantics of Infinitary Languages*, Springer, Berlin, 1968.
26. Becker. O., *Gröse und Grenze der Mathematischen Denkweiser*, Karl Alber, Freiburg-München, 1959.
27. Becker, O., *Grundlagen der mathematick in geschichtlicher Entwiklung*, Karl Alber, Freiburg-München, 1963.
28. Beju, A.E.; Beiu, I., *Compendium of Mathematics* (in Romanian), Ed. Şt., Bucureşti, 1983.
29. Behnke, H. et. al. (editors), *Grundlagen der Geometrie, Elementargeometrie*, Vadenhoeck und Rupert, Göttingen, 1967.
30. Bell, E.T., *Men of Mathematics*, Penguin Books, Harmond worth, 1953.
31. Bell, E.T., *The Development of Mathematics*, Mc. Graw-Hill, New York, 1945.
32. Belloc, J.C., *Mathematiques. Algebre*, Masson, Paris, 1992.
33. Benacerraf, P.; Putnam, H. (editors), *Philosophy of Mathematics*, Prentince Hall, New York, 1964.
34. Benedetti, L., *Fondamenti di geometria*, 2 vols., Hoepli, Milano, 1942.
35. Bernays, P.; Fraenkel, A., *Axiomatic Set Theory*, North Holland, Amsterdam, 1958.
36. Beth, E.W., *Les fondaménts logiques des mathematiques*, Gauthier-Villars, Paris, 1955.
37. Beth, E.W., *Mathematical Thought*, Reidel, Dordrecht (Holland), 1965.
38. Beth, E.W., *The Foundations of Mathematics*, North Holland, Amsterdam, 1959.
39. Beth, E.W.; Piaget J., *Épistémologie mathématique et psychologie*, Press Universitaire de France, Paris, 1961.
40. Birkhoff, G.D.; Bettley, R., *Basic Geometry*, Chelsea, New York, 1957.
41. Bishop, E., *Foundation of Constructive Analysis*, Mc Graw-Hill, New York, 1967.
42. Blaga, L., *Experiment and Mathematical Spirit* (in Romanian), Ed. Şt., Bucureşti, 1969.
43. Blattner, J.W., *Projective Plane Geometry*, Holden Day, San Francisco, 1968.
44. Blumenthal, L.M., *A Modern View of Geometry*, Freeman, San Francisco, 1961.
45. Blumenthal, L.M.; Menger, K., *Studies in Geometry*, Freeman, San Francisco, 1970.
46. Boltiansky, V.G; Yaglom J.M., *Transformations. Vectors* (in Russian), Prosv., Moscow, 1964.
47. Bolyai, J., *Appendix* (in Latin and Romanian), Ed. Acad., Bucureşti, 1954 (Available in English in [48]).
48. Bonola, R., *La geometria non euclidea* , (Zanichelli, Bologna, 1900) Dover, New York, 1955.
49. Bonola, R.; Liebemann, H., *Die nichteuklidische Geometrie*, Teubner, Leipzig-Berlin, 1908.
50. Born, M., *Einstein's Theory of Relativity*, Dover, New York, 1962.

51. Borsuk, K.; Szmielev, W., *Foundations of Geometry*, North Holland, Amsterdam, 1960.
52. Borş, C., *Complex Numbers* (in Romanian), Ed. Tehn., Bucureşti, 1962.
53. Borş, C., *Elements of Projective Geometry* (in Romanian), Ed. Tehn., Bucureşti, 1956.
54. Boole, G., *The Mathematical Analysis of Logic*, Univ. Press, Oxford, 1951.
55. Both, N., *The Algebra of Logic and Its Applications* (in Romanian), Dacia, Cluj, 1984.
56. Bourbaki, N., *Elémentes d'histoire des mathématiques*, Herman, Paris, 1948.
57. Borbaki, N., *Théorie des ensembles*, Chap. I-IV, Herman, Paris, (Available in English: *Theory of Sets*, Addison-Wesley, Reading, Masachusetts, 1968).
58. Brânzei, D., *Circumstantial Geometry* (in Romanian), Junimea, Iaşi, 1983.
59. Brânzei, D., *Foundations of Structural Mathematics* (in Romanian), University, Iaşi, 1976.
60. Brânzei, D., *Geometric models* (in Romanian), University, Iaşi, 1982.
61. Brânzei, D., *Introduction in Geometrie* (in Romanian), University, Iaşi, 1972-1977.
62. Brânzei, D., *Theory of Categories* (in Romanian), University, Iaşi, 1982.
63. Brânzei, D. et. al., *Bases of Geometric Reasoning* (in Romanian), Ed. Acad., Bucureşti, 1983.
64. Brânzei, D. et. al., *Competence and Performance in Geometry* (in Romanian), 2 vols., Minied, Iaşi, 1992-1993.
65. Brânzei, D. et. al., *Euclidean Plane and Space* (in Romanian), Ed. Acad., Bucureşti, 1986.
66. Brânzei, D. et. al., *Analogies Triangle-Tetrahedron* (in Romanian), To appear, Ed. Acad., Bucureşti.
67. Braunschviecg, L., *Les étapes de la philosophie mathématiques*, Alcan, Paris, 1922.
68. Brisac, R., *Exposé élémentaire des principes de la géométrie euclidiene*, Gauthier-Villars, Paris, 1955.
69. Brouwer, E.L.J., *Collected Works* (1 vol: *Philosophy and Foundations of Mathematics*), North Holland, Amsterdam, 1975.
70. Brumfiel, C.E. et.al., *Geometry*, Addison-Wesley, Reading, Massachusetts, 1975.
71. Bukreev, B.Ya., *Analitical Study of Lobatchevsky's Geometry* (in Russian), Gosud Izd., Moscow, 1951.
72. Bumcrot, R., *Modern Projective Geometry*, Holt, New York, 1969.
73. Burn, R.P., *Deductive Transformation Geometry*, Univ. Press, Cambridge, 1975.
74. Burn, R.P., *Groups, a Path to Geometry*, Univ. Press, Cambridge, 1985.
75. Busemann, H.; Keely, P.J., Projective Geometry and Projective Metrics, Acad. Press, New York, 1953.

76. Calude, C., *Complexity of Calculus. Qualitative Looks* (in Romanian), Ed. St. Enc., Bucureşti, 1982.
77. Câmpan, F.T., *Bolyai* (in Romanian), Albatros, Bucureşti, 1971.
78. Câmpan, F.T., *Lights in Depth. Talks on the Mathematics Foundation* (in Romanian), Albatros, Bucureşti, 1983.
79. Câmpan, F.T., *The Adventure of Non-Euclidean Geometries* (in Romanian), Albatros, Bucureşti, 1978.
80. Câmpan, F.T., *God and Mathematics* (in Romanian), To appear, Minied, Iaşi.
81. Campbell, D.M.; Higgins, J.C. (editors), *Mathematics-People-Problems-Results*, 3 vols., Wadsworth International, Belmont, California, 1984.
82. Carnap, R., *Einführung in die symbolische Logik*, Viena, 1960.
83. Carnap, R., *Meaning and Necesity. A Study in Semantics and Model Logic*, Univ. Press, Chicago, 1947
84. Cavailles, J., *Méthode axiomatique et formalisme*, 3 vols., Herman, Paris, 1938.
85. Cavailles, J., *Remarques sur la formation de la theorie abstraite des ensembles*, 2 vols., Herman, Paris, 1938.
86. Chetveruhin, N.F., *Projective Geometry* (in Russian), G.T.I., Moscow, 1953.
87. Chislom, J.S.R., *Vectors in Three Dimensional Space*, Univ. Press, Cambridge, 1978.
88. Chistyakov, V.D., *Foundations of Geometry* (in Russian), V.S., Minsk, 1961.
89. Choquet, G., *Geometry in a Modern Setting*, Kershaw, London, 1969.
90. Church, A., *Introduction to Mathematical Logic*, Univ. Press, Princeton, 1956.
91. Ciani, E., *Lezioni di geometria proection ed analition*, Spoeri, Pisa, 1919.
92. Cohen, P.J., *Independence of the Axiom of Choice*, Stanford Univ., 1963.
93. Cohen, P.J., *Set theory and the Continuum Hypothesis*, Benjamin, New York, 1966.
94. Colerus, E., *Vom Punkt zur vierten Dimension. Geometrie für jedermann*, Bischoff, Berlin, 1943.
95. Constantinescu, C., *Set Theory* (in Romanian), Ed. Acad., Bucureşti, 1962.
96. Coolidge, J.L., *The Geometry of the Complex Domain*, Clarendon Press, Oxford, 1924.
97. Copi, I.M.; Gould, J.A. (editors), *Contemporary Readings in Logical Theory*, Macmillan, New York, 1967.
98. 98. Courant, R.; Robbins, H., *What is Mathematics?*, Oxford University Press, London, 1941.
99. Coxeter, H.S.M., *Introduction to Geometry*, Willey & Sons, New York-London, 1961 and 1969.
100. Coxeter, H.S.M., *Non-Euclidean geometry*, Univ. Press, Toronto, 1968.
101. Coxeter, H.S.M., *Projective Geometry*, Blaisdell, New York, 1964.
102. Coxeter, H.S.M., *The Real Projective Plane*, Mc Graw-Hill, New York, 1949.

103. Coxeter, H.S.M.; Greitzer, S.L., *Geometry Revisited*, Random House, New York, 1947.
104. Craioveanu, M.; Albu, I.D., *Affine and Euclidean Geometry* (in Romanian), Facla, Timişoara, 1982.
105. Creangă, I., *Algebrical Introduction in Informatics* (in Romanian), Junimea, Iaşi, 1974.
106. Creangă, I., *Introduction in Number Theory* (in Romanian), EDP, Bucureşti, 1965.
107. Creangă, I., *Linear Algebra* (in Romanian), EDP, Bucureşti, 1962.
108. Creangă, I., *Textbook of Analitical Geometry* (in Romanian), Ed. Tehn., Bucureşti, 1965.
109. Cruceanu, V., *Elements of Linear Algebra and Analytic Geometry* (in Romanian), EDP, Bucureşti, 1973.
110. Cundy, M.; Rollet, A.P., *Mathematical Models*, Clarendon Press, Oxford, 1961.
111. Dalen, D. van, *Logic and Structure*, Springer, Berlin, 1980.
112. Daus, P., *College Geometry*, Prentince Hall, New York, 1941.
113. Davenport, H., *The Higher Arithemetic*, Hutchinson, London, 1952.
114. Davis, P.J.; Hersch, R., *The Mathematical Experience*, Birhhauser, Boston, 1981.
115. Deaux, R., *Introduction ŕ la géométrie des nombres complexes*, Bruxelles, 1947 (Available in English: *Introduction to the Geometry of Complex Numbers*, Ungar, New York, 1956).
116. Degen, W.; Profke, L., *Grundlagen der affinen und euklidishen Geometrie*, Teubner, Stuttgard, 1976.
117. Delachet, A., *Contemporary Geometry*, Dover, New York, 1962.
118. Devlin, K.J., *Aspects of Constructibility*, Springer, Berlin, 1973.
119. Devlin, K.J., *Sets, Functions and Logic. An Introduction to Abstract Mathematics*, Chapman & Hall, London, 1992.
120. Diek, W., *Nichteuklidiche Geometrie in der Kugeleben*, Teubner, Leipzig, 1918.
121. Dieudonné, J., *Algébre linéaire et géométrie élémentaire*, Herman, Paris, 1964 (Available in English: *Linear Algebra and Geometry*, Houghton Mofflin, Boston, 1969).
122. Dieudonné, J., *A Panorama of Mathematics*, Acad. Press, New York, 1982.
123. Dieudonné, J., *Foundations of Modern Analysis*, Acad. Press, New York, 1960.
124. Dingler, H., *Die Grundlagen der Geometrie ihre Bedeautung fur Philosophie*, Enke, Stuttgart, 1933.
125. Dingler, H., *Die Grundlagen der angewantden Geometrie*, Akad., Leipzig, 1911.
126. Doehlemann, G., *Geometrische Transformationen*, 2 vols., Goschen, Leipzig, 1902-1908.
127. Doehlemann, G., *Projective Geometrie in Synthetisher Behandlung*, 2 vols., Goschen, Leipzig, 1924.

128. Dorward, H.L., *The Geometry of Incidence*, Prentince Hall, New York, 1966.
129. Dragomir P.; Dragomir A., *Algebric Structures* (in Romanian), Facla, Timişoara, 1975.
130. Dubislav, W., *Die Definition*, F. Meiner, Hamburg, 1981.
131. Dubislav, W., Die Philosofie der Mathematik der Gegenwart, Berlin, 1968.
132. Duican, L.; Diucan, I., *Geometric Transformations* (in Romanian), Ed. Şt. Enc., Bucureşti, 1987.
133. Dumitriu, A., *Logical Machinery of Mathematics* (in Romanian), Ed. Acad., Bucureşti, 1968.
134. Dumitriu, A., *Logic History* (in Romanian), Ed. Acad., Bucureşti, 1975.
135. Dumitriu, A., *Logic Theory* (in Romanian), Ed. Acad., Bucureşti, 1973.
136. Dumitriu, A., *Polivalent Logic* (in Romanian), Ed. Şt. Enc., Bucureşti, 1968.
137. Dumitriu, A., *The Solution of Logico-Mathematical Paradoxes* (in Romanian), Ed. Şt., Bucureşti, 1966.
138. Durell, C.V., *Modern Geometry*, Macmillan, London, 1931.
139. Durell, C.V., *Projective Geometry*, Macmillan, London, 1931.
140. Durst, L.k., The Grammar of Mathematics, Addison-Wesley, Massachusets, 1969.
141. Eaves, J.C.; Robinson, A.T., An Introduction to Euclidean Geometry, Random House, New York, 1957.
142. Eccles, F.M., *An Introduction to Transformational Geometry*, Addison-Wesley, Reading, 1979.
143. Efimov, N.V., *Géométrie superieure*, Paix, Moscow, 1982.
144. Einstein, A., *Relativity, the Special and General Theory*, Crown, New York, 1961.
145. Egorov, I.P., *Geometry* (in Russian), Prosv., Moscow, 1979.
146. Enescu, Gh., *Logical Foundations of Thinking* (in Romanian), Ed. Şt., Bucureşti, 1980.
147. Enescu, Gh., *Logical Sistems Theory Metalogic* (in Romanian), Ed. Şt., Bucureşti, 1976.
148. Enescu, Gh., *Philosophy and Logic* (in Romanian), Ed. Şt., Bucureşti, 1973.
149. Enescu, Gh., *Semantic of Logic* (in Romanian).
150. Ershov, Yu.L.; Palyutin, E.A., *Mathematical Logic* (in Russian), Nauka, Moscow, 1975.
151. Eves, H.; Newsom, C.V., *The Foundations and Fundamental Concepts of Mathematics*, Rinehart, New York, 1958.
152. Ewald, G., *Geometry, an Introduction*, Wadsworth, Belmond, 1971.
153. Fano, G., *Complementi di geometria*, Felice Gili, Torino, 1935.
154. Fano, G., *Geometria non euclidea*, Zannichelli, Bologna, 1935.
155. Faulkner, T.E., *Projective Geometry*, Intercience, New York, 1952.
156. Faure, R.; Heurgon, E., *Structures ordonees et algebre de Boole*, Gauthier-Villars, Paris, 1971.

157. Feferman, S., *The Number Systems*, Addison-Weley, Reading, Mass., 1963.
158. Felix, L., *Initiation ŕ la géométrie*, Dunod, Paris, 1964.
159. Felix, L., *Exposé moderne des mŕthématique élémentaires*, Dunod, Paris, 1959.
160. Fenstad, J.E., *General Recursion Theory*, Springer, Berlin, 1980.
161. Fischback, W.T., *Projective and Euclidean Geometry*, J. Willey & Sons, New York, 1969.
162. Fischbein, A., *Concept and Image in the Developing of the Mathematical Thinking* (in Romanian), Ed. Şt., Bucureşti, 1965.
163. Fischbein, A., *Figurative Concepts* (in Romanian), Ed. Şt., Bucureşti, 1965.
164. Fladt, K., *Elementar geometrie*, (3 vols.), Teubner, Leipzig, 1928-1931.
165. Fletcher, T.I. (Ed.), *A Handbook on the Teaching of Modern Mathematics*, Cambridge Univ. Press, Lomdon, 1964.
166. Flohr, F., F. Raith, *Affine und Euklidische Geometrie*, (2 vols.),Vadenhoest-Rupert, Gotingen, 1971.
167. Fogarasi, B., *Logik*, Berlin, 1956.
168. Förder, H.G., *The Foundations of Euclidean Geometries*, Univ. Press, Cambridge, 1927.
169. Förder, H.G., *The Calculus of Extension*, Chelsea, New York, 1960.
170. Fraenkel, A.A., *Abstract Set Theory*, North Holland, Amsterdam, 1953.
171. Fraenkel, A.; Barhillel, Y., *Foundations of Set Theory*, North Holland, Amsterdam, 1958.
172. Frege, G., *Philosophical Writings (I-Foundations of Arithmetic)* (in Romanian), Ed. Şt. Enc., Bucureşti, 1977.
173. Freudenthal, H., *The Language of Logic*, Elsevier, Amsterdam, 1966.
174. Freudenthal, H. (editor), *Raumtheorie*, Wiss. Buchgesell, Darmstadt, 1978.
175. Frey, G., *Einführung in die philosophischen Gruntlagen der Mathemathik*, Shroedel, Hanovra, 1968.
176. Froda, A., *Error and Paradox in Matematics* (in Romanian), Ed. Enc., Bucureşti, 1971.
177. Galbură, Gh., *Algebra* (in Romanian), EDP, Bucureşti, 1972.
178. Galbură, Gh., *Cours of Geometry* (in Romanian), Univ., Bucureşti, 1973.
179. Galbură, Gh.; Radó, F., *Geometry* (in Romanian), EDP, Bucureşti, 1979.
180. Gautier, C. et. al., *Alpeh₀ Algebra*, Hachette, Paris, 1970.
181. Gautier, C., *Aleph₀ Géométrie*, 2 vols. Hachette, Paris, 1970.
182. Gergely, E., *Riemann's Hypothesis on the Bases of Geometry* (in Romanian), Ed. Tehn., Bucureşti, 1963.
183. Girard, G.; Thierce, C., *Aleph₀-Geometrie*, 4 vols., Hachette, Paris, 1970.
184. Gheorghiev, Gh. et. al., *Analytic and Differential Geometry* (in Romanian), 2 vols., EDP, Bucureşti, 1968.

185. Gheorghiu, Gh.Th., *Elements of Algebra and Analytic Geometry* (in Romanian), EDP, Bucureşti, 1961.
186. Godeaux, L., *La géométrie*, Herman, Paris, 1931.
187. Godeaux, L., *Lecons de géométrie projective*, Herman, Paris, 1933.
188. Godeaux, L., *Les géométries*, Armand Colin, Paris, 1937.
189. Gödel, K., *On Formally Undecidable Propositions*, Oliver and Boyd, Edinburg, 1962.
190. Gödel, K., *The Consistency of the Continuum Hypotheses*, Univ. Press, Princeton, New Jersey, 1940.
191. Gonseth, F., *Les Entretien de Zürich sur les Fondements et la Methode des Sciences Mathematiques*, S.A. Leeman, Zürich, 1941.
192. Gonseth, F., *Les fondements des Mathematiques*, A. Blanchard, Paris, 1922.
193. Goodstein, R.L., *Mathematical Logic*, Univ. Press, Leichester, 1957.
194. Goodstein, R.L., *Recursive Number Theory*, North Holland, Amsterdam, 1957.
195. Gorski, D.P., *Logic* (in Russian), Moscova, 1958.
196. Grassmann, H., *Projektive Geometrie der Ebene*, Teubner, Leipzig-Berlin, 1913.
197. Graustein, W., *Introduction to Higher Geometry*, Macmillan, New York, 1930.
198. Green, S.L., Algebraic Solid Geometry, Univ. Press, Cambridge, 1947.
199. Greenberg, M.I., *Euclidean and Non-Euclidean Geometries*, Freeman, San Francisco, 1974.
200. Griffits, M.B.; Hilton, P.J., *A Comprehensive Textbook of Classical Mathematics*, Springer, New York-Heidelberg-Berlin, 1978.
201. Grosche, G., *Projektive Geometrie*, 2 vols., Teubner, Leipzig, 1957.
202. Grotemeyer, K.P., *Annalytische Geometrie*, Gruyter, Berlin, 1964.
203. Gruenberg, K.W.; Weir, A.J., *Linear Geometry*, Springer, Berlin, 1977.
204. Grzegorczik, A., *An Outline of Mathematical Logic*, D. Reidel & Polish Sc. Publ., Warsaw, 1974.
205. Guggenheimer, H.W., *Plane Geometry and its Groups*, Holden Day, San Francisco, 1967.
206. Guillen, M., *Bridges to Infinity*, Rider, London, 1983.
207. Gurevich, G.B., *Projective Geometry* (in Russian), Fizmat., Moscow, 1960.
208. Hadamard, J., *Essai sur la psychologie de l'invention dans le domaine mathematique*, Gauthier-Villars, Paris, 1975.
209. Hadamard, J., *Lecons de geometrie elementaire*, 2 vols., Colin, Paris, 1932.
210. Haimovici, A., *Groups of Geometric Transformations* (in Romanian) EDP, Bucureşti, 1968.
211. Haimovici, A. et. al., *Elements of Plane Geometry* (in Romanian), EDP, Bucureşti, 1968.
212. Haimovici, A., Lectures on Elementar Geometry (in Romanian), Univ., Iaşi, 1975.
213. Hajos, G., *Einfuhrung in die geometrie*, Teubner, Leipzig, 1970.

214. Halmos, P.R., *Naive Set Theory*, Van Nostrand, Princeton, N.I., 1960.
215. Halsted, G.B., *On the Foundations and Technic of Arithmetic*, Court, Chicago, 1912.
216. Halsted, G.B., *Rational Geometry*, Willey, New York, 1904.
217. Hardy, G.H., *Pure Mathematics*, Cambridge Univ. Press, London, 1965.
218. Hartshorne, R., *Foundation of Projective Geometry*, Benjamin, New York, 1967.
219. Hasenlager, G,; Scholz, H., *Grundzuge der mathematischen Logic*, Springer, Berlin, 1961.
220. Hausdorff, F., *Grundzüge der Mengenlehre*, Leipzig, 1914 (Available in English, Chelsea, New York, 1949).
221. Heath, T.L., *The thirteen Books of Euclid's Elements*, 3 vols., Dover, New York, 1956.
222. Heijenoort, J. van, *From Frege to Gödel*, Harvard Univ. Press, Cambridge, Mass., 1967.
223. Heijenoort, J. van., *Frege and Gödel*, Harvard Univ. Press, Cambridge, Mass., 1967.
224. Hermes, H., *Enumerability, Decidability, Compatibility*, Springer, Berlin, 1969.
225. Heyting, A., *Axiomatic Projective Geometry*, Noordhoff, Groningen, 1963.
226. Heyting, A., *Constructivity in Mathematics*, North Holland, Amsterdam, 1959.
227. Heyting, A., *Intuitionism. An Introduction*, North Holland, Amsterdam, 1956.
228. Heyting, A. et. al., *Les fondements des mathématiques. Intuitionisme, Théorie de la démenstration*, Gauthier-Villars, Paris, 1955.
229. Hilbert, D., *Grundlagen der Geometrie*, Teubner, Berlin, 1930 (Available in English: *Foundation of Geometry*, Open Court, La Salle, Ilinois, 1971).
230. Hilbert, D.; Ackermann, W., *Principles of Mathematical Logic*, Chelsea, New York, 1950.
231. Hilbert, D.; Bernays, P., *Grundlagen der Mathematik*, 2 vols., Springler, Berlin, 1934-1939.
232. Hilbert, D.; Cohn-Vossen, S., *Geometry and the Imagination*, Chelsea, New York, 1952.
233. Hinman, P.G., *Recursion-Theoretic Hierarchies*, Springer, Berlin, 1978.
234. Hintikka, I. (editor), *The Philosophy of Mathematics*, Univ. Press, Oxford, 1969.
235. Hölder, O., *Die Matematische Methode*, Springer, Berlin, 1924.
236. Holland, G., *Geometrie fur Lehrer und Studenten*, 3 vols., Schroedel, Hanovra, 1974-1977.
237. Hughes, D.R., *Projective Planes*, Springer, Berlin, 1973.
    -For Iaglom see Yaglom.
238. Iacob, C. (editor), *Mathematics Today and Tomarrow* (in Romanian), Ed. Acad., Bucureşti, 1985.

239. Ioan, P., *Axiomatics, Morphological Study* (in Romanian), Ed. St. Enc., Bucureşti, 1980.
240. Ioan, P., *Logic and Metalogic* (in Romanian), Junimea, Iaşi, 1983.
  -For Jaglom see Yaglom.
241. Jacobs, H.R., *Geometry*, Freeman, San Francisco, 1974.
242. Jech, T.J., *The axiom of Choice*, North Holland, Amsterdam, 1973.
243. Jeger, M., *Transformation Geometry*, Allen and Unwin, London, 1966.
244. Johnson, R.A., *Avanced Euclidean Geometry*, Dover, New York, 1960.
245. Joja, A., *Studies of Logic* (in Romanian), 2 vols., Ed. Acad., Bucureşti, 1966-1968.
246. Jones, B.W., Elementary Concepts of Mathematics, Macmillan, New York, 1947.
247. Juel, C., *Vorlesungen uber projektive Geometrie*, Springer, Berlin, 1934.
248. Kagan, V.F., *Foundations of Geometry* (in Russian), GITTL, Moscow, 1949.
249. Kagan, V.F., *Lobatshevski*, Mir, Moscow, 1974.
250. Kamke, E., *Théorie des ensembles*, Dunod, Paris, 1964.
251. Karzel, M. et. al., *Einführung in die Geometrie*, Vadenhoeck and Rupert, Göttingen, 1973.
252. Kasner, E.; Newman, J., *Mathematics and Imagination*, Simon & Schuster, New York, 1940.
253. Keedy, M.L., *A Modern Introduction to Basic Mathematics*, Addison-Wesley, Massachusetts, 1963.
254. Keedy, M.L.; Nelson, C.W., *Geometry. A Modern Introduction*, Addison-Wesley, Reading, 1945.
255. Kerekjartó, B., *Les fondements de la géométrie*, 2 vols., Akad. Kiado, Budapesta, 1955-1966.
256. Klaua, D., *Allgemeine Mengenlehre. Ein Fundament der Mathematik*, Akad., Berlin, 1964.
257. Klaua, D., *Konstruktive Analysis*, DVW, Berlin, 1961.
258. Klaua, D., *Mengenlehre*, Walter de Gruyter, Berlin-New York, 1979.
259. Klaus, G., *Moderne Logik. Abriss der formalen Logik*, DVW, Berlin, 1973.
260. Kleene, S.C., *Introduction to Metamathematics*, Van Nostrand, New York, 1953.
261. Kleene, S.C.; Vesley, R.E., *The Foundations of Intuitionistic Mathematics*, North Holland, Amsterdam, 1965.
262. Klein, F., *Einleitung in die hohere Geometrie*, Teubner, Leibzig, 1907.
263. Klein, F., *Elementary Mathematics from an Advanced Standpoint*, Addison-Wesley, Reading, Mas., 1963.
264. Klein, F., *Famous Problems of Elementary Geometry*, Dover, New York, 1945.
265. Klein, F., *Le programme d'Erlagen*, Gauthier-Villars, Paris, 1974.
266. Klein, F., *Vorlesungen über die entwicklung der mathematik im 19 Jahrundert*, 2 vols., Springer, Berlin, 1926-1927.
267. Klein, F., *Vorlesungen über nicht-Euklidische Geometrie*, Springer, Berlin, 1928.

268. Klibanski, R. (editor), *Contemporary Philosophy* (1st vol.: *Logic and Foundations of Mathematics*), La nouva Italia Ed., Firenze, 1968.
269. Kline, M., *Mathematical Thought from Ancient to Modern Times*, Univ. Press, Oxford, 1972.
270. Kline, M., *Mathematics in Western culture*, Oxford Univ. Press, New York, 1953.
271. Kline, M. (editor), *Mathematics in the Modern World*, Freeman, San Francisco, 1969.
272. Klingenberg, W., *Grundlagen der Geometrie*, Teubner, Leipzig, 1957.
273. Klotzec, B., *Geometrie*, D.V.V., Berlin, 1971.
274. Kneebone, G.T. et. al., *Mathematical Logic and the Foundation of Mathematics*, Van Nostrand, London, 1963.
275. Koperman, R., *Model Theory and Its Application*, Allyn & Bacon, London, 1972.
276. Körner, S., *The Philosophy of Mathematics. An Introductory Essay*, Huthinson Univ., Library, London, 1960.
277. Kostin, V.I., *Bases of Geometry* (in Russian), Uciped., Moscow, 1948.
278. Kramer, E.E., *The Nature and Growth of Modern Mathematics*, Univ. Press, Princeton, 1970.
279. Kreisel, G.; Krivine, J.I., *Éléments de logiques mathématiques. Théorie des modeles*, Dunod, Paris, 1967.
280. Krivine, J.L., *Théorie axiomatique des ensembles*, Dunod, Paris, 1968.
281. Kuratowski, K.; Mostowski, A., *Set Theory*, North Holland, Amsterdam, 1967.
282. Kutuzov, B.V., *Lobachevsky's Geometry and Elements of Foundations of Geometry* (in Russian), UPGI, Moscow, 1955.
283. Lakatos, I., *Proof and Refulations. The Logic of Mathematical Discovery*, Univ. Press, Cambridge, 1976.
284. Lakatos, I. (editor), *Problems of the Philosophy of Mathematics*, North Holland, Amsterdam, 1972.
285. Lalescu, T., Geometry of Triangle (in Romanian), Schneider, Craiova, 1992.
286. Lang, S., *Algebra*, Addison-Wesley, Reading, Mass, 1965.
287. Lautman, A., *Essai sur l'unité des mathématiques*, UEG, Paris, 1977.
288. Leibniz, G.W., *Fragmente zur Logik*, ..., Berlin, 1960.
289. Lebesque, H., *Lecon sur les constructiones géométriques*, Gauthier-Villars, Paris, 1950.
290. Lebesque, H., *Les coniques*, Gauthier-Villars, Paris, 1942.
291. Lehmer, D.N., *An Elementary Course in Synthetic Projective Geometry*, Ginn and Co, Boston, 1917.
292. Lemmon, E.J., *Introduction to Axiomatic Set Theory*, Routledge & Kegan, London, 1968.
293. Lenz, H., *Grundlagen der Elementar Mathematik*, DVW, Berlin, 1967.

294. Lenz, M., *Nichteuklidische Geometrie*, DVW, Mannheim, 1967.
295. Lenz, M., *Vorlesungen über projective Geometrie*, DVW, Leipzig, 1965.
296. Levi, A., *Basic Set Theory*, Springer, Berlin, 1974.
297. Liebmann, W., *Nicht Euklidische Geometrie*, Goschen, Berlin, 1923.
298. Liebmann, W., *Pangeometrie von N.I. Lobacevskij*, Engelmann, Leipzig, 1912.
299. Liebmann, W., *Syntetische Geometrie*, Teuber, Leipzig, 1934.
300. Lietzmann, W., *Lustiges und Merkurwürgiges Von Zahlen Und Formen*, Hirt, Breslan, 1928.
301. Lietzmann, W., *Methodik des Mathematische unterrichts*, 3 vols., Quelle, Leipzig, 1919-1924.
302. Lingenberg, R., *Grundlagen der Geometrie*, ..., Manheim, 1969.
303. Lingenberg, R., *Metric Planes and Metric Vector Spaces*, J. Willey & Sons, New York, 1979.
304. Lionnais, F., *Les grandes courents de la pensées mathématiques*, Dunod, Paris, 1962.
305. Littlewood, *A Mathematician Miscellany*, Methuen & Co, London, 1963.
306. Lobachevski, N.I., *The Theory of Parallels*, In [48].
307. Lockwood, E.M; Macmillan, R.M., *Geometric Symmetry*, Univ. Press, Cambridge, 1978.
308. Lorenzen, P., *Métamathématique*, Gauthier-Villars/Mouton, Paris, 1967.
309. Lukasiewicz, J., *Elements of Mathematical Logic*, PSP, Wassaw, 1963.
310. Lukyachenco, S.J., *Elements of Non-Euclidean Geometry Lobachevsky-Boylay* (in Russian), Moscow, 1993.
311. Lyndon, R.C., *Notes on Logic*, Van Nostrand, Princeton, New Jersey, 1966.
312. Lyndon, R.C., *Groups and Geometry*, Univ. Press, Cambridge, 1985.
313. Malcev, A.I., *Algebraic Systems*, Springer, Berlin, 1973.
314. Malitz, J., *Introduction to Mathematical Logic. Set Theory. Computation Functions. Model Theory*, Springer, Berlin, 1974.
315. Malița, M., *Bases of Artificial Inteligence* (in Romanian), Ed. Tehn., București, 1987.
316. Manning., H.P., *Geometry of Four Dimension*, Dover, New York, 1955.
317. Manning, H.P., *Non-Euclidean Geometry*, Ginn, Boston, 1901 and Leipzig, 1951.
318. Marcus, S., *The Paradox* (in Romanian), Albatros, București, 1984.
319. Markwald, W., *Einführung in die formale Logik and Metamathematik*, Klebt, Stuttgard, 1972.
320. Martin, G.E., *Transformation Geometry*, Springer, Berlin, 1983.
321. Martin, G.E., *The Foundation of Geometry and the Non-Euclidean Plane*, Springer, New York, 1975.
322. Maxwell, E.A., *Geometry by Transformation*, Univ. Press, Cambridge, 1975.
323. Mayer, O., *Projective Geometry* (in Romanian), Ed. Acad., București, 1970.

324. Melcu, I., *Projective Geometry* (in Romanian), EDP, Bucureşti, 1963.
325. Mendelson, E., *Introduction to Mathematical Logic*, Van Nostrand, New York, 1964.
326. Mérö, L., *Ways of Thinking. The Limits of Rational Thought and Artificial Inteligenge*, World Scientific, Singapore, 1990.
327. Mesarosiu, I.V., *General Logic* (in Romanian), EDP, Bucureşti, 1971.
328. Meschkovski, M., *Non-Euclidean Geometry*, Academic Press, New York-London, 1964.
329. Meschkovski, M., *Unsolved and Unsolvable Problems in Geometry*, Oliver and Boyd, Edinburg-London, 1966.
330. Meserve, B.E., *Fundamental Concepts of Geometry*, Addison-Wesley, Cambridge, 1955.
331. Mihai, A., *Lectures in Geometry* (in Romanian), Univ., Bucureşti, 1957.
332. Mihăileanu, N.N., *Complements of Synthetic Geometry* (in Romanian), EDP, Bucureşti, 1965.
333. Mihăileanu, N.N., *Elements of Projective Geometry* (in Romanian), Ed. Tehn., Bucureşti, 1966.
334. Mihăileanu, N.N., *Analytic, Projective and Differential Geometry. Complements* (in Romanian), EDP, Bucureşti, 1972.
335. Mihăileanu, N.N., *Non-Euclidean Geometry* (in Romanian), Ed. Acad., Bucureşti, 1954.
336. Mihăileanu, N.N., *Complex Numbers in Geometry* (in Romanian), Ed. Tehn., Bucureşti, 1968.
337. Mihăileanu, N.N.; Neumann, M., *Foundations of Geometry* (in Romanian), EDP, Bucureşti, 1973.
338. Mihalescu, C., *Geometry of Remarkable Elements* (in Romanian), Ed. Tehn., Bucureşti, 1957.
339. Milman, R.S.; Parker, G.D., *Geometry*, Springer, Berlin, 1983.
340. Miron, R., *Analytic Geometry* (in Romanian), EDP, Bucureşti, 1976.
341. Miron, R., *Elementar Geometry* (in Romanian), EDP, Bucureşti, 1976.
342. Miron, R., *Vectorial Introduction to Plane Analytic Geometry* (in Romanian), EDP, Bucureşti, 1970.
343. Miron, R.; Brânzei, D., *Foundaments of Arithmetic and Geometry* (in Romanian), Ed. Acad., Bucureşti, 1983.
344. Modenov, P.S., *Analytic Geometry*, Univ., Moscow, 1955.
345. Modenov, P.S.; Parhomenko, A.S., *Geometric Transformations*, Acad. Press, New York, 1965.
346. Mohrmann, H., *Einführung in die nichteuklidische Geometric*, Akad., Leipzig, 1930.

347. Moise, E., *Elementary Geometry for an Advanced Standpoint*, Addison-Wesley, Reading, Messachusetts, 1963.
348. Moisil, Gh., *Elements of Mathematical Logic and set Theory* (in Romanian), Ed. Şt., Bucureşti, 1967.
349. Moorman, R.H., *Fundamental Concepts of Mathematics*, Burgess, Minneapolis, 1955.
350. Morley, F., Morley, F.V., *Inversive Geometry*, Bell, London, 1933.
351. Moschovakis, Y.N., *Elementary Induction on Abstract Structures*, North Holland, Amsterdam, 1974.
352. Moschovakis, Y.N., *Descriptive Set Theory*, North Holland, Amsterdam, 1982.
353. Mostowski, A., *Thirty Years of Foundational Studies*, Blackwell, Oxford, 1966.
354. Mousinho, M.L., *Conceitos fundamentais da geometria*, Bahaia Blanca, Buenos Aires, 1962.
355. Murgulescu, E. et. al., *Analytic and Differential Geometry* (in Romanian), EDP, Bucureşti, 1962.
356. Murgulescu, E. et. al., *Analytic Solid Geometry* (in Romanian), EDP, Bucureşti, 1972.
357. Murtha, J.A.; Wilord E.R., *Linear Algebra and Geometry*, New York, 1969.
358. Myller, A., *Analytic Geometry* (in Romanian), EDP, Bucureşti, 1972.
359. Năstăsescu, C., *Introduction in Set Theory* (in Romanian), EDP, Bucureşti, 1974.
360. Neumann, J. van, *Collected Works*, 1st vol., Pergamon Press, Oxford, 1961.
361. Neuman, M., *Bases of Geometry* (in Romanian), Lit. Inv., Timişoara, 1956.
362. Nevanlina, R., *Raum, Zeit und Relativität*, Birkhäuser, Basel, 1964.
363. Newmann, J.R. (editor), *The World of Mathematics*, 4 vols., Simon and Schuster, New York, 1956.
364. Newson, C.V.: Eves, N., *Introduction to College Mathematics*, Prentince Hall, New York, 1954.
365. Newton, I., *Principia*, Univ. of California Press, berkeley, 1970.
366. Nicod, J., *Foundations of Geometry and Induction*, Humanities Press, New York, 1950.
367. Nicod, J., *La geometrie dans le monde sensibles*, Presses Universitaires, Paris, 1962.
368. Niewenglowski, B., *Cours de geometrie analytique*, 4 vols, Gauthier-Villars, Paris, 1926-1929.
369. Norden, A.P., *Elementare Einfuhrung in die Lobatschewskische Geometrie*, DVW, Berlin, 1958.
370. Norden, A.P., *Basic Concepts of Geometry* (in Russian), GTI, Moscow, 1956.
371. Novikov, P.S., *Elemento of Mathematical Logic* (in Russian), Nauka, Moskow, 1973.

372. Nut, Iu.Iu., *Lobachevski's Analytically Treated* (in Russian) Akad. Nauk., Moskow, 1961.
373. Obădeanu, V., *Elements of Linear Algebra and Analitic Geometry* (in Romanian), Facla, Timişoara, 1981.
374. O'Hara, C.V.; Ward, D.R., *An Introduction to Projective Geometry*, Clarendon Press, Oxford, 1937.
375. Onicescu, O., *Principes de logique et de philosophie mathématique*, Ed. Acad., Bucureşti, 1971.
376. Oproiu, V., *Course of Geometry* (in Romanian), Univ., Iaşi, 1981.
377. Papelier, G., *Exercises de géométrie moderne*, 9 vols., Viubert, Paris, 1925-1927.
378. Papy, G.; Debaut, P., *Géométrie affine plane*, Press Universitaires, Bruxeles, 1967.
379. Parsons, Ch., *Logic Foundations of Mathematics and Computability Theory*, Canada, 1975.
380. Pârvu, Gh., *Semantic and Logic of the Science*, Ed. Şt., Bucureşti, 1974.
381. Pasch, M.; Dehn, M., *Vorlesungen über neuere Geometrie*, Springer, Berlin, 1926.
382. Păun, Gh., *The Show of Mathematics* (in Romanian), Albatros, Bucureşti, 1974.
383. Păun, Gh., *Actual Problems in the Formal Languages Theory* (in Romanian), Ed. Şt. Enc., Bucureşti, 1964.
384. Pedoe, D., *An Introduction to Projective Geometry*, Macmillan, New York, 1964.
385. Pedoe, D., *A Course of Geometry for Colleges and Universities*, Univ. Press, Cambridge, 1970.
386. Perron, O., *Nichteuklidische elementargeometrie der Ebene*, Stutgard, 1962.
387. Pic, Gh., *Algebra* (in Romanian), EDP, Bucureşti, 1966.
388. Pikert, G., *Projektive Ebenen*, Springer, Berlin, 1955.
389. Podehil, E.; Reidemeister, K., *Eine Begründung der Ebenen elliptischen Geometrie*, Leipzig, 1934.
390. Pogorelov, A.V., *Lectures on the Bases of Geometry* (in Russian), Univ., Harkov, 1959.
391. Pogorelov, A.V., *The Fundation of Geometry* (in Russian), Nauka, Moscow, 1968.
392. Poincaré, H., *Dernieres Pensees*, Flamarion, Paris, 1933.
393. Poincaré, H., *La Science et l'Hypothese*, Flamarion, Paris, 1925.
394. Poincaré, H., *La valeur de la Science*, Flamarion, Paris, 1925.
395. Poincaré, H., *Science et Methode*, Flamarion, Paris, 1927.
396. Polskiy, N.I., *About Distinct Geometries* (in Russian), Akad. Nauk, Kiev, 1962.
397. Pontrjagin, L.S., *Learning Higher Mathematics*, Springer, Berlin, 1984.
398. Pompeiu, D., *Mathematic Works* (in Romanian and French), Ed. Acad., Bucureşti, 1959.
399. Pop, I., *Algebra* (in Romanian), Univ., Iaşi, 1979.
400. Popa, C., *Theory of Definition* (in Romanian), Ed. Şt., Bucureşti, 1972.

401. Popescu, I.P., *Affine and Euclidean Geometry* (in Romanian). Facla, Timişoara, 1984.
402. Popovici, C.P., *Logic and Number Theory* (in Romanian), EDP, Bucureşti, 1970.
403. Prenowitz, W.; Jantosciak, I., *Join Geometries*, Springer, New York-Heildeberg-Berlin, 1979.
404. Prenowitz, W.; Jordan, M., *Basic Concepts of Geometry*, Ginn and Co. & Blaisdell, New York, 1964.
405. Prüfer, H., *Projektive Geometry*, Noske, Leipzig, 1935.
406. Quine, W.V., *From a Logical Point of View*, Univ. Press, Harvard, Mass., 1953.
407. Quine, W.V., *Mathematical Logic*, Univ. Press, Cambridge, Massachusetts, 1951.
408. Quine, W.V., *Methods of Logic*, Holt, London, 1952.
409. Quine, W.V., *Set Theory and Its Logic*, Harvard Univ. Press, Cambridge, Mass., 1971.
410. Quine, W.V., *The Ways of Paradox*, Random House, New York, 1969.
411. Rademacher, H.; Toeplitz, O., *The Enjoiment of Mathematics*, Univ. Press, Princeton, New Jersey, 1957.
412. Radó, F., *On* ⁻-linear Spaces, Unpublished MS.
413. Ramsey, F.P., *The Foundations of Mathematics*, Kegan, Trench, Trubner & London, 1931.
414. Rashevski, P.K., *Geometry and Its Axiomatic* (in Russian), FizMat., Moscow, 1960.
415. Rashevski, P.K., *Riemannsche Geometrie und Tensoranalyses*, Deutsche Ubersetzung, Berlin, 1959.
416. Rasiowa, H.; Sikorski, R., *The Mathematics of Metamathematics*, Panst. Wydaw. Naukowe, Warsaw, 1963.
417. Redei, L., *Foundation of Euclidean and Non-Euclidean Geometries*, Akad. Kiadó, Budapest, 1968.
418. Reghis, M., *Elements of Set Theory and Mathematical Logic* (in Romanian), Facla, Timişoara, 1981.
419. Reghis, M., *The Formal System of Mathematical Logic* (in Romanian), Univ., Timişoara, 1974.
420. Reichardt, H., *Gauss und die nicht-euklidische Geometrie*, Teubner, Leipzig, 1976.
421. Reid, C., *Hilbert*, Springer, New York, 1970.
422. Reidemeister, K., *Grundlagen der Geometrie*, Springer, Berlin-Heidelberg-New York, 1968.
423. Revuz, A., *Mathématique moderne, mathématique vivante*, O.C.D.L., Paris, 1968.
424. Richardson, M., *Fundamental of Mathematics*, Macmillan, New York, 1941.
425. Riemann, B., *Über die Hypothesen, welche der Geometrie zu Grunde liegen*, Springer, Berlin, 1923.
426. Rindler, W., *Essential Relativity*, Springer, New York, 1977.
427. Rob, A.A., *Geometry of Time and Space*, Univ. Press, Cambridge, 1936.

428. Robinson, A., *Definition*, Clarendon Press, Oxford, 1950.
429. Robinson, G. de B., *The Foundations of Geometry*, Univ. Press, Toronto, 1964.
430. Robinson, R., *Analytic Geometry*, Mc Graw-Hill, New York, 1949.
431. Roman, T., *Symmetry* (in Romanian), Ed. Tehn., Bucureşti, 1963.
432. Roman, T.G., *A Background to Geometry*, Univ. Press, Cambridge, 1967.
433. Rootselear, B. von; Stall, J.F. (editors), *Logic, Methodology and Phylosophy of Science*, North Holland, Amsterdam, 1968.
434. Rosen, J., *Symmetry Discovered*, Univ. Press, Cambridge, 1975.
435. Rosser, J.B., *Logic for Mathematics*, Mc Graw-Hill, New York, 1953.
436. Rouché, C.; Comberouse, Ch. de, *Éléments de géométrie*, Gauthier-Vollars, Paris, 1912.
437. Rozenfeld, B.A., *Many Dimensional Spaces* (in Russian), Nauke, Moscow, 1955.
438. Rozenfeld, B.A., *Noneuclidean Spaces* (in Russian), Nauka, Moscow, 1969.
439. Russel, B., *An Essay on the Foundations of Geometry*, Dover, New York, 1956.
440. Russel, B., *Introduction to Mathematical Philosophy*, Allen and Unwin, London, 1930.
441. Russel, B.; Whittehead, A.N., *Principia Matematica*, 3 vols., Cambridge Univ. Press, London, 1910-1913.
442. Russu, E., *Arithmetic and Number Theory* (in Romanian), EDP, Bucureşti, 1963.
443. Russu, E., *Bases of Number Theory* (in Romanian), Ed. Tehn., Bucureşti, 1953.
444. Russu, E., *Elementar Geometry* (in Romanian), EDP, Bucureşti, 1976.
445. Russu, E., *Methodical Problems in Collegium Mathematics* (in Romanian), EDP, Bucureşti, 1970.
446. Ruzsa, I.; Urban, I., *Mathematical Logic* (in Magyar), Tankönyvkiadó, Budapest, 1966.
447. Saaty, T.L., *Lectures on Modern Mathematics*, 3 vols., Wiley, New York, 1963-1966.
448. Sanger, R.G., *Synthetic Projective Geometry*, McGraw-Hill, New York-London, 1939.
449. Sâmboan, G., *Grounds of Mathematic* (in Romanian), EDP, Bucureşti, 1974.
450. Scherk, P.; Lingenberg, R., *Rudiments of Plane Affine Geometry*, Univ. Press, Toronto, 1975.
451. Schlaffi, L., *Gesammelte matematische Abhandlungen*, Birkhauser, Basel, 1950.
452. Schimidt, M., *Die Inversion und ihre Anwendungen*, München, 1950.
453. Schoenberg, T., *Mathematical Time Exposure*, Math. Asoc. of America, 1982.
454. Schreider, Iu. A., *Equality, Resemblance and Order*, Mir, Moscow, 1975.
455. Schreier, O.; Sperner, E., *Projektive Geometry of n Dimension*, Chelsea, New York, 1961.
456. Schutte, K., *Proof Theory*, Springer, Berlin, 1977.

457. Schwerdtfeger, H., *Geometry of Complex Numbers*, Univ. Press, Toronto, 1962.
458. Segre, B., *Lectures on Modern Geometry*, Cremonese, Roma, 1969.
459. Seidenberg, A., *Lectures in Projective Geometry*, Van Nostrand, Toronto, 1962.
460. Shervatov, V.G., *Hyperbolic Functions*, Heath, Boston, 1963.
461. Shirokov, P.A., *Essential of Lobachevsky's Geometry Foundation* (in Russian), Moscow, 1955.
462. Shirokov, P.A., *Selected Works on Geometry* (in Russian), Univ., Kazan, 1966.
463. Shirokov, P.A.; Kagan, V.F., *Construction of Non-Euclidean Geometries* (in Russian), GTI, Moscow, 1950.
464. Shively, L.S., *An Introduction to Modern Geometry*, J. Willey & Sons, New York, 1939.
465. Shoenfield, J.R., *Mathematical Logic*, Addison-Wesley, Reading, 1967.
466. Sikorsky, R., *Boolean Algebras*, Springer, Berlin, 1960.
467. Sindoons, A.W., Snell, K.S., *A New Geometry*, Univ. Press, Cambridge, 1938.
468. Simionescu, Gh.D., *Spherical Trigonometry* (in Romanian), Ed. Tehn., Bucureşti, 1965.
469. Simon, M., *Nichteuklidische Geometrie in elementarer Behandlung*, Teubner, Leipzig, 1925.
470. Singh, J., *Great Ideas of Modern Mathematics*, Dover, New York, 1959.
471. Sjostedt, C.E., *Le axiome de paralleles de Euclides ŕ Hilbert*, Interlingue Fundation, Upsala, 1968.
472. Skolem, Th., *Selected Works* (J.E. Fenstad-editor), Univ., Oslo, 1970.
473. Slupecki, J., Borkowski, L., *Elements of Mathematical Logic and Set Theory*, Pergamon Press & PWN, Warsaw, 1967.
474. Smaranda, D.; Dragomir, S., *Introduction in Axiomatic Geometry* (in Romanian), Univ., Bucureşti, 1979.
475. Smaranda, D.; Soare, N., *Geometric Transformations* (in Romanian), Ed. Acad., Bucureşti, 1988.
476. Smirnov, S.A., *Projective Geometry* (in Russian), Nedra, Moscow, 1976.
477. Smogorjevski, A.S., *Bases of Geometry* (in Russian), Radysanka Sk., Kiev, 1947.
478. Smogorjevski, A.S., *Lobachevsky's Geometry*, Nauka, Moscow, 1957.
479. Smullyan, R.M., *First-order Logic*, Springer, Berlin, 1968.
480. Snapper, E.; Trover, R.I., *Metric Affine Geometry*, Academic Press, New York, 1971.
481. Soare, N., *Course of Geometry* (in Romanian), Univ., Bucureşti, 1986.
482. Seminskiy, I.S., *Method of Mathematical Induction* (in Russian), FMI, Moscow, 1961.
483. Sommerville D.M.Y., *An Introduction to the Geometry of n Dimensions*, Methuen Co, London, 1929.

484. Sommerville, D.M.Y., *Bibliography of Non-Euclidean Geometry*, Chelsea, New York, 1970.
485. Sommerville, D.M.Y., *The Elements of Non-Euclidean Geometry*, Dover, New York, 1958.
486. Stabler, E.R., *An Introduction to Mathematical Thought*, Addison-Weslley, Cambridge, 1953.
487. Stati, S. et. al., *Language, Logic, Philosophy* (in Romanian), Ed. Şt., Bucureşti, 1974.
488. Steen, L.A. (editor), *Mathematics Today*, Springer, New York, 1978.
489. Steen, L.A. (editor), *Mathematics Tommorow*, Springer, New York-Heidelbery-Berlin, 1981.
490. Stein, S.K., *Mathematics: the Man-made Universe*, Freeman, San Francisco, 1962.
491. Stevenson, F.W., *Projective Planes*, Freeman, San Francisco, 1972.
492. Stewart, I., *Concepts of Modern Mathematics*, Penguin, Harmondsworth, 1975.
493. Stewart, I., *The Problems of Mathematics*, Oxford, Univ. Press, New York, 1978.
494. Stewart, I.; Tall, D., *The Foundations of Mathematics*, Univ., Press, Oxford, 1977.
495. Stoll, R.R., *Sets, Logic and Axiomatic Theories*, Freeman, San Francisco, 1962.
496. Stolyar, A.A., *Introduction to Elementary Mathematical Logic*, MIT Press, 1970.
497. Strawson, T., *Introduction to Logical Theory*, London, 1952.
498. Struik, D.J., *Lectures on Analitic and Projective Geometry*, Addison-Weslley, Cambridge, 1953.
499. Surdu, A., *Classical and Mathematical Logic* (in Romanian), Ed. Şt., Bucureşti, 1971.
500. Surdu, A., *Elements of Intuitionist Logic* (in Romanian), Ed. Acad., Bucureşti, 1976.
501. Surdu, A., *Neo-Intuitionism* (in Romanian), Ed. Acad., Bucureşti, 1977.
502. Takeuti, G.; Zaring, W.M., *Introduction to Axiomatic Set Theory*, Springer, Berlin, 1971.
503. Tarski, A., Einfuhrung in die matematische Logik und die Methodologie der Mathematic, Viena, 1937 (Available in English: New York, 1951).
504. Tarski, A., *Introduction a la logique*, Gauthier-Villars, Paris-Louvain, 1960.
505. Tarski, A., *Undecidable Theories*, North-Holland, Amsterdam, 1953.
506. Taylor, E.H.; Bartoo, G.C., *An Introduction to College Geometry*, Macmillan, New York, 1944.
507. Teleman, K. et. al., *Geometry* (in Romanian), 2 vols., EDP, Bucureşti, 1979.
508. Teodorescu, I.D., *Analytic Geometry and Elements of Higher Algebra* (in Romanian), EDP, Bucureşti, 1972.
509. Teodorescu, N., *Vectorial Methods in Mathematical Physics* (in Romanian), Ed. Tehn., Bucureşti, 1954.

510. Thurston, H.A., *The Number System*, Interscience, New York, 1954.
511. Timerding, H.E., *Aufga bensammlung zur projektion Geometrie*, Gruyter, Berlin, 1933.
512. Ţiţeica, Gh., *Selected Problems of Geometry* (in Romanian, 6th edition), Ed. Tehn., Bucureşti, 1981.
513. Trajnin, Ja.L., *Foundations of Geometry* (in Russian), UPGI, Moscow, 1961.
514. Tresse, A., *Théorie élémentaire des géométries non euclidiennes*, Gauthier-Villars, Paris, 1957.
515. Troelstra, A.S. (editor), *Mathematical Investigation of Intuitionistic Analysis, Arthimetic and Analysis*, Springer, Berlin, 1973.
516. Ţurlea, M., *Philosophy and the Foundations of Mathematics* (in Romanian, English abstract), Ed. Acad., Bucureşti, 1981.
517. Tuller, A., *A Modern Introduction to Geometries*, Van Nostrand, Toronto, 1982.
518. Udrişte, C., *Problems in Linear Algebra, Analytic and Differential Geometry* (in Romanian), EDP, Bucureşti, 1981.
519. Udrişte, C.; Tomuleanu, V., *Analytic Geometry* (in Romanian), EDP, Bucureşti, 1990.
520. Udrişte, C. et. al., *Problems in Algebra, Geometry and Differential Ecuations* (in Romanian), EDP, Bucureşti, 1976.
521. Udrişte, C. et. al., *Mathematical Problems and Methodological Remarks* (in Romanian), Facla, Timişoara, 1980.
522. Uspensky, V.A., *Un Completeness Godel's Theorem* (in Russian), Nauka, Moscow, 1982.
523. Vaisman, I., *Foundations of Mathematics* (in Romanian), EDP, Bucureşti, 1968.
524. Vaisman, I., *Foundations of Three-Dimensional Euclidean Geometry*, Marcel Dekker, New York-Basel, 1980.
525. Vaismann, F., *Introduction of Mathematical Thinking*, Fr. Ungar, New York, 1951.
526. Vasilache, G., *Elements of Set Theory and Algebric Structures* (in Romanian), Ed. Acad., Bucureşti, 1956.
527. Vasiu, A., *Affine and Metric Geometry* (in Romanian), Univ., Cluj, 1993.
528. Vasiu, A., *Foundations of Geometry* (in Romanian), Univ., Cluj, 1978.
529. Weber, H.; Wellstein, *Encyklopädie der Elementar Mathematic*, 3 vols., Druk und Teubner, Leipzig, 1904.
530. Veblen, O.; Young, J.W., *Projective Geometry*, 2 vols., Blaisdell-Waltham, Massachusetts, 1960.
531. Verriest, G., *Introduction á la géométrie non euclidienne par la methode élémentaire*, Gauthier-Villars, Paris, 1951.
532. Vinogradov, I.M., *Bases of Number Theory* (in Russian), GTL, Moscow, 1953 (in Romanian: Ed. Acad., Bucureşti, 1954).
533. Volberg, O.A., *Leading Ideas of Projective Geometry* (in Russian), UPGI, Moscow-

Leningrad, 1944.
534. Vrănceanu, G., *Analytic and Projective Geometrie* (in Romanian), Ed. Tehn., Bucureşti, 1954.
535. Vrănceanu, G.; Teleman, K., *Euclidean Geometry, Non-Euclidean Geometries, Relativity Theory* (in Romanian), Ed. Şt., Bucureşti, 1965.
536. Vrănceanu, G. et. al., *Elementary Geometry from a Modern Stand Point* (in Romanian), Ed. Tehn., Bucureşti, 1967.
537. Wade, T.L.; Howard, H.E., *Fundamental Mathematics*, McGraw-Hill, New York, 1956.
538. Wang Hao, *A Survey of Mathematical Logic*, Science Press, Peking, 1962.
539. Wang Hao; Naughton Mac, *Le systĕmes axiomatiques de la théorie des ensembles*, Gauthier-Villars, Paris, 1963.
540. Ward, M.; Hardgrave, C.E., *Modern Elementary Mathematics*, Addison-Wesley, New York, 1964.
541. Watson, E.E., *Elements of Projective Geometry*, Heath, Boston, 1935.
542. Weszely, T., *Farkas Bolyai. The Man and the Mathematician* (in Romanian), Ed. Şt., Bucureşti, 1974.
543. Weyl, H., *Philosophy of Mathematics and Natural Science*, Univ. Press, Priceton, 1949.
544. Weyl, H., *Space, Time, Matter*, Dover, New York, 1962.
545. Whitehead, A.N., *The Axioms of Projective Geometry*, Univ. Press, Cambridge, 1906.
546. Wilder, R.L., *Evolution of Mathematical Concepts*, Willey, New York, 1968.
547. Wilder, R.L., *Introduction to the Foundation of Mathematics*, J. Willey & Sons, New York, 1952.
548. Winger, R.M., *An Introduction to Projective Geometry*, Heath, Boston, 1935.
549. Woolfe, H.E., *Introduction to Non-Euclidean Geometry*, Driden, New York, 1945.
550. Woods, F.S., *Higher Geometry*, Ginn and Co, Boston, 1922.
551. Wylie, C.R., *Foundations of Geometry*, Mc Graw-Hill, New York, 1964.
552. Yaglom, I.M., *A Simple Non-Euclidean Geometry and Its Physical Basis*, Springer, New York-Heideberg-Berlin, 1979.
553. Yaglom, I.M., *Complex Number in Geometry*, Academic Press, New York, 1968.
554. Yaglom, I.M., *Geometric Transformations*, 3 vols., Random House, New York, 1962-1968-1973.
555. Yaglom, I.M.; Ashkinuze, V.G., *The Ideas and Method of Affine and Projective Geometry* (in Russian), UPGI, Moscow, 1963.
556. Yale, P.B., *Geometry and Symmetry*, Holden Day, San Francisco, 1968.
557. Young, J.V., *Lectures on Fundamental Concepts of Algebra and Geometry*, Macmillan, New York, 1925.

558. Young, J.V., *Monographs on Topics of Modern Mathematics*, Dover, New York, 1955.
559. Young, J.V., *Projective Geometry*, Open Court, La Salle, 1933.
560. Zacharias, M., *Einführung in die projektive Geometrie*, Teubner, Leipzig, 1912.
561. Zacharias, M., *Elementargeometrie der Ebene und des Raume*, Gruyter, Berlin, 1930.
562. Zacharias, M., *Das Parallelenproblem und seine Losung*, Teubner, Leipzig, 1951.
563. Zamorzaev, A.M. et. al., *Theory of Discret Groups of Simmetry* (in Romanian), Univ. Moldova, Chişinău, 1991.
564. Zippin, L., *Uses of Infinity*, Random House, New York, 1962.

Section B: Selected Papers

565. Brânzei, D., *On some Categories of Miron Spaces*, Works of national Conf. of Geometry and Topology, Timişoara, 28-30 apr. 1977.
566. Brânzei, D., *Structure affines et opérations ternaires*, An. Şt. Univ. "Al.I. Cuza" Iaşi, XXIII (1977), s. I-a, t. 1, p. 33-38.
567. Cohen, I.P., *The Independence of the Continuum Hypothesis*, Proc. Nat. Acad. of Sci., 50 (1963), p. 1143-1148, 51 (1961), p. 105-110.
568. Gheorghiev, Gh., *Geometrical Groups and the Evolution of the Idea of Space* (in Romanian), Probleme actuale de matematică, EDP, Bucureşti, 1963.
569. Gheorghiev, Gh.; Miron, R., *Octav Mayer and Erlangen Program* (in Romanian), National Sympos. "Gh. Ţiţeica", Craiova, 21-22 sept. 1978, p. 71-81.
570. Golab, S., *Über die Grundlagen der affinen Geometrie*, Jber. Deutsch. Math Verein, 71 (1969), 3, p. 139-155.
571. Liebeck, H., *The Vector Space Axiom 1 · v = v*, The Math Gazette, LVI (395) 1972, p. 30-33.
572. Miron, R., *Almost Affine Spaces over a Field*, An. Şt. Univ. "Al.I. Cuza" Iaşi, XXIV, s. I, 2 (1978), P. 327-344.
573. Miron, R., *On the Almost Linear Spaces*, Matematica 18 (41), 2 (1976), p. 187-190.
574. Miron, R., *Sur quelques strustures affines*, Dem. Math. VI (1973), P. 289-294.
575. Miron, R., *The Minimality of Weyl's System of Axioms for the Affine Geometry over an Unitary Ring*, An. Şt. Univ. "Al.I. Cuza" Iaşi, XXIV, s. I (1978), p. 15-19.
576. Miron, R.; Opaiţ, Gh., *Sur les espaces affines généralises*, Rev. Roumaine Math. Pures et Appl., 14 (1969), P. 653-660.
577. Miron, R.; Radó, F., *On the Definition of a Module and Vector Space by Independent Axioms*, Mathematica, 18 (41), 2 (1976), p. 179-186.
578. Rugină, A., *Skew Modules. Applications in Informatics* (in Romanian), These for Doctorate, Univ. :Al.I. Cuza" Iaşi, september 1982.
579. Vaisman, I., *An Axiomatic for the Plane Euclidean Geometry* (in Romanian), Gaz. Math., s. A, vol. LXXI (1966), nr. 9, p. 321-329.

Index

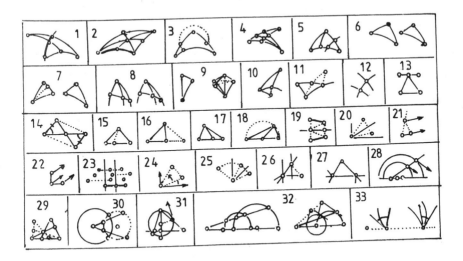